PLANT GENETIC ENGINEERING

Volume 4

Improvement of Commercial Plants - II

The Editors

Dr Pawan K. Jaiwal is a Reader in the Department of Biosciences, M. D. University, Rohtak, India. He is teaching Molecular Biology, rDNA technology and Plant Biotechnology courses to graduate and postgraduate students since 1986. He has throughout first division academic career with Ph.D. in Botany (Plant Physiology) from Kurukshetra University, Kurukshetra, India. He has more than 70 research publications including original research papers, reviews and book chapters in reputed International research journals and books. He has edited three books published by International publishers. Dr Jaiwal has been awarded DST Young Scientist Project, INSA Visiting Fellowship, New Delhi and DBT Overseas Associateship (Long Term) for advance research at laboratory of Prof Ingo Potrykus, ETH, Switzerland. Dr Jaiwal is a member of several academic bodies and on the Editorial Board of four International research journals. His has guided many students for Ph.D., M. Phil. and M. Sc. degrees. He has completed several research projects on Biotechnology for improvement of important grain legumes. His group has developed gene transfer protocols in *Vigna radiata* and *V. mungo* for the first time. His current research interests are metabolic engineering for resistance to abiotic and biotic stresses, improvement of nutrient use efficiency and nutritional quality of grain legumes.

Dr Rana P. Singh is a Reader in the Department of Biosciences, M. D. University, Rohtak, India. He is teaching courses in Plant Biology and Biotechnology to graduate and postgraduate students since 1986. He earned his first class M. Sc. in Botany from DDU Gorakhpur Univ., Gorakhapur and Ph.D. in Life Sciences (Plant Biochemistry) from Devi Ahilya University, Indore, India. He has more than 70 research publications including original research papers, reviews and book chapters in reputed International research journals and books. He has edited 5 books published by International publishers. He is fellow and member of various academic societies. Dr Singh has been awarded R. D. Asana Medal from Indian Soc. Plant Physiol., New Delhi and JEB Young Scientist award of The Academy of Environmental Biology, India for significant research contributions. He is recipient of INSA, New Delhi and UNESCO, Paris Biotech. Fellowships and visited twice at University of Guelph, Canada for advance research. He has visited Yunnan Agri. Univ., Kunming, China and hosted three Senegalese Scientists for research collaborations. He has guided many students for Ph.D., M. Phil. and M. Sc. degrees and completed many projects in areas of his interest. His group has contributed significant work in nitrogen relation to plants. Dr Singh is Editor-in-chief of an International research journal, Physiology and Molecular Biology of Plants and on editorial board of three other journals. His current research interests are metabolic engineering for nutrient use efficiency, stress tolerance and nutritional improvement in plants.

Other Volumes in this series:

Volume 1 **Applications and Limitations :** *Eds.* Rana P. Singh & Pawan K. Jaiwal

Volume 2 **Improvement of Food Crops :** *Eds.* Pawan K. Jaiwal & Rana P. Singh

Volume 3 **Improvement of Commercial Plants - I :** *Eds.* Rana P. Singh & Pawan K. Jaiwal

Volume 5 **Improvement of Vegetables :** *Eds.* Rana P. Singh & Pawan K. Jaiwal

Volume 6 **Improvement of Fruits :** *Eds.* Pawan K. Jaiwal & Rana P. Singh

PLANT GENETIC ENGINEERING

Volume 4

Improvement of Commercial Plants-II

Editors

Pawan K. Jaiwal **Rana P. Singh**

Department of Biosciences
M. D. University,
Rohtak-124001 India

2003

SCI TECH PUBLISHING LLC, U.S.A.

ISBN : 1-930813-05-8
SERIES ISBN : 1-930813-17-1

First Published 2003

Published by
SCI TECH PUBLISHING LLC.
9207, Country Creek Drive
Houston, Texas - 77036 (U.S.A.)
Tele. : 713-541-94-- Fax : 713-541-9401
E-mail : sci@hotmail.com

Printed by
RAINBOW PROCESSORS & PRINTERS
New Delhi-110059 India

PREFACE

Plants have a major share in the commercial sector which is increasing day by day due to the recent advances in bio-techniques and a tilt in people's liking towards the products of herbal origin. *In vitro* cell culture and gene technologies have created possibilities for obtaining novel products in commercially viable quantity from plants which was never realized before. The major plant products of market value are edible oils, pharmaceuticals and fragrances, agrochemicals, beverages, fibres, wood, pulp and spices etc. Advances in genetic engineering has revolutionized biotechnology industries dealing with these products.

The companies which have developed GM plants want to protect the exclusive right of their products and are keen to develop gene protection technologies within the GM plants (Chapter 1). A debate has been generated on such issues globally to ascertain the rights of farmers for the conservation of gene pool traditionally. *Agrobacterium rhizogenes*-mediated gene transfer induces hairy roots which regenerate into transgenic plants with desirable traits. The hairy root system has also been used for the production of valuable metabolites with wide ranging applications i.e. from natural products of pharmaceutical importance to antibodies and valuable foreign proteins. Besides, this system has been used for the enhanced bioaccumulation of heavy metals and other environmental toxicants and in studies on plant-microbe interaction. The recent advances in applications and limitations of transgenic hairy roots have been discussed in chapter 2. Cotton is one of the most important natural plant fibres used in textile industry. Transgenic cotton has been developed for improved agronomic traits using *Agrobacterium*-mediated genetic transformation (see chapter 1, *Plant Genetic Engineering Vol. 3* in this series), but many elite commercial cotton varieties are recalcitrant to regenerate from callus and therefore are not directly amenable to *Agrobacterium*-mediated gene transfer. Particle bombardment-mediated transformation of cotton is independent of genotype and large number of genes have been transferred to improve such varieties quantitatively and qualitatively using particle bombardment (Chapter 3). Barley is a major raw material for brewing industry which requires alterations in structural grain constituents or the enzyme activities that mobilize storage reserves of the seed during the mashing. In order to produce improved malting barley several alterations must be made. Improved malting barley has been developed by genetic engineering which otherwise, is very difficult through conventional breeding (Chapter 4).

Many plants are cultivated for their oil contents which is used for cooking and industrial purposes. The production of oilseeds has not improved significantly due to their susceptibility to pests, pathogens and sensitivity to abiotic stresses and low nutrient use efficiency etc. Manipulation of fatty acid composition and their contents can improve nutrition or industrial qualities of the oil as per requisite. *In vitro* culture and gene transfer techniques have played a significant role in improving the desired quantitative and qualitative characters in sunflower (Chapter 5), safflower (Chapter 6), niger (Chapter 7), sesame (Chapter 8), linseed (Chapter 9), oilpalm (Chapter 10) and salicornia (Chapter 11). *Linum* is also cultivated as a natural fibre plant as the fibres are used in textile and other industries. The woody part of stem and root of linseed after yielding fibres is also a valuable raw material for fibre board and paper industries. *Salicornia* has been domesticated as a new crop of costal areas which is irrigated by sea water due to its halophytic nature. Chapter 11 deals with biology and biotechnology of *Salicornia* which is emerging as a significant commercial crop due to its multipurpose products e.g. food source, seed oil and pharmaceutical metabolites etc.

We are grateful to all the contributors for providing updated insightful account of the various plants of their interest. The response from the colleagues was so overwhelming that the title on the commercial plants could cover two volumes in this series. Thanks to our family members, friends and students for their patience and understanding during the planning and preparation of this volume and the series. We appreciate the publishers especially Mr. Anil Jain and Mr. Arvind Jain of Sci-Tech Pub. LCC, Houston, USA and designer Ms. Reema and Mr. Ashok Datta of *LaPrints*, New Delhi, India for their keen interest in bringing out this volume with quality work.

March, 2002
Rohtak, India

Pawan K Jaiwal
Rana P Singh

CONTENTS

Contents of other volumes in the series

PLANT GENETIC ENGINEERING

SERIES ISBN : 1-930813-17-1

ISBN : 1-930813-02-2

Volume 1 : **Applications and Limitations**
Editors : Rana P. Singh & Pawan K. Jaiwal

ISBN: 1-930813-03-1

Volume 2 : **Improvement of Food Crops**
Editors : Pawan K. Jaiwal & Rana P. Singh

ISBN: 1-930813-04-X

Volume 3 : **Improvement of Commercial Plants - I**
Editors : Rana P. Singh & Pawan K. Jaiwal

ISBN : 1-930813-14-7

Volume 5 : **Improvement of Vegetables**
Editors : Rana P. Singh & Pawan K. Jaiwal

ISBN : 1-930813-15-5

Volume 6 : **Improvement of Fruits**
Editors : Pawan K. Jaiwal & Rana P. Singh

Chapter 1

GENE PROTECTION TECHNOLOGIES AND THEIR IMPLICATIONS

AK Singh★ and K Hariprasanna

Division of Genetics, Indian Agricultural Research Institute, New Delhi – 110 012

Summary

Increasing area under transgenic crops globally reflects growing acceptance of transgenic technology by the farmers in both industrial and developing countries. The private sector companies expect maximum profits from the commercialization of transgenic crops, as high investments are required for developing these transgenics. The operational difficulties in implementing the patents and PBR system for protection of transgenes necessitated the development of genetic use restriction technologies of gene protection technologies. The 'Control of Plant Gene Expression' or 'Terminator' technology based on the transfer of a combination of three genes aroused much global debate. The other such gene protection technologies, which are at various stages of development, are 'Verminator', 'Traitor' etc. The terminator technology may have far reaching consequences especially in developing countries where small and resource-poor farmers make up the major portion of farming community. The implications may also be serious on transgenic research, and growth and development of science of molecular biology as a whole, in developing countries. However, some beneficial aspects of this technology are also being highlighted. In spite of the global opposition, many private sector seed companies are actively engaged in perfecting the further advancement or refinement of the terminator technology.

Keywords : Gene protection technology, terminator, traitor, transgenics, verminator.

★Corresponding author

1. INTRODUCTION

Since early nineties concerted efforts are being made on development of transgenic crops genetically engineered for various agronomic and quality traits. The transgenic crops are developed by incorporating foreign gene(s) by methods of genetic engineering. With the commercialization of transgenic crops, biotechnology has attained a firm place in agriculture. The isolation, modification and development of appropriate transgenes for expressing a given trait is quite expensive, thus the private sector companies as well as public sector institutions involved in transgenic research usually have to make substantially high investment in order to develop transgenic crops. Therefore, in turn, the investors expect to make as much profit as they can from the commercialization of transgenic crops. Patents and Plant Breeders' Rights (PBR) have been used to protect the interests of multinational companies in past. However, there are certain operational difficulties in implementing the patents and PBR system for protection of transgenes particularly in developing countries. This is so because of small average farm size and a large number of small and marginal farmers. Therefore development of 'genetic use restriction technologies' (GURTs) or 'gene protection technologies' (GPTs) have been sought by certain multinational seed companies in association with public sector institutions. These efforts have led to development of technologies like 'Terminator', 'Verminator' and 'Traitor', which are at various stages of development. Use of these technologies is likely to provide a foolproof protection mechanism to transgenic crop varieties having novel transgene(s). Obviously therefore, the development of GPTs is in relation to the commercialization of transgenic crops in past one decade.

2. TRANSGENIC CROPS : A GLOBAL PERSPECTIVE

The global area under transgenic crops has increased significantly in the last decade. During the five-year period 1996 to 2000, the global area of transgenic crops increased by more than 25-fold *i.e.*, from 1.7 million hectares in 1996 to 44.2 million hectares in 2000 (Table 1). Global adoption rates of transgenic crops are unprecedented and are the highest for any new technology by agricultural industry standards. High adoption rate of transgenic crops reflects grower satisfaction, with the products that offer significant benefits ranging from more convenient and flexible crop management, higher productivity or net returns per hectare and a safer environment through decreased use of conventional pesticides, which collectively contribute to more sustainable agriculture.

Table 1. Global area of transgenic crops (1996-2000) (James, 1999; 2000).

Year	Hectares (million)	Acres (million)
1996	1.7	4.3
1997	11.0	27.5
1998	27.8	69.5
1999	39.9	98.6
2000	44.2	109.2
2001	45.3	125

The high rate of adoption reflects the growing acceptance of transgenic crops by farmers using the technology in both industrial and developing countries. The number of countries growing transgenic crops has increased from six in 1996 to 13 in 2000.

As evident form Table 1, the global area of transgenic crops increased from 27.8 m ha in 1998 to 39.9 m ha in 1999, recording an increase of 12.1 m ha or 44%. The increase in area between 1999 and 2000 is 11%, equivalent to 4.3 m ha. This increase of 4.3 m ha between 1999 and 2000 is about one quarter of the corresponding increase of 12.1 m ha between 1998 and 1999. Out of the 13 countries that grew transgenic crops in 2000 four countries contributed 99% of the global transgenic crop area (Table 2). It is noteworthy that they are two industrial countries, USA and Canada, and two developing countries, Argentina and China. Consistent with the pattern since 1996, the USA grew the largest transgenic crop hectarage in 2000 with 30.3 m ha followed by Argentina with 10 m ha, Canada with 3 m ha and China 0.5 m ha. In 2000 area increases were reported for the USA, Argentina and China, with a decrease in area in Canada. Transgenic crops were grown commercially in all six continents of the world-North America, Latin America, Asia, Oceania, Europe (Eastern and Western), and Africa. Of the top four countries that grew 99% of the global transgenic crop area, the USA grew 68%, Argentina 23%, Canada 7% and China 1%. The other 1% was grown in the remaining nine countries (James, 2000).

The predominant transgenic crops grown during 2000 included soybean, corn, cotton and canola (Table 3) engineered for herbicide tolerance, insect resistance (Bt) or a combination of the two. The growth in the area under genetically engineered crops gives an indication of development

Table 2. Distribution of transgenic crops by country (area in millions of hectares)[1, 2]

Country	1998	%	1999	%	2000	%
USA	20.5	74	28.7	72	30.3	68
Argentina	4.3	15	6.7	17	10.0	23
Canada	2.8	10	4.0	10	3.0	7
China	<0.1	<1	0.3	1	0.5	1
Australia	0.1	1	0.1	<1	0.15	<1
South Africa	<0.1	<1	0.1	<1	0.2	<1
Mexico	<0.1	<1	<0.1	<1	<0.1	<1
Spain	<0.1	<1	<0.1	<1	<0.1	<1
France	<0.1	<1	<0.1	<1	<0.1	<1
Portugal	0.0	0	<0.1	<1	0	0
Romania	0.0	0	<0.1	<1	<0.1	<1
Ukraine	0.0	0	<0.1	<1	0	0
Germany	0.0	0	0.0	0	<0.1	<1
Bulgaria	0.0	0	0.0	0	<0.1	<1
Uruguay	0.0	0	0.0	0	<0.1	<1
Total	**27.8**	**100**	**39.9**	**100**	**44.2**	**100**

Table 3. Distribution of transgenic crops by crop (area in millions of hectares) (James, 1999; 2000).

Crop	1998	%	1999	%	2000	%
Soybean	14.5	52	21.6	54	25.8	58
Corn	8.3	30	11.1	28	10.3	23
Cotton	2.5	9	3.7	9	5.3	12
Canola	2.4	9	3.4	9	2.8	6
Potato	<0.1	<1	<0.1	<1	<0.1	<1
Squash	0.0	0	<0.1	<1	<0.1	<1
Papaya	0.0	0	<0.1	<1	<0.1	<1
Total	**27.8**	**100**	**39.9**	**100**	**44.2**	**100**

in transgenic crops and their adoption. Given its potential and not withstanding fact that the debate about its advantage and disadvantage still going on, biotechnology will be a significant component of sustainable agricultural development and global food security.

The foregoing discussion about transgenic crops is relevant in context to GPTs, which prevent the use of farm saved seed for planting the next crop by one or the other mechanism including control of seed germination, which will be discussed in this chapter.

3. WHY GENE PROTECTION TECHNOLOGIES?

The development of transgenic crops requires a heavy investment on the part of the multinational seed companies and other research organizations. Once the finished product (transgenic crop) is developed using these technologies the investors want to make as much profit as they can from the sale of the transgenic seed. In case of pure line varieties of self-pollinated crops (like rice, wheat, pulses) and open-pollinated varieties of often cross-pollinated crops (like cotton, pigeon pea, sorghum etc.), farmers usually save seed from their harvest for planting next crop at least for 3-4 years after buying the seed once. This reduces the profit margin for the company. Although, patents and plant breeders' rights have been implemented in developed countries for protection of plant varieties, there are inherent problems in these systems of protection requiring a close watch and legal litigation etc. Therefore, the multi-national companies (MNCs) have devised fool-proof technologies based on sound genetic principles using genetic engineering methods to ensure the maximum benefit out of the investment made on developing such expensive technologies, which is perfectly on line with the business ethics. However, development of gene protection technologies has attracted worldwide criticism. Our effort in this chapter is to provide a balanced view on the excellent science used in developing such technologies and pros-and cons of the technology there of.

4. THE TERMINATOR TECHNOLOGY

On March 3, 1998, the U.S. Department of Agriculture (USDA) and Delta and Pine Land Co. (D&PL), a subsidiary of the seed and the agrochemicals multinational Monsanto/American Home Products, received a US patent for a technology that changes the genetic makeup of a plant cell. Plants regenerated from this cell will develop seeds, which will not germinate. It is claimed that the genetic system described in the patent has worked well with tobacco and cotton, although not completely proven yet. It is believed that the tobacco and cotton system should become workable completely in a very short time. The first commercial crop (cotton) utilizing this technology is expected to be marketed by the year 2004 (Lehmann, 1998). The patent work was mainly done by Melvin Oliver, a scientist with USDA-ARS in Lubbock, Texas, through cooperative research with D&PL. The patent claims a very broad protection and is valid for plant cells, tissues, seeds and whole plants of any species containing the combination of genes described in the patent. USDA and D&PL have announced that they will make this technology ideally available to many seed companies through licensing agreements.

The technology is based on the transfer of a combination of three genes and the patent is titled "Control of plant gene expression" (US patent No. 5,723,765) (Anonymous, 1998a). The patented technology enables a seed company to genetically alter seed so that it will not germinate if re-planted. Hence, the technology has been dubbed as "Terminator" by the Canadian non-governmental organization Rural Advancement Foundation International (RAFI) which plunged into an international campaign against the patent. According to RAFI, it took four years and an investment of US$ 720,000 to develop the technique (Cully, 1998). The USDA spent $ 190,000 and D&PL $ 275,000 on in-house research expenses and the two jointly spent $ 255,000. While D&PL has been given exclusive licence to sell the seeds, the USDA is to get a royalty of five percent on the net sales. The USDA and D&PL have applied for licence in 78 countries.

There are two basic purposes of inventing this anti-farmer technology (Singh, 1998). Firstly, to prevent farmers from using saved seeds for sowing next crop particularly, in case of transgenic crops having transgene(s) for desirable traits, for example insect resistance, disease resistance, herbicide resistance etc. From this angle, the technology is quite relevant to pure lines, open pollinated varieties (OPVs) and F_2 hybrids but not so relevant to hybrid varieties because in case of hybrids, farmers otherwise also change seed fresh each year.

The 1991 Act of International Union for the Protection of New Plant Varieties (UPOV) also diluted the provision of the farmers' privilege, which earlier allowed them to save seeds from their harvest for re-plantation. With the terminator technology the rights of farmers to save seed for re-use are further curtailed and they will be forced to buy seed fresh each year even in the case of pure lines and OPVs, there by increasing the turnover of the company. In fact, the American Company

Monsanto, which has developed transgenic soybean seed with resistance to herbicides, requires its customers to sign a licence agreement not to use the seeds obtained from transgenic crops to plant next crop. The company has even employed Pinkerton investigators (private hired police) to identify unauthorized growers and taken legal action against them. Now the use of terminator technology will fully prevent such unauthorized growing and this will increase the sale of the company and ultimately its profit. This technology can also be used in high-yielding varieties developed by companies, even if they do not have transgenes to increase the sale.

The second objective of inventing this technology is to prevent the mobilization of traits of interest from transgenic crop variety to another or in other words to protect the transgene(s) from being pirated by researches elsewhere. Plant breeding in most of the (developing) countries has been under the purview of public sector institutions for many decades and no patents or plant breeders' rights (PBRs) were ever available to protect the cultivars developed through plant breeding. This was envisaged to encourage free flow of seeds to researchers and resource-poor small farmers. Only recently, considerable activity for seeking patents/PBRs has been witnessed in the private sector, particularly in the post GATT period. Under the Trade Related Intellectual Property Rights (TRIPs) it is mandatory for all the signatories, to allow some kind of protection of the rights on plant variety. Many countries have brought about amendments in their patent laws to allow patents on living organisms. This has become necessary in crop plants also due to the advent of genetically modified plant varieties, which are entering the global agricultural scenario on a faster rate.

The identification, isolation, modification and transformation or the use of transgene(s) in developing transgenic crops is a time consuming and highly expensive process. Once these transgenic crop plants are developed and reach farmers, the transgene(s) present in these plants can easily be transferred by plant breeders through hybridization to a normal crop variety following ordinary breeding procedures. Thus the monopoly of inventors on transgene(s) is lost. This concern has necessitated the provision of patents/PBRs to allow fair returns for the heavy investments made by the private companies. With the use of terminator technology the seed obtained after hybridization will also not germinate and thus the transgene(s) will be protected from being pirated and the rights of corporate plant breeders will be further strengthened.

From this angle the technology is relevant to all crop varieties propagated by seed (PLs, OPVs and hybrids) which carry transgene(s).

4.1. About the genes and promoters

Variously quoted as the 'Neutron bomb for seeds' or 'Trojan seeds' (Gopinathan, 1998) the science behind the terminator provides an excellent manifestation of combination of theory and practice of controlling the gene expression patterns. This makes use of the classical information of the basic principles of gene regulation by operator-repressor interactions (the well-known paradigm of control of gene expression proposed by the Nobel laureates Jacob and Monad in the early 1960s), the current developments in transgenesis technology which permits the introduction of foreign genes into living organisms unrelated to the donor, as well as the application of advanced molecular knowledge in differential gene expression and transposition of DNA sequences through the recombination process.

The lethal gene or 'terminator' can be any gene encoding for a protein or an enzyme, or any other toxic molecule which when expressed should result in an altered plant phenotype (as claimed in the patent) or in the death of the organism. For instance, the uncontrolled presence of suicidal enzyme like protease which can degrade a vital protein essential

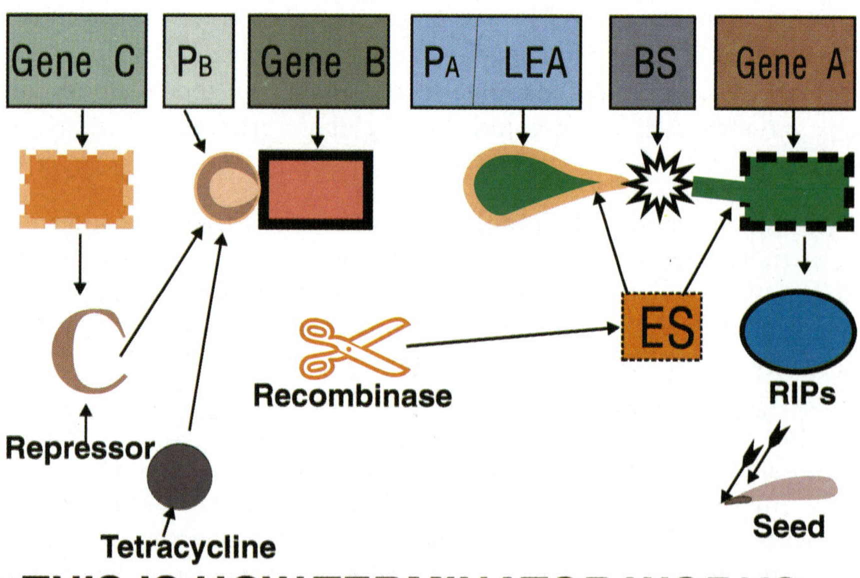

THIS IS HOW TERMINATOR WORKS

for the survival, or a nuclease which can chew up the nucleic acids (DNA or RNA) may result in the death of the cell. The results will be similar if the gene product is a toxin.

A schematic representation of genes, promoters and functioning of terminator technology is given in the figure. This package consists of three genes, *viz.,* DNA sequence I, DNA sequence II and DNA sequence III which will be referred to as genes A, B and C respectively. Gene A produces the toxin, ribosomal inactivating proteins (RIPs-see box), which kill the embryo, thereby making seed incapable of germination. According to claim no. 3 of the patent this gene is selected from the group comprising a lethal gene, an insecticidel gene, a fungistatic gene, a fungicidal gene, a bactericidal gene, a drought resistance gene, a protein product gene or a gene that alters secondary metabolism. Gene A is hooked with its promote PA, which is a transiently-active promoter called Late Embryogenesis Abundant (LEA) because it is active only at late embryogenesis stage of seed development. The claim no. 3 also covers the promoter as the transiently-active promoter selected from the group comprising a promoter active in late embryogenesis, in seed development, in flower development, in leaf development, in root development, in vascular tissue development, in pollen development, after wounding, during heat or cold stress, during water stress, or during or after exposure to heavy metals. The gene A and the LEA promoter are being operably linked to one another, but separated by a blocking sequence (BS) that is flanked by specific excision sequences (ES), such that the presence of the BS prevents the expression of gene A. The specific signal sequences (*i.e.*, ES) are selected from the group comprising LOX sequences and sequences recognizable by either flippase, resolvase, FLP, SSV1-encoded integrase or transposase.

The gene B encodes a specific recombinase selected from the group comprising CRE, flippase etc., which cleaves DNA at ES, removing thereby the BS and ligating PA with gene A and making it functional. The promoter of gene B (PB) is a repressible promoter selected from the group comprising a 35S promoter modified to contain one or more tet operons, a modified ubiquitin promoter, a modified MAS promoter and a modified NOS promoter. The PB is repressed by the repressor produced by gene C.

The gene C which encodes the Tn 10 *tet* repressor is selected from the group comprising the Tn 10 *tet* repressor and the *lac* operator-repressor system. The master control of the entire process is through an

FUNCTION OF RIBOSOME-INACTIVATING PROTEINS (RIPs)

Many plants produce ribosome-inactivating proteins (RIPs) - the enzyme that acts on ribosome in a highly specific way thereby inhibiting protein synthesis. Some RIPs can bind to and enter cells, making them among the **most toxic substances** known. Most commonly, however, RIPs are unable to enter healthy cells, and are therefore poorly cytotoxic. Their role in plants is currently a matter of debate, although it has been suggested that they may have defensive functions, or an involvement in programmed senescence, in certain organs. The RIPs are frequently produced in large quantity-up to 10 percent of the total protein-that are in far excess of the amount required to inhibit protein synthesis. In most RIPs producing plants, the ribosomes are sensitive to enzyme, and the two are physically separated (Martin *et al.*, 1996).

antibiotic tetracycline, which acts as an inducer to PB by interacting with the repressor molecule bound to PB.

4.2. How terminator technology works?

As the first step a transgenic plant is developed by stably transforming a plant cell or cell culture (claim no. 55 of the patent) with DNA sequences comprising PA+BS+gene A; PB+gene B and gene C followed by regenerating whole plant from the plant cell or cell culture. This plant is allowed to flower and produce first generation seed. Since, in this plant gene A is separated from its promoter PA by BS, the seed produced on this plant will germinate. Using these seeds and their progenies, the seed is multiplied in the required quantity. A part of the multiplied seed is kept separately as nucleus seed for further multiplication and remaining seeds are treated with tetracycline. The treated seeds are then sold to farmers.

When these seeds germinate and seedlings grow, right from the beginning tetracycline interacts with repressor molecule attached to PB and pulls it off. As a result PB becomes free and gene B is expressed producing recombinase. Recombinase removes BS by cleaving at ES. Therefore, the cells entering megasporogenesis and microsporogenesis do not have BS between gene A and PA, and so is the case with egg cells and pollen grains and zygotes resulting from fertilization. Zygote divides and embryo develops normally till the late embryogenesis stage is reached. At late embryogenesis, LEA promoter PA becomes active and gene A is expressed resulting into formation of RIPs, which are lethal to embryos. Therefore, seeds harvested by farmers although can be used for consumption but lack the ability to germinate, hence cannot be used as seed for raising next crop.

4.3. Technology for purelines and hybrid seed production

In self-pollinated crops where purelines are used as cultivars, the above mentioned mechanism can be used as such along with the development of transgenics. In actual commercial production and sale of seeds of purelines, the crop for commercial seed production will be grown by the breeder or seed company without tetracycline treatment. The harvested seeds from this crop will be treated with tetracycline before it is sold to farmers.

In case of hybrid seed production a different strategy utilizing only two genes (gene A and gene B), one in each of the two parents of the hybrid can be used and gene C is not needed. One of the parental lines contains the recombinase gene, which becomes active only after germination and the other parent contains the lethal (terminator) gene separated from its promoter by BS. The hybrid progeny, which is the technology protected hybrid seed bought and planted by the farmer, thus contains both the elements of the system in every cell. The recombinase expressed right after germination, excises the BS, thus bringing together gene A and PA. Since the promoter is embryo specific, the gene expresses once the seed development starts, resulting in embryo mortality. Thus the seeds harvested from the first generation hybrid crop will be normal in all respects, except that they will not germinate if sown as a crop (Gupta, 1998).

4.4. Implications of terminator technology

The terminator technology may have much wider implications than being foreseen presently, particularly in relation to research and development in public sector institutions. Most of these implications are against farmers and pro-private sector seed industry, though some are welcome from the point of view of large scale commercialization of transgenic technology. Some of the important implications will be discussed here under.

4.4.1. Threat to farmers' independence

As already discussed the basic purpose of the invention seems to prevent farmers from using saved seeds. In case of self-pollinated crops like rice and wheat where pureline varieties are being used the seed replacement rate is very low (less than 10%), this means rest 90 percent seed used for sowing next crop comes from saved seed from previous crop. If this technology is used, farmers will be prevented from their age-old privilege of using saved seed and thus will terminate farmers' independence and threaten the food security of over a billion resource-poor farmers in developing countries. Overall in these countries, farm-saved seed accounts for an estimated 80 percent of the total seed requirement (Lehmann, 1998). In this context it will become virtually impossible for seed producing agencies to meet the seed demand of farmers.

On the other hand according to Harry B. Collins from D&PL, farmers in developing countries will still have the choice between saving seeds of traditional variety and newly developed cultivars protected by this technology. The inventors claim that it will stimulate plant breeders' interest to develop new varieties of crops for which hybrid development is difficult or cumbersome such as wheat, rice and soybean. Farmers will profit from this new development because they will gain broad access to continuing agricultural improvements and more productive varieties. Furthermore, the incentives to breed new variety will enhance genetic diversity in many important crops. The Europe based NGO Genetic Resources Action International (GRAIN), however, rejects this view because biodiversity cannot be reduced to packages in breeding lines. GRAIN maintains that seed saving is necessary for farmers to adapt the seeds to their own needs, thereby generating and nurturing biodiversity in their fields.

4.4.2. Access to transgenes and influence on transgenic development

In the 1990s many developing countries including India, have made considerable progress in developing infrastructure and standardizing protocols for development of transgenic plants in different crops like brinjal, tomato, potato, cotton etc., particularly with respect to insect resistance using '*Bt*' gene. However, the '*Bt*' gene and many other transgene(s) (see box) on which scientists are currently working, have been obtained from other developed countries or MNCs. Use of such gene(s) by researchers other than inventors, only for research purposes is ensured through signing in proper transfer agreements. But for commercializing the transgenic crops, the researchers may be forced to hook-up terminator gene with the novel transgene(s) so that the inventor of the novel transgene(s) may harvest a handsome royalty on annual sale of transgenic seeds. The rejection to do so may lead the researchers to complete denial of transgenes, *per se.* Therefore, this technology is likely to give a serious set back to transgenic research in developing countries. For avoiding the use of terminator and transgene(s) from MNCs and other countries, the developing countries have to make their

TRANSGENICS COMMERCIALISED BY MNCs

- 'Flavr Savr' Tomato
- 'IMI' Corn
- 'SR' Corn
- 'Bollgard' Cotton
- 'Liberty Link' Soybean
- 'STS' Soybean
- 'Bt' Cotton
- 'Liberty Link' Corn
- 'BXN' Cotton
- 'Roundup Ready' Cotton
- 'Roundup Ready' Soybean
- Barnase/Barstar Canola

research and development programmes more sound than the existing, so that researchers can develop useful transgene(s) themselves and use them at will. For this there is a need to strengthen the public sector institutions both in terms of infrastructure and human resource than ever before rather than gradually handing over R&D to private seed sector.

4.4.3. Impact on future growth and development of science in developing countries

Another concern is with respect to the basic/applied molecular biology research in the area of plant gene expression and/or regulation. Though considerable information is being generated, further research efforts regarding constitutive or regulated (time/ tissue/ developmental stage specific) gene expression patterns are very much essential. As previously mentioned the terminator patent covers a wide range of genes and promoters involved in gene expression patterns. For example, the claim no. 3 covers the different transiently-active promoters like a promoter active not only in late embryogenesis stage but also in seed development, in flower development, in leaf development, in root development, in pollen development, after wounding, during heat or cold stress or during water stress etc. Restrictions on the identification, isolation, characterization or use of such specific promoters by other researchers worldwide will have a serious negative impact on the advancement of science of molecular biology in those countries. Hence development of transgenic technologies independent of MNCs will become a trance in developing countries.

4.4.4. Use of the technology by public sector institutions

In many third world countries the lack of funds and proper infrastructure is always a serious concern in the case of public sector institutions involved in research activities. In view of the increasing pressure to generate resources, these institutions may decide to use terminator technology in the crop varieties developed by them, so that they can sustain themselves using the revenue generated by sale of these varieties. This will have a drastic impact on the resource-poor farmers who entirely depend on government institutes and machineries for their requirements and who cannot knock the doors of private seed agencies and financing establishments.

4.4.5. Threat to seed security

Farmers who have to rely on the commercial seed sector for seed supply become more vulnerable to all sorts of disturbances of, for instance, transportation or markets to raise capital for purchasing seeds. Further more, farmers would no longer be able to adapt their crops to environmental variation. Control over seed material is the baseline of agricultural practice for most farmers in developing countries. Any technology that undermines this prerequisite is not likely to have a positive impact on food security.

Terminator seeds may be used as a weapon against developing world. This technology will be a boon in the prolific growth of private seed sector in many developing countries where already several MNCs have set their foot in the agrochemical sector. Moreover, the size of a patent holder can have a major influence on the introduction of a new technology as acquisitions and merger of private seed companies is a recent feature. The American giant Monsanto has purchased another company Kargil seeds and it has also taken about 26 per cent share of the leading Indian seed company Mahyco. Monsanto also made attempts to acquire the D&PL and the main motive behind this acquisition is the patent held on sterile seed production by the latter. It is obvious that Monsanto's interest in protecting its own varieties against seed saving is likely to benefit from this technology. Other private seed companies will also buy this technology from inventors or owners for use in their crop varieties.

In the longer run this process will become a grave threat to the seed security of the country. For instance, if the terminator technology along with a useful transgene is used for developing new crop varieties and these varieties do extremely well on farmers' field, farmers would like to use these seeds in spite of high cost. This way, the new package will become extremely popular with farmers and the companies will also be happy due to increased turnover. However, this will lead to complete dependency on MNCs for seeds. If at the peak sowing time of, say rice or wheat, the MNCs decide to withdraw the seed supply or they are forced to do so, the farmers will not get seeds to sow the crops. The government's godowns will be full of seeds that will not germinate, as they are terminator seeds. This will lead to complete failure of crop and ultimately to mass starvation, enormous sterility and loss of human life. This is how seeds may be used as weapon.

4.4.6. Terminator and nutritional quality

The product of so-called terminator gene is a protein called ribosomal inactivating protein. As the name indicates RIPs inactivate ribosome, the site of protein synthesis. Hence expressing RIPs in seed embryo may have deleterious effect on seed protein, particularly in non-endospermic seed, and seed protein quantity or quality may be reduced. In seeds where the embryo is the major supplier of nutrients (protein), if the terminator technology is used, it may have a profound impact on the protein nutrition of millions of poor people who cannot afford the non-vegetarian source of protein. This will further deteriorate the nutritional imbalance or malnourishment scenario prevalent in the developing countries.

4.4.7. Threat to biodiversity

The terminator gene threatening entire biodiversity as one of the most serious implications as expressed by media and press, needs to be analyzed critically. The frequency of gene flow through pollen from a terminator crop to a normal crop would vary depending on mode of reproduction or pollination; in case of self-pollinated crops like rice and wheat the extent of out crossing is less than five percent, which may go up to 50 percent in case of often cross-pollinated crops like cotton, sorghum and sunflower. In case of cross-pollinated crops like maize and pearlmillet out crossing may be still greater. However, this gene flow to wild and weedy relatives of respective crops will be very low (may be one in thousand). In any case the seeds formed by pollination with terminator pollen will not germinate because of terminator effect. Thus the terminator gene is responsible for its own death and there is no way by which this gene will perpetuate in environment. If one seed out of a thousand seeds of a wild or weedy species does not germinate, the species will not face the danger of extinction. Because otherwise also, these species have mechanism of over production in order to survive. In case of normal crop varieties, the loss in seed germination resulting from pollination with terminator pollen can easily be compensated by increasing the seed rate proportionate to out crossing, if farmer is using saved seed for raising the next crop. Therefore, the fear expressed earlier about enormous sterility and loss of the entire biodiversity needs to be re-examined.

The inventors of the technology claim that it can be applied as a safety technology. In D&PL's view, the escape of transgenic traits to wild or other non-targeted plants becomes impossible because seed produced from unwanted pollination will not germinate and therefore be non-viable (as previously discussed). But GRAIN has criticized this effect on biosafety grounds. Since the genetic information for sterility is also contained in the pollen, cross-pollination and gene transfer to adjacent crops could accidentally spread sterility. A farmer trying to save seed from a conventional variety grown next door to a sterile seed variety might find that the yield would be drastically reduced, due to non-germination of seeds (Lehmann, 1998).

4.4.8. Effects on soil biodiversity

Large scale use of tetracycline treated seeds may adversely affect soil health by killing beneficial microorganisms, for example, the nitrogen fixing bacteria. Biological nitrogen fixation plays a significant role in fixing atmospheric nitrogen and thereby improving the soil productivity.

4.5. Terminator vs. apomixis

As discussed earlier, there is no need to change seed fresh each year in case of purelines and OPVs but with use of terminator technology it will become necessary to do so. On contrary, in case of hybrid varieties it is necessary to change fresh seed each year but with use of apomixis approach farmers will not be required to do so and at the same time they will get the benefit of heterosis. Several laboratories worldwide are working on the phenomenon of apomixis. Through genetic engineering approach it should be possible to use the apomictic gene(s) in hybrid cultivars to make them breed true. Plant breeders and molecular biologists have successfully transferred the genes that confer apomixis from a wild grass species, *Tripsacum dactyloides*, to maize. In developing terminator technology the science of genetic engineering has been used purely from business angle. On contrary use of genetic engineering for cloning, characterization and utilization of apomictic gene(s) in hybrid seed production is pro-farmer and pro-poor.

4.6. Beneficial aspects of terminator technology

A brief description of some of the good/ beneficial aspects of the technology are given below:

4.6.1. Private sector participation in development of purelines and OPVs

This technology will induce private sector to make more investment in research and development of pureline varieties and OPVs, which they were not doing so far, because in these varieties farmers did not change seed, fresh each year. With the use of terminator technology, farmers will have to change seed fresh each year in these varieties as well; therefore companies will get more profit. This will also result into a stiff competition between public and private sector institutions to show a better performance and ultimately farmers will be benefited.

4.6.2. Higher seed replacement rate in purelines and OPVs

As previously discussed good quality seed is a key factor for getting good harvest. Since the seed replacement rate in many developing countries including India is very low, the national production suffers for the want of good quality seeds. With terminator technology farmers will be forced to buy fresh seed each year, therefore, seed replacement rate will automatically go up and this will directly increase the production at national level. However, as mentioned earlier, it will be a great challenge to meet the seed demands of the farmers.

4.6.3. Prevention of pre-harvest sprouting

Pre-harvest sprouting of grains due to rains at maturity is a serious problem in many crops. Many a times this will have a serious impact on the total grain production and quality of the harvested produce. Use of the terminator technology can prevent this sprouting.

4.6.4. Prevention of transgene(s) escape

Escape of transgene(s) for example, transgene conferring resistance to herbicide, to wild and weedy relatives of the crops through cross-pollination, may lead to development of resistance in weeds against the herbicide. Hooking the transgene(s) with terminator gene can effectively control this gene flow and risk of possible development of super weeds can be checked. Similarly in future plants are likely to be used as bioreactors for producing vaccines, industrial oil and plastics etc. and it will be necessary to prevent the escape of pollen from such transgenic plants to normal crop varieties. Terminator technology can be effectively used for this purpose.

5. SECOND GENERATION TERMINATOR TECHNOLOGIES

Even when the original 'terminator' patent aroused a global debate and concern, the MNCs are busy in perfecting the further advancement or refinement of the terminator technology. Some of such technologies are being dubbed as 'Verminator' or 'Traitor' (negative trait) by RAFI. RAFI reports that every gene giant multinational has patented, or admits that it is working on genetically sterilized or chemically dependent seeds (Anonymous, 1999). RAFI's report provides details and analysis on over two dozen such patents obtained by 12 institutions. The patents seek to exploit -or could exploit- new genetic engineering techniques that use inducible promoter to disable critical plant functions governing reproduction, disease resistance and seed viability. If commercialization of such seeds proceeds, farmers worldwide will be tangled in an expensive web of chemicals, intellectual property and disabled germplasm that leads to bioserfdom.

The latest batch of patent claims connect the terminator's 'suicide sequence' to enhanced herbicide or fertilizer applications thereby transferred the costs of sterilization of seeds (to prevent their being saved and used from the harvested crops) from the company to the farmer. Some of these terminator type patent claims seem to reach beyond plants to insects and animals, while others seem to be aimed at expressively weakening a plant's pest and disease resistance capacity as part of the genetic sterilization process (Chakravarthi Raghavan, 1999). "The ultimate goal appears to be, not to force farmers to buy corporate seeds each year, but to force them to pay for their seeds every year". This would capture to the companies' enormous cost savings and render the commercial merit of aggressive new plant breeding moot.

The patents describe the use of external chemicals to turn on and off specific genes or possibly multi-gene traits in plants and go well beyond D&PL's original 'Terminator' patent. They are techniques to control a wide variety of 'input' and 'output' (production and processing) traits by spraying with proprietary herbicides or fertilizers.

5.1. Killer genes

The AstraZeneca's (UK based seed company) patent aim to switch the plant's germination on or off. The technology makes use of a gene from fat tissue of a rat, which it calls 'killer genes', which will block the normal plant growth, unless the blocking process is deactivated by a chemical. The technology has been dubbed as 'Verminator' by RAFI (Anonymous, 1998b). AstraZeneca's new verminator II patent can create plants that need repeated application of proprietary chemicals not only for germination, but also for continued healthy growth. The precise chemical necessary to avoid plant death depends upon the particular genes involved (AstraZeneca has at least three different promoter systems under patent claim); but the chemically dependent plant must have it in order to survive.

5.2. Genetic mutilation

An especially disturbing feature of some of the new patents profiled in RAFI's report is the deliberate disabling of natural plant functions that help to fight disease. Swiss biotech giant Novartis is most advanced in this aspect of 'Traitor' technology. Novartis blandly refers to it as "inactivation of endogenous regulation" so that "genes which are natively regulated can be regulated exclusively by the application to the plant of a chemical regulator". Since spraying to de-activate negative or 'traitor' traits can be a great incentive for the farmer, it will be particularly attractive for the company. Among the traits proposed for control in Novartis patent claims are:

1. **Input traits :** Germination, Flowering, Herbicide resistance, Insect resistance
2. **Output traits :** Nutritional qualities, Flavour qualities

Among the genes, which Novartis can control in this manner are patented SAR (systemic acquired resistance) genes, which are critical to plant's ability to fight off infections from many viruses and bacteria. Thus, Novartis has patented techniques to create plants with natural healthy functions turned off.

Although all the 'traitor' patents uncovered by RAFI involve external chemical inducers, they do not all confine their targets to plants. One patent, issued to the University of Texas (US patent no. 5,846,768) suggests that the inventors could activate a dormant suicide trait in insect pests by later spraying the crop with almost any chemical they can link to an inducible promoter. Indeed, the sequence could even trigger suicide through 'natural causes' - changed climatic conditions, for example. The Texas researchers refer to their invention - actually a gene from a fly, as the "GRIM Protein".

Syngenta, the world's largest agrochemical enterprise and the third largest seed corporation, was formed on 13 November 2000 with the merger of AstraZeneca and Novartis. The next day the company won its newest Terminator patent, US patent no. 6,147,282, 'Method of controlling the fertility of a plant'. (The patent was issued to Novartis – but the company's intellectual property goes to Syngenta.) The patent describes a complex system for chemical control of a plant's fertility. The application of a chemical inducer can be used to either abolish or restore a plant's fertility. The technique allows companies to engineer crops that depend on the external application of a chemical in order to develop into fertile or healthy plants (Anonymous, 2001).

At one level, traitor offers the opportunity to load a number of commercial characteristics onto a plant variety (or animal breed) which the company can choose to either activate or de-activate at or after the point of sale. This turns the 'traitor' into a launching pad or platform technology upon which proprietary traits are placed. Farmers can buy seed and depending on what traits the farmer can afford - or what traits the company wants to disclose - external chemical sprays or soaking could activate the purchased qualities in the 'platform' seed. The Gene Giants want to tie traitor seed to their proprietary chemicals so that one is useless without the other. Economic and commercial realities point the way clearly to how the technology will evolve.

6. CONCLUSIONS AND PROSPECTS

The increase in the global area planted to transgenic crops in the last five years has demonstrated that the early promises of transgenic crops are meeting expectations of large and small farmers planting transgenic crops in both industrial and developing countries barring all the skepticism about the appropriateness of transgenic crops. The confidence and trust farmers have placed in transgenic crops can make a vital contribution to global food, feed and fiber security.

The development of GPTs or GURTs along with the commercialization of transgenic crops has aroused global debate, especially in the developing countries where small and resource-poor farmers make up the major portion of farming community. Terminator technology, the genetic engineering of plants to produce sterile seeds, is universally considered the most morally offensive application of agricultural biotechnology, because over 1.4 billion people depend on farm-saved seeds. The Terminator and the subsequent patents on GURTs have attracted almost total global opposition from the United Nations, scores of national governments, scientific institutions and non-governmental organizations. The Director General of the United Nations Food and Agriculture Organization (FAO) Dr. Jacques Diouf had declared his opposition to Terminator. In publicly rejecting Terminator, he has come to the defense of the people who depend upon farm-saved seed for their survival.

Among the national governments that have announced their intention to oppose Terminator technology are Panama, India, Ghana and Uganda. Canada based RAFI, and other NGOs are calling for a global ban on the use of the Terminator technology and are actively involved in educating the farmers about the consequences of Terminator technology and pressurizing the concerned authorities to publicly denounce the technology as a threat to food security. India, one of the first government to publicly reject Terminator, explicitly prohibits Terminator genes in a draft bill now before the Indian Parliament. Because of the mounting opposition from the farming community, NGOs and national government, the MNCs including seed and agrochemical giant Monsanto have announced publicly that they will not introduce any seed with terminator technology in India. Provisions are being made to check thoroughly the presence of terminator gene sequence(s) in introductions (seed, propagule, gene sequences etc.) by MNCs.

REFERENCES

Anonymous (1998a) http://www.patents.ibm.com/claims?patent_ number = 5723765.

Anonymous (1998b) Seed industry consolidation: Who own whom? *RAFI Communique*, July/August, 1998: 1-4.

Anonymous (1999) Traitor technology: "Damaged goods" from the gene giants. *RAFI News Release,* 29 March 1999.

Anonymous (2001) New Terminator Patent goes to Syngenta. *RAFI News Release*, 12 March 2001.

Chakravarthi Raghavan (1999) From 'Terminator' to Traitor' (technologies). *South-North Development Monitor* (SUNS), 28 April 1999.

Cully DE (1998) The terminator technology. *RAFI Communique,* March/April, 1998: 1-5.

Gopinathan KP (1998) The terminator saga. *Curr. Sci.,* **75** : 416-419.

Gupta PK (1998) The terminator technology for seed production and protection: Why and how? *Curr. Sci.,* **75** : 1319-1323.

James C (2000) Global Status of Commercialized Transgenic Crops: 2000. *ISAAA Briefs No. 21: Preview.* ISAAA, Ithaca, NY.

James C (1999) Global Review of Commercialized Transgenic Crops: 1999. *ISAAA Briefs No. 20 : Preview.* ISAAA, Ithaca, NY.

Lehmann V (1998) Patent on seed sterility threatens seed saving. *Biotechnology and Development Monitor,* **35**: 6-8, June 1998.

Martin RH *et al.* (1996) The structure and function of ribosome-inactivating proteins. *Trends Plant Sci.,* **1** : 254-259.

Singh AK (1998) Terminator seeds and their implications: issues ahead. *Indian farming,* **48** : 13-16.

Chapter 2

RECENT ADVANCES IN APPLICATIONS OF TRANSGENIC HAIRY ROOTS

Archana Giri[1], CC Giri[2] and M Lakshmi Narasu[1]★

[1]*Centre for Biotechnology, Jawaharlal Nehru Technological University, Mahaveer Marg, Hyderabad - 500 028. India*
[2]*Centre for Plant Molecular Biology, Department of Genetics, Osmania University, Hyderabad - 500 007 India*

Summary

Agrobacterium rhizogenes causes hairy root disease in plants. These roots are characterized by their fast growth rate, biochemical and genetic stability. Genetically transformed roots exhibit profuse growth and branching which results in high biomass accumulation. In view of their growth characteristics and cellular differentiation, hairy root cultures provide a suitable alternative as a production system. They offer promise for production of valuable metabolites with wide ranging applications i.e. from natural products to foreign proteins. The main constraint for commercial exploitation of hairy root cultures is their scaling up, as there is a need for developing specially designed bioreactors that permit the growth of interconnected tissues unevenly distributed throughout the vessel. Development of bioreactor models for hairy root cultures is still a recent phenomenon. It is also necessary to develop computer-aided models for different parameters involving medium components, aeration and product removal. Further, transformed roots are able to regenerate genetically stable plants as transgenics or clones thus utilizing A. rhizogenes as a gene delivery system. The property of rapid growth and high plantlet regeneration frequency allows clonal propagation of elite plants. In vitro transformation and regeneration from hairy roots facilitates

★Corresponding author : E-mail : mangamoori@rediffmail.com

application of biotechnology to tree species. The ability to manipulate trees at cellular and molecular level shows great potential for clonal propagation and genetic improvement. In addition, the altered phenotype of hairy root regenerants (hairy root syndrome) proves to be useful in plant breeding programs with plants of ornamental interest. Genetic engineering of secondary metabolism in hairy roots may prove to be a strategy with prospects. Transgenic root system offers tremendous potential for application of metabolic engineering principles to plants leading to overproduction of a given metabolite and introduction of newer pathways in a given plant system. This can be achieved by completing partial pathways, overexpression of cloned genes, blockage of pathways for channeling precursors for the synthesis of desired product etc. Hairy roots can also act as an alternative expression system for production of antibodies, valuable animal proteins and products for pharmaceutical importance. Hairy roots grown in vitro are suitable model for studies of heavy metal uptake and genetic manipulations to enhance bioaccumulation. They also offer a means to simplify the study of complicated plant microbe interaction.

Keywords : *Agrobacterium rhizogenes,* bioreactor, genetic manipulation, hairy roots, metabolic engineering, phytoremediation, secondary metabolites, transgenics.

I. INTRODUCTION

Plants are the potential source for discovery of new products and fine chemicals for drug development. Compared to the field production of plants, *in vitro* production of elite and valuable secondary metabolites by cultured plant cells is considered to be a suitable standardized alternative. Despite considerable efforts, a few commercial processes have been achieved using cell cultures (*e.g.* shikonin, berberine, gensenosides, anthroquinones etc.). The major constraint with cell cultures is that they are genetically unstable and cultured cells tend to produce lower yield of secondary metabolites. *Agrobacterium rhizogenes* a gram-negative soil bacterium, is a natural plant pathogen responsible for adventitious root formation (hairy roots) at the site of infection (Hooykaas, 1988). This hairy root phenotype is due to the integration and expression of T-DNA of Ri plasmid in the plant cell genome. Genetically transformed hairy roots produced by infection of plants with *A. rhizongenes,* offers a promising system for secondary metabolite production. Hairy roots are unique for their genetic and biosynthetic stability. Besides, their fast growth in hormone free media offers an additional advantage promoting its exploitation in plant biotechnology. These fast growing hairy roots can be used as a continuous source for the production of valuable secondary metabolites. Moreover, transformed roots are able to regenerate whole viable plants and maintain their genetic stability during root culture and plant regeneration. It has generally been accepted that a single hairy root line that arises from the explant is a clone (Chilton *et al.*, 1982). However, a wide range of somaclonal variation between these clones is also evident. Altered phenotype of hairy root regenerants proves to be useful for plant breeding programs. *In vitro* transformation and regeneration shows a great potential for manipulating tree species in terms of clonal propagation both at cellular and molecular levels. The technology of transformed root culture provides the advantage of a built in genetic vector for introducing foreign DNA which proves to be beneficial for genetic manipulations. As a heterologus antibody expression system hairy roots are capable of production of assembled antibody proteins. Hairy roots also offer a means to simplify the study of root-organism interaction and phytoremediation. Recent developments in plant metabolic engineering, and the transgenic hairy root system can be exploited for production and yield enhancement of valuable secondary metabolites.

2. *AGROBACTERIUM RHIZOGENES*

Agrobacteria are gram negative rod shaped bacteria that are abundant in soil. They were classified as *Agrobacterium tumefaciens* (Crown gall causing), *A. rhizogenes* (hairy roots causing), *A. rubi* (causing cane galls) and *A. radiobacter* (the non-pathogenic *Agrobacteria*). *Agrobacterium rhizogenes,* a natural plant pathogen was identified as the aetiological agent for the neoplastic outgrowth of fine roots symbolyzing the symptoms of "Hairy root disease". Subsequently it was conclusively proved that *A. rhizogenes* is responsible for adventitious root formation at the site of infection (Hooykaas, 1988). *Agrobacterium* possess a complex genetic system comprising of main chromosomal genes, virulence genes and T-DNA genes.

2.1. Chromosomal virulence genes

Eleven chromosomal virulence genes have so far been identified and the function of these genes have also been studied. *Psc A, Chv A* and *Chv B* are responsible for biosynthesis, modification and export of specific extracellular polysaccharide necessary in the plant microbe attachment process. *Chv D, Chv E, mia A* and *ros* products provide regulation of vir expression. *Chv G* and *Chv I* provide an additional two-component system required for virulence and *Acv B* participates in the T-complex formation.

2.2. Vir Region

In *vir* region at least seven transcriptional units have been identified. Some loci comprise polycistronic operons and four of them (*vir A, vir B, vir D* and *vir G*) are essential for hairy root formation. Three loci (*vir C, vir E* and *vir F*) are important for some plants and thus determine host specificity (Zambryski, 1988). Two component regulatory system of *vir A* and *vir G* is constitutively expressed. Although *vir G* together with *vir B, vir C, vir D* and *vir E* is also inducible by specific low molecular weight compounds released from wounded plant cells such as acetosyringone and hydroxy acetosyringone or with a similar efficiency, by a dilute extract of plant phenolics (Stachel *et al.*, 1985; Bolton *et al.*, 1986). The induction takes upto 16 h to reach a maximum within which time two polypeptides (*vir D1* and *vir D2*) coded by *vir D* locus are induced. They are endonucleases with specific affinity for 25 bp direct imperfect repeats confining the borders flanking the T-DNA region. The endonucleases introduce nicks at the site of the 25 bp repeats which leads to the polar synthesis of single stranded free DNA molecule, T-strands (Stachel *et al.*, 1986; Veluthumbi *et al.*, 1988). Eleven proteins probably coded by the *vir B* locus, play an active role in mediating the transfer of T-strand molecule into plant cells, perhaps by directing and catalyzing DNA transfer (Thompsom *et al.*, 1988). Virulence of *A. rhizogenes* is coded by Ri or root inducing plasmids. Vir A detects phenolic compounds and acts as a signal sensor and transmits extracelluar signal to intracellular *vir G* gene, leading to its expression at increased level. Vir G protein acts as an inducer for rest of the *vir* genes. Vir C interacts with overdrive for T region transfer. On the other hand Vir D1 and D2 produce single strand cleavage between the third and fourth nucleotide in the bottom strand of each border, Vir D2 remains closely associated. Vir D4 mediates interaction between Vir B complex and T-complex. *vir E* encodes single stranded nucleic acid binding proteins. *vir B* codes for proteins necessary for forming a membrane associated apparatus. It produces approximately 24 proteins, which help in transfer of T-DNA.

2.3. Ri T-DNA genes

The hairy root phenotype is due to the integration and expression of T-DNA of Ri plasmid in the plant cell genome. Total genomic constitution of Ri-plasmid of *A. rhizogenes* is illustrated in Fig. 1. The rhizogenic strains contain a single copy of a large Ri plasmid. In Agropine Ri plasmid T-DNA is referred to as left T-DNA (TL-DNA) and right T-DNA (TR-DNA). TR-DNA contains genes homologous to Ti plasmid tumor inducing genes, genes involved in agropine synthesis are also located in the TR-DNA regions. T-DNA is transferred to wounded plant cells and stably integrates to its genome. Genes encoded in T-DNA are of bacterial origin but have eukaryotic regulatory sequences enabling their expression in infected plant cells. Synthesis of auxins can be ascribed to the TR-DNA. However, even in the absence of TR-DNA directed auxin synthesis as in mannopine type which lack *tms* loci, root induction occurs. Genes of Ri TL-DNA direct the synthesis of a substance that

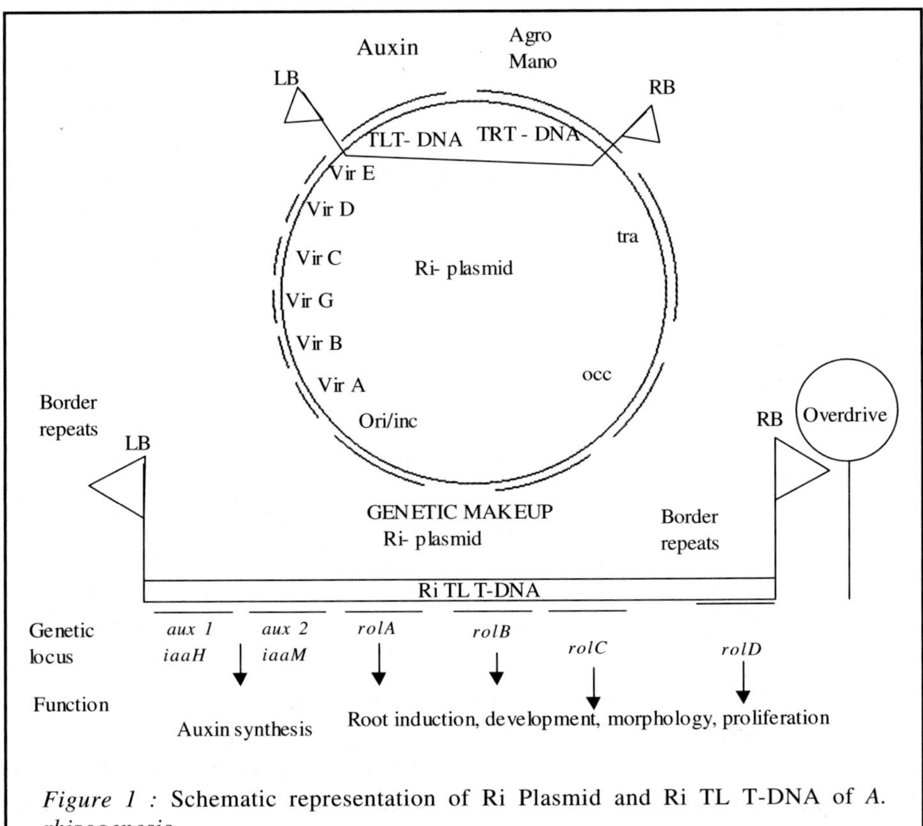

Figure 1 : Schematic representation of Ri Plasmid and Ri TL T-DNA of *A. rhizogenesis*

recruits the cells to differentiate into roots under the influence of endogenous auxin synthesis (Ooms *et al.*, 1986b; Shen *et al.*, 1988).

2.4. Transfer and integration of Ri-T DNA

Agrobacterium recognizes some signal molecules exuded by susceptible wounded plant and becomes attached to it (chemotactic response). Vir region is induced to express genes necessary for T-DNA transfer. It has been demonstrated that with the exception of border sequences, none of the other T-DNA sequences are required for the transfer. Transfer of T-DNA is mediated by virulence genes that form *vir* region of Ri plasmid and *chv* genes found on bacterial chromosomes. Transcription of *vir* region is induced by various phenolic compounds released by wounded plant such as acetosyringone and α-hydroxy-actosyringone. Recalcitrant plant species for transformation can be transformed by inducing the *vir* genes of the bacteria by signal molecules or it can be achieved *in vitro* by co-cultivating *Agrobacterium* with wounded tissues or in media that contains signal molecules (Stachel *et al.*, 1986).

Expression of T-DNA genes upon introduction and integration into plant nuclear genome considerably affects plant cell biology. First T-DNA encodes several proteins influencing cell growth of which the most notable are enzymes enhancing endogenous auxins, secondly it encodes genes for opine synthesis.

Medium has a significant effect on root formation, high salt medium like LS (Linsmaier and Skoog, 1965), MS (Murashige and Skoog, 1962) favour hairy root formation in some plants. Low salt containing medium like B5 (Gamborg *et al.*, 1976) favours bacterial multiplication on the medium, so that explant needs to be transferred several times to the fresh antibiotic containing medium before incubation as it inhibits hairy root formation. Acetosyringone or related compounds have been reported to increase *Agrobacterium*-mediated transformation frequencies in a number of plant species *e.g. Salvia, Antirrhinum majus, Allium cepa, Brassica campestris, Glycine max* and *Lycopersicon esculentum* (Hu and Alfernan, 1993; Vanhala *et al.*, 1995). Various sugars also act synergistically with acetosyringone to induce high levels of *vir* gene expression (Delmotte *et al.*, 1991). Bacterial concentration also plays an important role for the production of transformed roots, concentration below optimum may result in lower availability of bacteria for transforming plant cells while supra optimal concentrations may decrease it by competitive inhibition (Kumar *et al.*, 1991). More than 450 species of many different genera and families are known to be susceptible to *A. rhizogenes* infection (Porter, 1991). Hairy roots are fast growing and plagiotropic and require no external supply of growth hormones, this plagiotropic characteristic proves to be advantageous as it increases the aeration in liquid medium. Since roots grow in air resulting in an elevated accumulation of biomass.

3. SECONDARY METABOLITE PRODUCTION

3.1. Use of hairy root system for production secondary metabolites

Hairy root cultures are characterized by a high growth rate, phytohormone independence and their ability to synthesize higher amounts of root derived secondary metabolites, as compared to the phytohormone dependence, slow growth and low levels of secondary metabolites of control root cultures. However, use of hairy root cultures has added a new dimension to the role of plant tissue culture for secondary metabolite production (Giri and Lakshmi Narasu, 2000). These hairy roots are unique in their genetic and biosynthetic stability. Their fast growth, low doubling time, easy maintenance and ability to synthesize similar range of secondary metabolites offer an additional advantage as a continuous source for the production of valuable secondary metabolites. Hairy roots are a valuable source of root derived phytochemicals that are useful as pharmaceuticals, cosmetics, and food additives. More than one type of product can be obtained in these transformed root cultures thereby enhancing the commercial prospects of the process. Establishment of transformed root cultures has proved to be critical for the production of secondary metabolites. Transformed roots of many medicinal plants have been widely studied for the *in vitro* production of secondary metabolites as depicted in Table 1 (Hamill *et al.*, 1986; Mano *et al.*, 1986; Signs and Flores, 1990; Bengamin *et al.*, 1993; Bajaj and Ishimaru, 1999). Transformed root lines can provide a source for a constant and standardized production of secondary metabolites. Hairy root cultures produce secondary metabolites over successive generations with out losing its genetic and biosynthetic stability. This property can be exploited by genetic manipulations to increase biosynthetic capacity.

Some advantages and limitations of transgenic hairy root cultures are listed below.

Table 1. Applications of transgenic hairy roots for secondary metabolite production

Plant	Secondary metabolite	Reference
Aconitum heterophyllum	Aconites	Giri *et al*., 1997
Ajuga reptans var. atroparpurea	Phytoecdysteroids	Matsumoto and Tanaka, 1991
Ambrosia sps.	Polyacetylenes and Thiophenes	Flores *et al*., 1987b
Amsonia elliptica	Indole alkaloids	Sauerwein *et al*., 1991a
Anisodus luridus	Tropane alkaloids	Jobanovic *et al*., 1991
Armoracia laphthifolia	Peroxidase, Isoperoxidase, Fusicoccin	Taya *et al*., 1989b; Saito *et al*., 1991; Babakove *et al*., 1995
Artemisia absinthum	Essential oils	Nin *et al*., 1997
Artemisia annua	Artemisinin	Jaziri *et al*., 1995; Teoh *et al*., 1996; Weathers *et al*., 1994; Rao *et al*., 1998
Astragalus mongholicus	Cycloartane Saponin	Ionkava *et al*., 1997
Atropa belladonna	Atropine	Christen *et al*., 1989; Lee *et al*., 1998
Azadirachta indica A. Juss.	Azadirachtin	Allan *et al*., 1999
Beta vulgaris	Betalain pigments	Hamill *et al*., 1986; Taya *et al*., 1992
Bidens sps.	Polyacetylenes and Thiophenes	Flores *et al*., 1987b
Brugmansia candida	Tropane alkaloids	Giulietti *et al*., 1993; Sandra *et al*., 1999
Calystegia sepium	Cuscohygrine	Jung and Tefper, 1987; Bhadra *et al*., 1993

Table 1. Continued

Plant	Secondary metabolite	Reference
Campanula medium	Polyacetylenes	Tada *et al.*, 1996
Carthamus	Thiophenes	Flores *et al.*, 1987b
Cassia obtusifolia	Anthraquinone, Polypeptide pigments	Asamizu *et al.*, 1988; Ko *et al.*, 1995
Catharanthus roseus	Indole alkaloids, Ajmalicine	Parr *et al.*, 1988; Toivonen *et al.*, 1989, 1990
Catharanthus tricophyllus	Indole alkaloids	Davioiud *et al.*, 1989
Centranthus ruber	Valepotriates	Granicher *et al.*, 1995b; Cristen, 1999
Chaenatis douglasis	Thiarubrins	Constable and Towers, 1988
Cinchona ledgeriana	Quinine	Hamill *et al.*, 1989
Coleus forskohlii	Forskolin	Sasaki *et al.*, 1998
Coluria geoides	Eugenol	Olszowska *et al.*, 1996
Coreopsis	Polyacetylene	Marchant, 1988
Datura candida	Scopolamine , Hyoscyamine	Christen *et al.*, 1989
Datura stramonium	Hyoscyamine , Sesquiterpene	Payne *et al.*, 1987; Furze *et al.*, 1991
Daucus carota	Flavonoids, Anthocyanin	Bel-rhlid *et al.*, 1993; Kim *et al.*, 1994

Table 1. Continued

Plant	Secondary metabolite	Reference
Digitalis purpurea	Cardioactive glycosides	Saito *et al.*, 1990
Duboisia myoporoides	Scopolamine	Deno *et al.*, 1987b
Duboisia leichhardtii	Scopolamine	Muranaka *et al.*, 1992
Echinacea purpurea	Alkamides	Mukundan and Hjortso, 1990; Trypsteen *et al.*, 1991
Fagra zanthoxyloides Lam.	Benzophenanthridine, Furoquinoline alanine	Couellerot *et al.*, 1999
Fagopyrum	Flavanol	Trotin *et al.*, 1993
Fragaria	Polyphenol	Motomori *et al.*, 1995
Geranium thubergee	Tannins	Ishimaru and Shimomura, 1991
Glycyrrhiza glabra	Flavonoids	Asada *et al.*, 1998
Gynostemma pentaphyllum	Saponin	Fei *et al.*, 1993
Hyoscyamus albus	Tropane alkaloids, Phytoalexins	Sauerwein *et al.*, 1992; Kuroyanagi *et al.*, 1998
Hyoscyamus muticus	Tropane alkaloids, Hyoscyamine, Proline	Vanhala *et al.*, 1995; Sevon *et al.*, 1997; Halperin and Flores, 1997
Hyoscyamus niger	Hyoscyamine	Jaziri *et al.*, 1988
Hyssopus officinalis	Rosmarinic acid and related phenolics	Murakami *et al.*, 1998

Table 1. Continued

Plant	Secondary metabolite	Reference
Lactuca virosa	Sesquiterpene lactones	Kisiel *et al.*, 1995
Lawsonia inermis	Lawsone	Bakkali *et al.*, 1997
Leontopodium alpinum	Anthocyanins and Essential oil	Hook, 1994
Linum flavum	Lignans (5–methoxy podophyllotoxins)	Oostdam *et al.*, 1993
Lippia dulcis	Sesquiterpenes (hernandulcin)	Sauerwein, 1991a
Lithospermum erythrorhizon	Shikonin Benzoquinone	Shimomura *et al.*, 1991b; Fukui *et al.*, 1998
Lobelia cardinalis	Polyacetylene glucosides	Yamanaka *et al.*, 1996
Lobelia inflata	Lobeline, Polyacetylene	Yonemitsu *et al.*, 1990
Lotus corniculatus	Condensed Tannins	Carron *et al.*, 1994
Nicotiana hesperis	Nicotine, Anatabine	Parr and Hamill, 1987
Nicotiana rustica	Nicotine, Anatabine	Hamill *et al.*, 1986
Nicotiana tabacum	Nicotine, Anatabine	Flores and Filner, 1985
Panax ginseng	Saponins	Yoshikawa and Furuya, 1987; Yoshimatsu *et al.*, 1996
Panax hybrid (P. ginseng × P. quinqifolium)	Ginsenosides	Washida *et al.*, 1998

Table 1. Continued

Plant	Secondary metabolite	Reference
Papaver somniferum	Codein	Yoshimatsu and Shimomura, 1992; Willium and Ellis, 1993
Paulownia elongata		Ben *et al.*, 1998
Perezia cuernavcana	Sesquiterpene quinone	Arellano *et al.*, 1996
Pimpinella anisum	Essential oils	Santos *et al.*, 1998
Platycodon grandiflorum	Polyacetylene glkucosides	Tada *et al.*, 1995b; Ahn *et al.*, 1996
Psoralea sps.	Flavonoids	Bougaud *et al.*, 1999
Rauwolfia serpentina	Reserpine	Sato *et al.*, 1991; Benjamine *et al.*, 1994
Rubia peregrina	Anthraquinones	Lodhi *et al.*, 1996
Rubia tinctorum	Anthroquinone	Sato *et al.*, 1991
Rudbeckia sps.	Polyacetylenes and Thiophenes	Flores *et al.*, 1987b
Salvia miltiorhiza	Diterpenoid	Hu and Alfermann, 1993
Scoparia	Methoxybenzoxazolin-on	Hayashi *et al.*, 1994
Scopolia japonica	Hyoscyamine	Mano *et al.*, 1987
Scutellaria baicalensis	Flavonoids and phenylethnoids	Zhou *et al.*, 1997
Serratula tinctoria	Ecdysteroid	Delbecque *et al.*, 1995

Table 1. Continued

Plant	Secondary metabolite	Reference
Sesamum indicum	Naphthoquinone	Ogasawara *et al.*, 1993
Solanum aculeatissi	Steroidal saponin	Ikenaga *et al.*, 1995
Solanum aviculare	Steroidal alkaloids	Yu *et al.*, 1996; Kittipongpatana *et al.*, 1998
Solanum lacinialum	Steroidal alkaloids	Hamill *et al.*, 1987
Swainsona galegifolia	Swainsonine	Ermayanti *et al.*, 1994
Swertia japonica	Xanthons	Ishimaru *et al.*, 1990
Tagetus erecta	Xanthons	Shin and Hjortso, 1998
Tagetus patula	Thiophenes	Flores *et al.*, 1987b; Arroo *et al.*, 1995
Tanacetum parthenium	Sesquiterpene coumarin ether	Kisiel and Stojakawashs, 1997
Tricosanthes kirilowii maxim var japonicum	Defense related proteins	Savary and Flores, 1994
Trigonella foenum graecum	Diosgenin	Merkli *et al.*, 1997
Valeriana officinalis L.	Valepotriates	Granicher *et al.*, 1994
Vinca minor	Indole alkaloids (vincamine)	Tanaka *et al.*, 1994
Withania somnifera	Withanoloides	Banerjee *et al.*, 1994

3.2. Advantages

1. Genetic stability with predictable growth pattern and product yield
2. Production coupled with growth
3. Simple growth medium without growth regulators
4. Generally no light requirement
5. Secretion of product may occur naturally or can be induced.
6. Inoculum density not very crucial
7. Naturally immobilized system for biotransformation
8. No change in medium viscosity

3.3. Limitations

1. Linear and branched network of root biomass is non-pumpable
2. Amenable for batch cultivation
3. Long growth/production periods
4. High and uneven absoption surface leading to disturbed mass transfer
5. Sensitive to oxygen stress
6. Not amenable to biomass sampling
7. Formation of callus due to hairy root rupture

It has been generalized that a single hairy root that arises from the explant tissue is a clone. A wide range of somaclonal variations between these clones is also evident. Different strains of *Agrobacterium rhizogenes* vary in their transforming ability.

Transformed green roots have been obtained in a few species belonging to Asteraceae, Solanaceae, Cucurbitaceae (Sauerwein *et al.*, 1992). Green hairy roots are known to produce certain metabolites that are normally synthesized in green parts of the plant (Saito *et al.*, 1992). Chloroplast dependent reactions are a vital part of certain metabolic pathways and could result in a novel pattern of compounds produced by roots. This aspect has been studied recently using Soybean hairy roots by functional analysis of the toabcco Rubisco large subunit AN-methyl transferase promoter and its light controlled reactions (Mazarei *et al.*, 1998). The site of accumulation in plants may not be the site of synthesis. Certain metabolites that are normally produced using shoot cultures or whole plants may be produced in hairy roots. Accumulation of lowsone, a naphthoquinone derivative is restricted to aerial parts and not in roots of wild type henna, however, lawsone is found in significant quantities in hairy root cultures (Bakkali *et al.*, 1997). Hairy roots produced by different bacterial strains may vary in their morphology. Differences in virulence and morphology could be explained by the plasmids harbored by the strains (Nguyen *et al.*, 1992). To obtain a high-density culture of roots, the culture conditions should be maintained at the optimum level. Hairy root cultures follow a definite pattern, however, the metabolite production may not be growth related. Secondary metabolite biosynthesis in transformed roots is genetically controlled but it is influenced by nutritional and environmental factors. Composition of culture medium affects growth and production. Sucrose level, exogenous growth hormone, nature of nitrogen source and their relative amount, light, temperature and presence of chemicals affect growth and total biomass yield of the culture and secondary metabolite production (Smith *et al.*, 1997). Sucrose is the best source of carbon and is hydrolyzed into glucose and fructose by plant cells by assimilation. Their rate of uptake varies in different plant cells (Srinivasan *et al.*, 1995). In hairy roots the source of new cells are in the tips so proliferation occurs only at the apical meristem and laterals form behind the elongation zone. Such a defined growth pattern leads to steady accumulation of biomass root cultures. To obtain a high-density culture of roots, the culture conditions should be maintained at the optimum level. Hairy root cultures are able to synthesize significant amounts of alkaloids but the desired compounds are poorly released into the medium and their accumulation in the roots can be limited by feedback inhibition. Growth and productivity of hairy roots can be improved by optimization of inoculam age and length of subculture cycle as some secondary metabolites are growth associated (Weathers *et al.*, 1997). Addition of phytohormones, elicitors and precursor feeding can also result in increased levels of secondary metabolites (Rjhwani and Shanks, 1998; Sim *et al.*, 1994). Media manipulations are reported to aid in release of metabolites. Release varies from one clone to another. Betacyanin release from hairy roots of *Beta vulgaris* was achieved by oxygen starvation. Short exposure of hairy roots in low pH media can be utilized to facilitate product release into the culture medium (Shanks and Morga, 1999). Permeabilization treatment using Tween 20 (Polyoxy ethylene sorbilane monolaurate) released high yield of hyoscyamine from roots of *Datura innoxia* without any detrimental effects (Boitel *et al.*, 1996). Addition of X A D-2, liquid paraffin stimulated production of shikonin (Shimomura *et al.*, 1991b). Lee *et al.* (1998) reported that treatment with 5 mM H_2O_2 induced a transient release of tropane alkaloids from transformed roots without affecting its viability.

3.4. Scaling-up of hairy roots

Hairy roots once established can be grown in a medium with low inoculum and high growth rate. Optimization of inoculum age and subculture cycle can improve growth rate as some secondary metabolites are growth related (Rijhwani and Shank, 1998). Productivity of valuable metabolites can be improved by manipulating the extracellular environment (Bhadra and Shank, 1997). The main constraint for commercial exploitation of hairy root cultures is the scaling up at industrial level. Hairy roots are complicated biocatalysts when it comes to scaling up. Root cultures present unique challenges, mechanical agitation causes wounding and leads to callus formation. With a product of sufficiently high value it is feasible to use batch fermentation, harvest the roots and extract the product. For less expansive products it may be desirable to establish a packed bed of roots and to run the fermenter in a continuous mode for extended periods collecting the product from the effluent stream. The scale up becomes complicated due to limitations in providing nutrients from both liquid and gas phases simultaneously. Meristem dependent growth of root cultures in liquid medium results in a root ball with young growing roots on the periphery and a core of older tissue inside. Restriction of nutrients and oxygen delivery to the central mass of tissue gives rise to a pocket of senescent tissue. Due to branching, the roots form an interlocked matrix that exhibits a resistance to flow. The main problem with hairy roots is supply of oxygen. The ability to exploit hairy root culture as a source of bioactive chemicals depends on development of suitable bioreactor system. It must fulfill the conditions like sterility, pH, temperature, dissolved oxygen etc. For large-scale production in bioreactors several physical and chemical parameters are to be taken into consideration.

Nutrient supply is the major chemical factor involved in scaling up. For large-scale cultivation in a bioreactor several aspects play an important role. Periodical examinations of specific nutrients at different periods provide information regarding biomass and alkaloids production in bioreactors. Carbon, Nitrogen, Oxygen and Hydrogen depletion in the medium with regard to biomass and alkaloid production has been studied in *Atropa belladonna* by Kwok and Doran (1995). This type of studies can be done for plant sps. where product leaches into the medium and can be recovered by adsorbents. The medium can be rejuvenated to maintain the supply of nutrients. By leaching of secondary metabolite synthesized by hairy roots the uptake of nutrients gets altered so leachets need to be removed regularly. Leaching of phenolics by hairy roots and their oxidation leads to inhibition of uptake of other nutrients which can be avoided by passing the spent medium through adsorbents or metabolite traps.

Mass transfer is also an important factor that influences the uptake of nutrients by hairy root cultures. The availability of water and nutrients to any region of hairy root network in a bioreactor at different periods is known as mass transfer. Hairy root bioreactor chambers become more heterogeneous owing to continuous growth of culture. Oxygen is the most important chemical that needs to be supplied continuously to a bioreactor, judicious mixing leads to efficient O_2 transfer. During the initial stages of growth in a bioreactor O_2 transfer is not difficult as the medium contains enough dissolved O_2 to support low inoculum. Mixing is a very important factor because it serves the dual purpose of supplying dissolved O_2 and driving away carbon dioxide. The rate of uptake of O_2 by a unit biomass in a unit time is known as O_2 transfer co-efficient. Other dissolved gaseous metabolites namely carbon dioxide and ethylene also affect the overall productivity. A high biomass transfer resistance by hairy roots will result in development of stagnant zones and non-uniform gaseous metabolite concentration. Sampling of gases can be done by studying the inlet and exit gases by passing through rotameter and then to a mass spectrophotometer interphased to a p c. Few attempts have been made for scaling up hairy root cultures for secondary metabolite production. Several bioreactor designs have been reported taking into consideration their complicated morphology and shear sensitivity.

These features call for a specially designed bioreactor that permit the growth of interconnected tissue unevenly distributed throughout the culture vessel. While designing a bioreactor it has to be taken into consideration that roots need to be fixed on some form of support, the root mass can restrict the flow by lodging in some part of the bioreactor. Moreover, for optimal biomass mode of exploitation in a bioreactor, the product must be a part released from the roots and must be able to be maintained in a high density packed beds without loss of viability.

Several bioreactor designs have been developed in the recent times for the mass scale cultivation of transgenic hairy root cultures. The models of bioreactor are depicted in Fig. 2 (A-I).

- **Airlift or Submerged bioreactors**

Basic configuration of airlift or submerged bioreactors are similar to stirred tank reactors but lack impeller. Plants cells have large vacuoles and slow growth. So similarly hairy roots require comparatively low O_2 supply of about 0.05-0.4 vol. of air/vol. of liquid/min. Humidified air is passed through glass grid that functions as aerators. These are found successful for hairy roots (Kwok and Doran, 1995; Tescione *et al.*, 1997) (Fig. 2 A,B).

- **Modified Stirred tank reactor**

The impeller or turbine blades facilitate mass transfer. These are not suitable for hairy root cultures because of damage that they cause resulting in callus formation because of shear stress due to impeller rotation. (Taya *et al.*, 1989a; Hilton and Rhodes, 1990). However recently some modified stirred tank bioreactors have been developed. This has large impellers and battles, this is agitated at a very low speed (Fig 2 C,D).

- **Turbine blade reactor**

This is combination of airlift/stirred tank reactor. Here cultivation space is separated form agitation space by stainless steel mesh (Fig. 2 D). So that hairy roots do not come in contact with impeller and the air is introduced from the bottom and dispersed by 8-blade impeller that stirs the medium. This is efficient for hairy roots (Kondo *et al.*, 1989).

- **Bubble column reactor**

This reactor is a modified form of airlift bioreactors. Air bubbles create very less shear stress, so it can be useful for organized structures like hairy roots. Bubbling rate needs to be gradually increased with the growth of hairy roots. Division of bubble column into segments and installation of multiple spargers or intermediate seive plates increase the mass transfer (Buitellar *et al.*, 1991) (Fig. 2 E).

- **Gas sparged bioreactor**

In this reactor humidified air is introduced from the bottom of the reactor through a sintered glass sparger. This is useful for mixing and

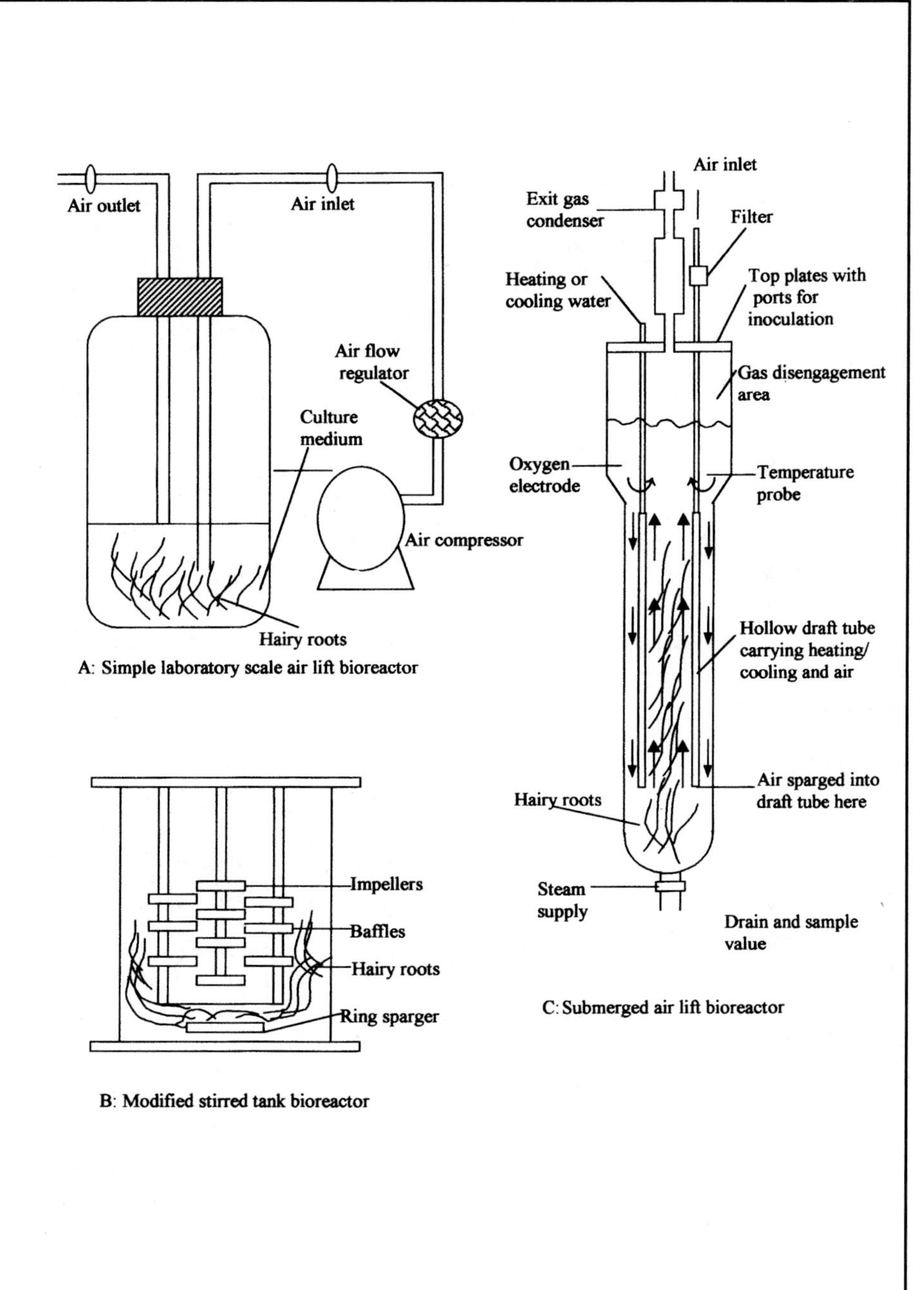

Figure 2 : (A-C) Schematic diagram of different models of bioreactors for culture of transgenic hairy root cultures.

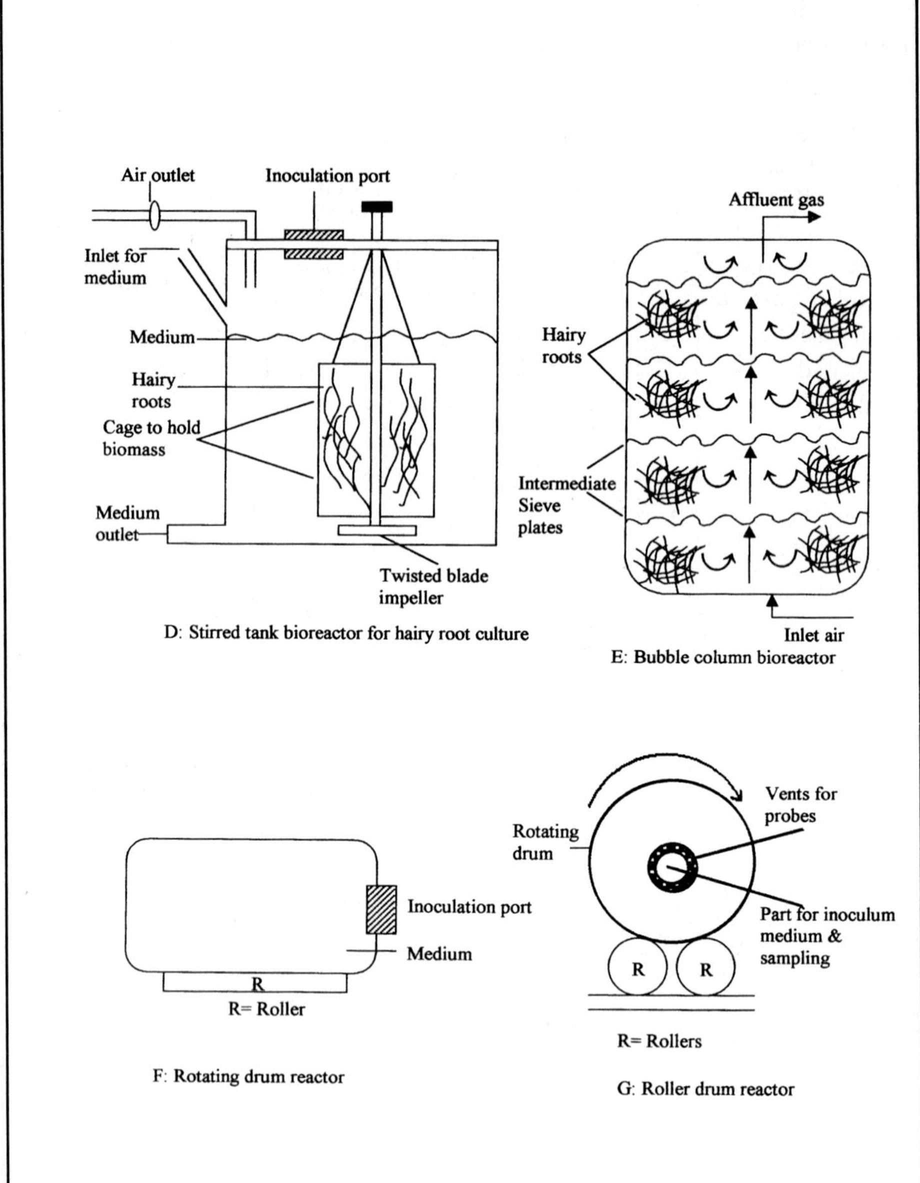

Figure 2 : (D-G) Schematic diagram of different models of bioreactors for culture of transgenic hairy root cultures.

oxygenation of medium needed for mass culture of hairy roots. The Model needs further modifications for its wider applications for scaling up hairy root cultures.

- **Mist or droplet bioreactor (Trickle bed reactor)**

Here the medium is trickled over a whatman paper where biomass is kept and spent medium is drained from the bottom of the bioreactor to a reservoir and recirculated at a specific rate. The degree of distribution of liquid varies according to the mechanism of liquid delivery at the top of the reactor chamber. For better dispersion, spraying is done by mixing humidifid air that creats the mist (Dilorio *et al.*, 1992: Whitney *et al.*, 1992). The knowhow used in the droplet or mist bioreactor is essentially to supply nutrients through a droplet or misting dispersion system from the top of the reactor (Fig. 2 F). The mist will further flow over the hairy root cultures thus providing nutrients and maintaining a mist environment in the vessel. Below the vessel there is a balancing device which would measure the biomass production and the spent medium. The spent medium will further become part of recycle nutrient feed system.

- **Rotating drum bioreactor**

This consists of a drum shaped container mounted on rollers for support and rotation. Rotating speed of the drum is kept as low as 2-6 rpm to exert minimal shear pressure on the hairy roots. Kondo *et al.* (1989) used it initially for the hairy roots of carrot. They observed that the roots adhered to the walls of the reactor during rotation leading to breakage. To overcome this, polyurethane foam sheet was fixed onto the inner surface of the drum, to which the hairy roots attached and showed higher growth without detachment (Fig. 2 H,I).

In a gas sparged reactor oxygen is delivered by local transfer from gas bubbles that rise through the reactor and the inoculum gets distributed evenly in the vessel and circulates. In addition to free roots cultivated in stirred tank reactor and an air lift column, the growth of hairy roots was also tested after immobilization in polyurethane foam. Buitelaar *et al.* (1989) tested three fermenter types for growth and thiophene production by *Tagetets patula* hairy roots and found the best productivity with a bubble column. Shimomura *et al.* (1991) used airlift reactor connected to a column containing a polymorphic adsorbent for continuous production of shikonin by hairy root cultures of *Lithospermum erythrorhizon*. Yoshikava and Furuya (1987) used airlift type of reactor

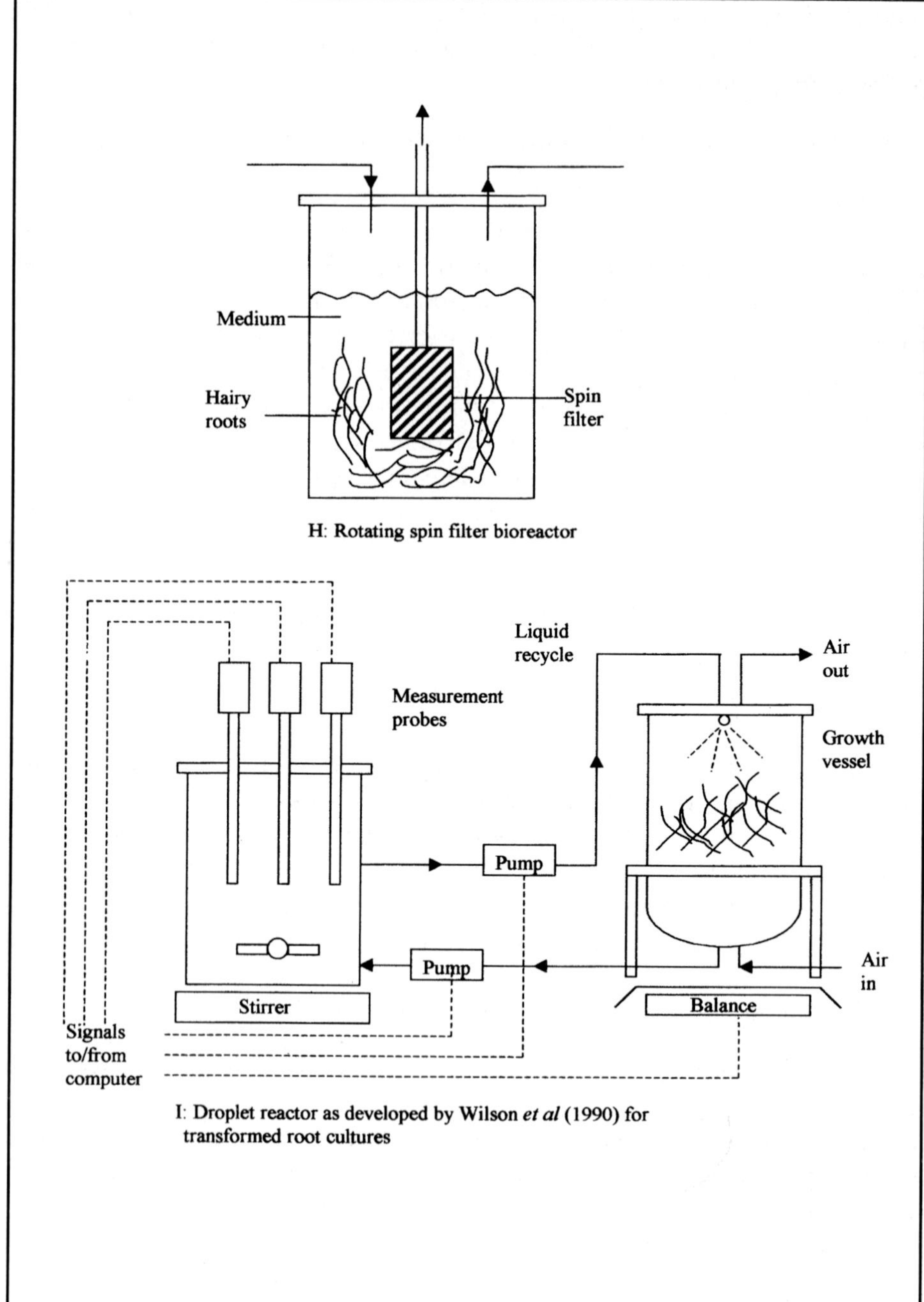

Figure 2 : (H-I) Schematic diagram of different models of bioreactors for culture of transgenic hairy root cultures.

successfully in *Panax ginseng* hairy roots and for hernandulin production in *Lippia dulicis* (Sauerwin *et al.*, 1991a).

- **Spin filter bioreactor**

In this bioreactor the rotating filter mildly mixes the cultures and simultaneously allows spent medium to be removed by adsorption into the filter (besides facilitating addition of fresh medium). In this type of reactor the shear stress to the hairy roots is due to friction caused by the spin filter.

Basically roots do not require additional illumination but certain hairy roots produce higher levels of metabolites in presence of light. Bioreactors can be illuminated externally or internally.

Temperature also plays an important role, Yu *et al.* (1996) studied the effects of temperature on *Solanum aviculare* hairy roots and found 25°C optimal.

Root morphology is an parameter for scaling-up. Shear sensitivity is another parameter. Hairy root system is of special interest because their rheology changes continuously because of their indefinite proliferation. Their cell walls are relatively weak and rupture easily which makes them more sensitive towards shear stress. Asepsis is another parameter that plays a very important role and contamination should be completely checked, which can be achieved through system design, effective operating procedure, scheduled checks and maintenance (Dilorio *et al.*, 1992). Table 2 depicts different bioreactor types used for the growth and secondary metabolite production from hairy roots.

The successful operation of a bioreactor becomes complicated due to the above mentioned parameters and variables. Some additional engineering aspects are also involved in it. Computer aided models can help in the efficient product formation and recovery. Kim *et al.* (1995) developed hairy root models based on branching pattern that helps to monitor shear stress and stoichiometry. Albiol *et al.* (1995) used artificial neural network for plant cell cultures. It can be adopted for hairy roots. Wyslouzyl *et al.* (1997) found good agreement between experimental models and predicted values. For complex hairy root cultures modeling involves multiple factors such as rheology, O_2 consumption and product excretion. Padmanabhan *et al.* (1998) have done computer analysis of the somatic embryos for assessing their ability to be converted to the

Table 2. Advancements in bioreactor models for scaling up of hairy roots for secondary metabolite production

Bioreactor	Volume	Plant sps		Secondary metabolite	Reference
Air-sparged vessel	880 ml.	*Nicotiana rustica*		Nicotine	Rhodes *et al.*, 1987a
Stirred tank	300 ml	*Armoracia rusticana*			Taya *et al.*, 1989b
	1.0 L	*Atropa belladonna*		Tropane alkaloids	Jung and Tepfer, 1987b
		Atropa belladonna	Stainless steel net	Tropane alkaloids	Lee *et al.*, 1998
	1.0 L	*Calystegia sepium*		Tropane alkaloids	Jung and Tepfer, 1987b
Stirred tank with impeller isolated	1.0 L	*Atropa belladonna*	The impeller separated by a mesh from roots	Tropane alkaloids	Jung and Tepfer, 1987b
	1.0 L	*Calystegia sepium*		Tropane alkaloids	Jung and Tepfer, 1987b
	12.0L	*Datura stramonium*		Tropane alkaloids	Hilton and Rhodes, 1990
	1.0 L	*Duboisia leichhardtii*		Scopolamine	Muranaka *et al.*, 1992
Fermenter with mechanical stirring		*Catharanthus tricophyllus*		Indole alkaloids	Davioud *et al.*, 1989
Air lift	300 ml	*Armoracia rusticana*	Roots immobilized in reticulate polyurethane foam		Taya *et al.*, 1989b

Table 2. Continued

Bioreactor	Volume	Plant sps		Secondary metabolite	Reference
	9.0 L	*Trigonella foenum-graceum*	Draft tube	Diosgenin	Rodriguez-mendiole et al., 1991
	9.0 L	*Trigonella foenum-graceum*	Nylon mesh replacing draft tube	Diosgenin	Rodriguez-mendiole et al., 1991
		Panax ginseng		Saponins	Yoshikava and Furuya, 1987
		Lippia dulcis		Hernandulcin	Sauerweins *et al.*, 1991a
Concentrically arranged three sparged set–up was used to provide air bubbles	2.0 L	*Lithospermum erythrorhizon*	Reactor connected with a column containing polymeric adsorbent for continuous production of shikonin		Shimomura *et al.*, 1991
	15 L	*Solanum tuberosum*			Tesicone *et al.*, 1997
Bubble column	2.5 L	*Atropa belladonna*		Tropane alkaloids	Sharp and Doran, 1990
	1.0 L	*Catharanthus roseus*		Indole alkaloids	Toivonen *et al.*, 1990
	6.0 L	*Tagetes patula*		Thiophene	Buitelaar *et al.*, 1991

Table 2. Continued

Bioreactor	Volume	Plant sps	Secondary metabolite	Reference
	2.5 L	*Atropa belladonna*	Atropine	Kwok and Doran, 1995
Trickle bed nutrient mist rotating drum	2.0 L	*Hyoscyamus muticus*	Tropane alkaloids	Flores and Curtis, 1992
	1.4 L	*Beta vulgaris*	Betacyanins	Dilorio *et al.*, 1992
	1.0 L	*Daucus carota*	Anthocyanins	Kondo *et al.*, 1989
Nutrient mist reactor		*Artemisia annua*	Artemisinin	Wyslouzil *et al.*, 2000

plants. The same type of analysis for hairy roots may be beneficial for assessing the growth, genetic and biosynthetic stability.

3.5. Co-culture of hairy roots and shooty teratomas

In most cases the compounds formed by hairy roots are similar to those produced by roots of the parent plant. Few hairy root cultures have been found that are capable of producing significant levels of compounds synthesized in the aerial parts of the plants. For some secondary compounds, production in plants require participation of the roots and leaves. Organ specific enzymes in the root may produce a metabolic precursor, which is translocated to the aerial parts of the plant for conversion to another product by leaves or vice versa. If these enzymes involved in this process retain their organ specificity *in vitro* synthesis of final product will be difficult. Considering the complexity of secondary metabolic pathway and involvement of specific organelles like chloroplast in the biosynthesis, a solution to this problem is the root shoot co-culture using hairy roots and their genetically transformed shoot counterparts called shooty teratomas (Subroto *et al.*, 1996).

Genetically transformed shooty teratomas are produced by infection of plant material with specific strains of *Agrobacterium tumefaciens.* These can be grown in the absence of exogenous phytohormones. Intergeneric co-culture of genetically transformed hairy roots and shooty teratomas proves to be effective for improving tissue specific secondary metabolites. It resembles the whole plant by localized metabolite synthesis and translocation of compounds between organs for further bioconversion. Developments of transgenic organs make the co-culture feasible by sharing the common medium requirement without any hormone supplement. Subroto *et al.* (1996) reported significant improvement in the production of scopolamine by co-culture of *Atropa belladonna* hairy roots and *A. belladonna* and *Duboisia* hybrid shooty teratomas. The ratio of roots: shoots used to inoculate the co-culture had a pronounced influence on alkaloid synthesis. Mahagamasekera and Doran (1998) reported co-culture of transgenic hairy roots and shooty teratomas for production of scopolamine.

4. PLANT REGENERATION

Plants can be regenerated from transformed hairy root cultures with high efficiency. The plants regenerated from hairy are genetically stable. However, transgenic plants exhibit altered phenotype compared to controls. These hairy roots can be maintained as organ cultures for long time and subsequent shoot regeneration can be obtained without any cytological abnormality.

Rapid growth of hairy roots on hormone free medium and high plantlet regeneration frequency allows clonal propagation of elite plants. It has generally been accepted that a single hairy root line that arises from an explant is a clone (Chilton *et al.*, 1982). In *in vitro* cultures, the hairy root regenerants show rapid growth, increased lateral bud formation and rapid leaf development. These regenerants are useful for micropropagation of plants that are difficult to multiply.

4.1. Tree improvement

Other plants that need more attention to exploit the applications of hairy roots are trees. Major limitation with the tree improvement program is their long generation cycle. Classical breeding program in trees are slow and tedious because of their long generation cycle and difficulty in introducing specific genes for genetic manipulation by crossing parental lines. *Agrobacterium rhizogenes*-mediated transformation can prove to be a useful alternative, as a rapid and direct route for introduction and expression of specific traits. Transformation of tree sps. and subsequent regeneration of transgenic plants has been reported only for a few genera. The ability to manipulate tree species at cellular and molecular level shows great potential, *in vitro* transformation and regeneration from hairy roots facilitates application of biotechnology to tree sps. This significantly reduces the time necessary for tree improvement and gives rise to new gene combinations that cannot be obtained using traditional breeding methods. Fully transformed transgenic trees are obtained in many species as depicted in Table 3. In some tree sps. root initiation limits vegetative progapagation. Rooting of cuttings from recalcitrant woody sps. have been improved by using *A. rhizogenes*-mediated transformation. Roy (1989) demonstrated this for some fruit trees *viz.* peach, apple, cherry, olive etc. It has also been reported for *Pinus* and *Larix* sps. (McAffe *et al.*, 1993; Huang *et al.*, 1991). Rugini and Mariotti, (1991) demonstrated successful rooting of some tree sps. These methods have potential to increase the efficiency of plant progagation in crops where propagation is difficult. *A. rhizogenes*-mediated transformation has the potential to introduce foreign genes specifically into root systems *e.g.* resistance to pathogens or pests, resistance to heavy metals.

Altered phenotypes produced from hairy root regenerants are useful in plant breeding programs (Giovanini *et al.*, 1997). Morphological traits

Table 3. Use of *Agrobacterium rhizogenes* and transgenic hairy roots as gene delivery system

Plant	Gene introduced	Reference
Anthyllis vulneraria	*nptII, ipt*	Stiller *et al.*, 1992
Antirrhinum majus	*rol*	Hoshino and Mii, 1998
Artemisia annua⋆	Recombinant farnesyl diphosphate synthase gene	Hua Chen *et al.*, 1999
Atropa belladonna	*Bar, 6 β H*	Saito *et al.*, 1992; Hashimoto *et al.*, 1993
Brassica napus	*uidA, nptII, als*	Christey and Sinclair, 1992
B. napus	*nptII*	Boulter *et al.*, 1990
B. campestris	*uidA, nptII, als*	Christey and Sinclair, 1992
B.oleracea	*nptII, uidA*	Christey *et al.*, 1997
B.oleracea	*uidA, nptII, als*	Christey and Sinclair, 1992
B. campestris	*nptII*	Christey *et al.*, 1997
B. napus	GS	Down *et al.*, 1994
Brassica sps.	*nptII*, *Bt*, *uidA*, 35 S-EFE5 7 gene	Christey *et al.*, 1997
Citrus aurantifolia (Christm.) swing	Nos-*nptII*, Cab-*uidA*	Perez-Molphe-Balch and Ochoa-Alejo, 1998

Table 3. Continued

Plant	Gene introduced	Reference
Chinchona officinalis Ledgeriana*	Tryptophan decarboxylase and Strictosidine synthase from *Catharanthus roseus*	Geerling *et al.*, 1999
Crotalaria juncea L.	*rol*	Ohara *et al.*, 2000
Cucumis satives	*nptII*	Trulson *et al.*, 1986
Gladiolus sps.	*uidA*	Kamo and Blowers, 1999
Glycine argyrea	*nptII*	Kumar *et al.*, 1991
G. canescens	*nptII*	Rech *et al.*, 1989
Ipomoea batatus	*nptII, uidA*	Otani *et al.*, 1993
Larix decidua	*nptII, aroA, Bt*	Shin *et al.*, 1994
Lotus corniculatus	GS from *Phaseolus vulgaris*	Forde *et al.*, 1989
Lotus corniculatus	Dihydroflavanol reductase from *Antirrhinum*	Bavage *et al.*, 1997
Lycopersicon esculentum	*nptII*	Shahin *et al.*, 1986
L. peruvianum	*nptII*	Morgan *et al.*, 1987
Medicago truncatula	*nptII*	Thomas *et al.*, 1992

Table 3. Continued

Plant	Gene introduced	Reference
M. arborea	*hpt*	Damiani and Arcioni, 1991
Nicotiana debneyi	*nptII*	Davey *et al.*, 1987
Nicotiana plumaginifolia, N. tabacum	*nptII*	Hatamoto *et al.*, 1990
N. rustica	*ods*	Hamill *et al.*, 1990
Nicotiana sps.	*rol*	Palazon *et al.*, 1998
Panax ginseng		Yang and Choi, 2000
Peganum harmala	*tds*	Berlin *et al.*, 1993
Pepino sps.	*rol*	Atkinson and Gardner, 1991
Populus tricocarpa × *P. deltoides*	*nptII*	Pythoud *et al.*, 1987
Prunus avium × *P. seudocerasus*	*rol*	Gutirrez-Pesce *et al.*, 1998
Robinia pseudoacasia	*nptII*	Han *et al.*, 1993
*Rubia peregrina**	*ics*	Lodhi *et al.*, 1996

Table 3. Continued

Plant	Gene introduced	Reference
*Scutellaria baicalensis**	*uidA*	Nishikawa and Ishimaru, 1997
Solanum aviculare	*Hmgr* from *Artemisia annua*	Argolo *et al.*, 2000
S. nigrum	*nptII*	Davey *et al.*, 1987
S.tuberosum	*nptII, uidA*	Visser *et al.*, 1989b
S. dulcamara	*nptII, rol*	Mc Inns *et al.*, 1991
Stylosanthes humilis	*nptII*	Manners and Way, 1989
Verticordia grandis	*nptII, uidA*	Stummer *et al.*, 1995
Vinca minor	*nptII, uidA*	Tanaka *et al.*, 1994
Vitis vinifera	*nptII, uidA*	Nakano *et al.*, 1994

als : Acetolactate synthase, *aro* A :5-enolpyruvylshikimate –3- phosphate synthase, *bar* : Phosephinothicin acetyltransferase, 6βH : 6 β hydroxylase from *Hyoscyamus muticus,* BT : *Bacillus thuringiensis* protein, GS : Glutamine synthase from soybean, *uidA* : β-glucuronidase, *hpt* : Hygromycin phosphotransferase, *ipt* : Isopentenyl transferase, *nptII* : Neomycin phsphotransferase, *ods* : Ornithine decarboxylase from yeast, 35 S-*EFE5* 7 : Coding region of ethylene forming enzymes from tomato in antisense orientation, *rol* : Root loci genes, H* : Only upto hairy root stage with relevant ornamental value are abundant adventitious root formation, reduced apical dominance and altered leaf and flower morphology. Dwarfing, altered flowering wrinkled leaves or increased branching may be useful for enhancing the ornamental value. Dwarf phenotype is an important characteristic for breeding flower crops such as *Eustoma grandiflorum* and *Dianthus* (Giovanini *et al.*, 1997). Higher levels of some target metabolites have been found in the leaves of plants regenerated from hairy roots. So plant regeneration is an important aspect for production of these chemicals. Pellegrineschi *et al.* (1994) improved ornamental quality of scented *Pelargonium* sps., this plant has pleasant odour but its long internodes and ungainly growth makes it unattractive, hairy root regenerants are of shorter stature. In snapdragon, flower number was increased (Handa *et al.*, 1995). Some perennial forage legumes became annual (Damiani and Arcioni, 1991). However, a wide range of somaclonal variation between the so-called clones is evident. There is possibility of initial root line being chimeric. Thus increasing the variation among regenerated plants (Yukimure *et al.*, 1994).

By isolating protoplasts from hairy root cultures and regenerating plants, this variation can be studied in detail. Sevon *et al.* (1997) reported plant regeneration via protoplasts from *Agrobacterium rhizogenes* transformed hairy root cultures of *Hyoscyamus muticus*. They reported strong differences in phenotypic characters and secondary metabolite production in different clones and reported considerable somaclonal variation among the regenerants. Plants regenerated from Ri transformed roots displace "hairy root syndrome", combined expression of *rol* (A, B and C) loci Ri plasmid is responsible for this expression. Each locus is responsible for a typical phenotypic alternation : *rol A* is associated with internode shortening and leaf wrinkling; *rol B* is responsible for protruding stigmas and reduced length of stamens and *rol C* causes internode shortening and reduced apical dominance (Cardarlli *et al.*, 1987; Van Altovorst *et al.*, 1992).

4.2. Artificial seed production

Higher levels of desired secondary metabolites have been found in the leaves of some plants. Plant regeneration from hairy roots is necessary for production of these chemicals. Micropropagation of plants regenerated from hairy roots is a suitable alternative for production of these secondary metabolites. Artificial seed is prepared by encapsulating the tissue in polysaccharide gels, such as alginate. These form a reliable deliver system for clonal propagation of elite plants with the potential for genetic uniformity, high yield and low cost of production (Redenbaugh and Walker, 1990).

Hairy roots used for artificial seed production must have the ability to regenerate. The hairy roots used to produce artificial seeds can be classified into

1. Root fragments of hairy roots
2. Adventitious shoot primordia produced from hairy roots
3. Plantlets produced from hairy roots
4. Direct somatic embryo production from hairy roots

Schematic sketch has been depicted for the applications of synthetic seeds from hairy roots in Fig 3. The transgenic hairy roots provide multidimensional applications for clonal propagation, transport and conservation of valuable germplasm.

Micropropagation can be done from hairy roots using artificial seeds. In *Ajuga reptans* GUS transformed hairy roots were used for producing artificial seeds (Uozumi *et al.*, 1996). Uozumi and Kobayashi (1997) reported regeneration procedure for *Ajuga* hairy roots and its usefulness for artificial seed production at industrial level. In many plants hairy roots have been induced with properties superior to the original plant. Artificial seeds are strongly required for mass propagation of these elite hairy root lines. Artificial seed using hairy roots has further potential for mass propagation. Bioreactor design, image analysis with computers and robotics can improve the process. Root tips of hairy roots of *Panax ginseng* (Yoshimatsu *et al.*, 1996) and shoot tips of hairy root regenerants have been cryopreserved in horseradish (Nakshimada *et al.*, 1995; Phunchindawan *et al.*, 1997). It can be regenerated and cultured when needed. Hairy roots in the form of transformed plant organs provide promising means for biotechnological exploitation of plant cells. Plant cells used for artificial seed production must have the ability to regenerate.

4.3. Genetic manipulation

Genetic manipulation using *Agrobacterium*-mediated transformation is a good alternative to the conventional and existing methods of gene transfer. Plants can be regenerated from hairy root cultures either spontaneously (direct from roots) or by transferring roots to hormone containing medium. The advantage of Ri plasmid based gene transfer is

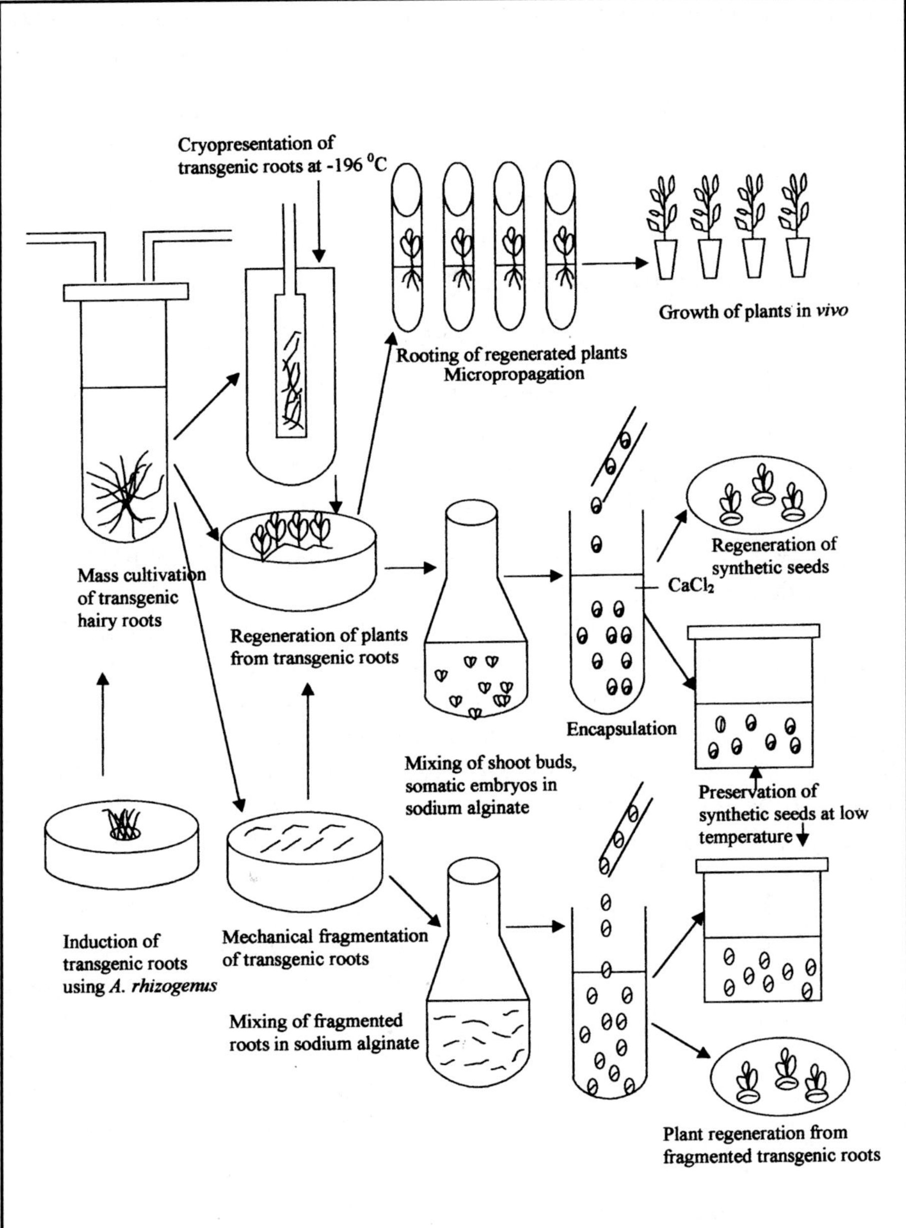

Figure 3 : Application of transgenic roots for micropropagation, germplasm preservation and synthetic seed production.

that spontaneous shoots regeneration is obtained. It offers several advantages over other methods of transformation such as:

- Lack of callus phase
- Higher rates of transformation and co-transfer of genes,
- Effective regeneration of transgenic plants,
- No requirement for selection agent,
- Avoiding chemicals that inhibit shoot regeneration.

Further, *Agrobacterium tumefaciens*-mediated transformation results in high frequency of escapes whereas in *Agrobacterium rhizogenes* mediated transformation only transformed cells can be obtained after several cycles of root tip cultures.

Application of *A. rhizogenes*-mediated transformation has further perspective for genetic modification of introducing additional genes along with the Ri T DNA (Plasmid as vector). *A. rhizogenes*-mediated transformation has been used to produce transgenic hairy root cultures as a source material for regenerating transgenic plants. It has been demonstrated that with the exception of the border sequences, none of the other T DNA sequences are required for the transfer. Entire T DNA can be replaced with the foreign DNA and introduced into cells from which whole plants can be regenerated. These foreign DNA sequences introduced into plants have shown to be stably inherited in Mendelian manner (Stougaard *et al.*, 1987; Zambryski *et al.*, 1989). This characteristic makes *Agrobacterium* the most widely used vector for plant genetic engineering. The *Agrobacterium rhizogenes*-mediated transformation has the advantage that any foreign gene of interest placed in binary vector can be simultaneously transferred to the transformed hairy root clone. Foreign genes can also be introduced through the use of co-integrate vectors. Co-integrate vectors contain the foreign gene within Ri T DNA.

Agrobacterium rhizogenes transfers T-DNA of binary vectors in trans enabling the production of transgenic plants containing foreign genes after regeneration from hairy roots. This method has been used for production of transgenic plants in several species. Transgenic plants produced by transformed hairy roots have been depicted in Table 3. Transgenic plants with genes from binary vector have been obtained in 12 cultivars of *Brassica*. The plants showed hairy root phenotype to varying degrees and were fertile. Segregation analysis confirmed the transmission of traits to the progeny (Chirstey *et al.*, 1997). Due to independent insertion of Ri T-DNA and binary vector T-DNA in subsequent generations, phenotypically normal transgenic plants were produced in tobacco (Hatamoto *et al.*, 1990) and in *Brassica napus* (Boulter *et al.*, 1990). Rapid growth of hairy roots on hormone free medium and their high plantlet regeneration frequency allows clonal propagation of elite plants. Bacterial isochorismate synthase coding gene was cloned in a binary vector and mobilised into *Agrobacterium rhizogenes*. Down *et al.* (1994) reported transgenic hairy roots in *Brassica napus* containing glutamine synthase gene from soyabean, these transformed hairy roots showed a 3 fold increase in GS activity. Greater tolerance to PPT was also seen in shoots regenerated from hairy roots with elevated GS activity.

A number of genes (that are readily available) coding for a wide range of desirable traits can be introduced into numerous plants including genes for herbicide resistance, pest resistance, disease resistance and environmental stress. This can be applied for producing different plant species with a range of agronomically useful traits.

5. METABOLIC ENGINEERING AND HAIRY ROOT SYSTEM

5.1. Genetic engineering of metabolite pathways

The current thrust is on achieving higher yields of secondary metabolites through recombinant DNA technology. This concept will not only enable us to maximise product synthesis but also may add novel economically feasible botanicals. The strategy of manipulation of metabolic pathway includes strengthening of partial pathway for its completion, amplification of a regular pathway by overexpression of cloned genes, interdiction of regular pathways, maximizing response cascades and revising metabolic regulation etc. Completion of pathways by insertion of heterologus genes for synthesis of enzymes that enhance natural product synthesis has been accomplished (Yun *et al.*, 1992). A schematic representation depicting different possibilities of metabolic pathway manipulation has been illustrated in Fig. 4. Amplification of pathways or activation of rate limiting enzymes by insertion and overexpression of heterologus or homologus genes can amount to increased production of secondary metabolites (Fig. 4) (Kutchan, 1993). Secondary metabolites are synthesized via multiple enzymatic steps that happen to be part of complex

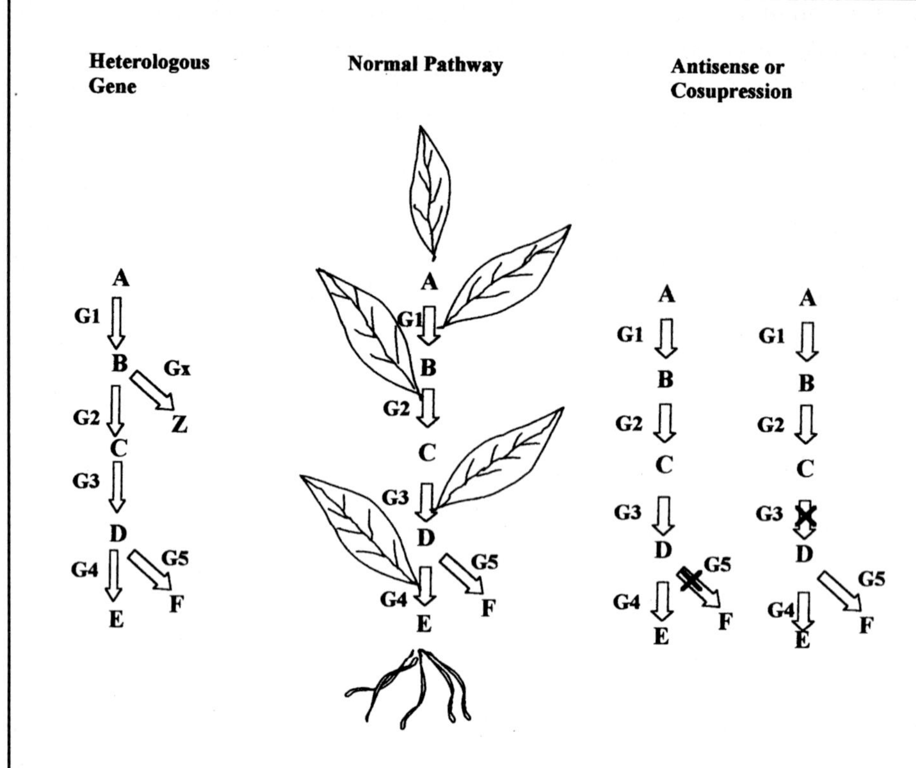

Figure 4 : Diagrammatic representation of options for metabolic pathway engineering using transgenic hairy root cultures. A, B, C, D, E and F are the intermediate products in the multiple enzymatic secondary meabolite pathway. G1, G2, G3, G4 and G5 are the respective genes to excel the metabolic process. Gx: heterologus gene. Z: Genetically engineered new product.

branched biosynthetic pathways. To enhance the yield of desired product a specific channeling of precursors into specific pathways would be desirable. Such channeling can be accomplished by blocking unwanted side branches by competitive pathway blockade and down regulation of genes by antisense technology (Fig. 4) (Chavadej *et al.*, 1994; Dwivedi *et al.*, 1994). The metabolic expression revision will permit us to target the cell components for genetic manipulation much more specifically than before. This approach will open up means to study regulatory genes that control the flux of compounds along a pathway (Nessler, 1994). The cascade of events following introduction of a heterologus gene may bring surprises rather than expected outcome. Therefore it is necessay to minimize unwanted response cascades (Bailey, 1991). For better specificity and expression a homogenic promoter sequence should be used. The promoter sequence may direct gene activation in a tissue specific manner for increase or interdiction of secondary metabolite production without undesirable phenotypes (Nessler, 1994). Metabolic engineering offers immense potential for the manipulation of secondary metabolites *in vivo* as well as *in vitro*. Transgenic hairy root system can act as a suitable candidate to extrapolate the recent developments in metabolic engineering for the production of secondary metabolites of high economic value.

Hairy roots have proved to be useful for testing developing and applying metabolic engineering principles to plants. Foreign genes ligated into binary vectors can be introduced and incorporated into the plant genome along with Ri-T-DNA at the time of transformation. Coding sequences can be fused to strong constitutive promoter, tissue specific or inducible promoter for overexpression of a gene for accumulation or synthesis of secondary metabolites in hairy roots. Down regulation of a particular enzymatic step can be done by expression of that gene in antisense orientation.

A number of genes which encode enzymes controlling key reactions at the interface between primary and secondary metabolism in plants have been isolated from different organisms. Full length plant cDNAs have also been isolated which encode enzymes in a number of secondary metabolic pathways. It seems to be feasible for alteration of secondary metabolites in a particular way by introducing certain hydroxylation, methylation, glycosylation for which specific enzymes exist. It may be possible in certain cases to make the cells actively secrete secondary metabolites in analogy with opine secretion by natural *Agrobacterium* T-DNA system, to achieve easier harvest and purification. Genes of interest with regard to secondary metabolism can be introduced into hairy roots. Experiments have been designed to increase the availability of substrate for a secondary metabolic pathway by increasing the levels of enzymes at the interface between primary and secondary metabolism in hairy roots *in vivo*. Hamill *et al.* (1990) overexpressed yeast ornithine decarboxylase gene in the hairy roots of *Nicotiana rustica* and recovered some lines with increased ornithine decarboxylase activity. Transgenic tissues expressed an increased putrescine content and elevated activity of putrescine methyl transferase which catalyses first step in the pyridine alkaloid biosynthesis. Overexpression of yeast ornithine decarboxylase gene in the hairy roots of *Nicotiana rustica* resulted in 3-4 fold increase in nicotine content. Overexpression of tryptophan decarboxylase from *Catharanthus roseus* in hairy roots of *Peganum harmala* resulted in increased tryptophan decarboxylase activity, thereby increasing the serotonin levels 10-20 fold in some hairy root lines. Yields of serotonin can be further enhanced by feeding L-Tryptophan to hairy root lines overexpressing the *C. roseus* tryptophan decarboxylase gene (Berlin *et al.*, 1993). Overexpression of a bacterial gene coding for lysine decarboxylase in hairy roots of *Nicotiana tabacum* led to an increased activity along with 10 fold increase in cadaverine content and a three fold increase in anabasine levels (it is derived from cadaverine) in transgenic lines (Fecker *et al.*, 1993). Hyoscyamine 6-β- hydroxylase catalyses conversion of hyoscyamine to scopolamine, which is more valuable. Constitutive expression of the gene encoding hyoscyamine 6-β-hydroxylase from *Hyoscyamus niger* in hairy roots of *Atropa belladonna* resulted in the root tissues of *A. belladonna* converting most of their hyoscyamine to scopolamine (Hashimoto *et al.*, 1993).

One possible outcome may be that the supply of metabolites from primary metabolism may become a limiting factor for secondary metabolite biosynthesis. Use of hairy roots will be advantageous rather than using full plant for feeding precursors for harvesting high levels of secondary metabolites (Lidgett *et al.*, 1994). A working model of the *Catharanthus roseus* indole alkaloid biosynthetic pathway in hairy roots has been developed using light, elicitors, precursors and enzyme inhibitors (Shank *et al.*, 1998; Bhadra *et al.*, 1998; Rijhwani and Shank, 1998). The gene for green fluorescent protein (GFP) was successfully used as a reporter gene in *Hyoscyamus* hairy roots (Shanks and Margon, 1999). Damiani *et al.* (1998) reported that use of maize *Sn* gene (a maize anthocyanin transactivator) for induction of pigmentation in hairy roots of dicot plants. The *Sn* gene product activates the anthocyanin biosynthetic pathway, thus producing a red pigment. Maize *Sn* gene can be used as a reporter gene.

6. RECENT DEVELOPMENTS

6.1. Production of antibodies

Transgenic hairy root system can be used as an alternate system for production of antibodies. Wongsamuth and Doran (1997) reported eight antibody producing hairy root clones initiated from the tobacco seedlings expressing Guy's 13 antibody. Guy's 13 is a murine Ig G1 monoclonal antibody which binds to a surface protein of *Streptococcus mutans,* a causative agent of dental caries in humans. Hairy roots were induced using *Agrobacterium rhizogenes* strain 8196 and 15834. As a heterologous antibody expression system hairy roots are capable of production of assembled antibody proteins. In B_5 medium most of the antibody remained associated with the biomass rather than being secreted extracellularly. However, medium additives such as gelatin, polyvinylpyrollidone and nitrate improved total antibody accumulation and allowed upto 43% of the product to be recovered from the medium. Antibody degradation occurs to a greater extent in hairy roots than hybridoma systems. Further work is needed to stabilize plantibody production in hairy root system (Wongsamuth and Doran, 1997).

6.2. Phytoremediation

Environmental contamination with heavy metals comes from a variety of sources. Hairy roots play an important role in delineating fundamental processes of phytoremediation. This is a technology where plants are used to remediate metal and organic contaminants *in situ*. Roots are the main parts of this remediation action. Hairy root cultures are valuable system to elucidate the fate of contaminants without interference from microbes.

Heavy metals *e.g.* Cd, Hg or Pb are toxic at higher concentrations and their accumulation in the environment is a health hazard. Use of vegetation for *in situ* treatment of contaminated soils and sediments is a promising and effective technology. Conventional treatment of soils contaminated with toxic metals cost US $ 1000 per ton of soil, while treating sites with phytoremediation costs approximately US $ 15-20 per ton (Jones, 1994). Roots are the plant parts, which facilitate primary contact between a plant and soil. Transformed roots grown *in vitro* are most suitable model for these studies. Their main advantages are their similarity with normal roots and high growth rate. These hairy roots can be a promising tool for selection of resistant clone, and introduction of new features by molecular biology technique. Metzger *et al.* (1992) reported bioaccumulation or biosorption of heavy metals by transformed root cultures of different sps. They also evaluated bioavailability of Cd from sewage sludge. Macek *et al.* (1994) and Koturba *et al.* (1995) studied Cd bioaccumulation using *Solanum nigrum* and *Solanum aviculare* hairy root cultures. These cultures were used for the studies of biotransformation of polychlorinated biphenyls (Macek *et al.*, 1995).

Hairy roots are also useful as a model system in the are of phytoremediation. *Solanum nigrum* hairy roots removed polychlorinated biphenyls from culture medium. significant progress on transformation processes and fate of nitroaromatic explosive 2,4,6- Trinitrotoluene (TNT) has been made in mass balance experiments using *Catharanthus roseus* hairy roots as model system. TNT was reduced to monoaminodinitrotoulenes, then conjugated at the amine groups with a six–carbon unit and then incorporated into unextractable bound residues in the cell wall material (Shank and Morgan, 1999).

6.3. Mycorrhizal interaction

In vitro culture of hairy root offers a means to simplify the study of complicated root-organism interaction. VAM (Vesicular arbuscular mycorrhizae) has beneficial association with land plants. The study of their interaction is difficult because of the inability to culture this fungi. Mugnier *et al.* (1982) reported the suitability of hairy root cultures as substrates for the symbiotic culturing of *Glomus mosseae* Gerd. and Trappe and *Gigaspora margarita* Backer and Hall (fungi capable of association with a number of crop species). Hairy root cultures can be used for culturing rhizosphere microorganisms, which require roots for the growth. It offers an opportunity to culture and study biotrophic fungi, nematodes and root colonizing bacteria such as *Rhizobia* and *Frankia* (Diouf *et al.*, 1995). Phytopathogenic nematodes can be divided into migratory ectoparasites and endoparasites, and sedentary endoparasites. Migratory ectoparasites and endoparasites can be maintained *in vitro* on plant callus tissues, however, sedentary endoparasties can not be maintained on undifferentiated callus. Hairy roots have been used to propagate these sedentary endoparasties (Paul *et al.*, 1990). A bicompartmental petri dish can be employed so that factors influencing root physiology and the process of infection could be regulated independently. By application of hairy root cultures the study of root feeding insect pests and other soil organisms which interact with roots but are not biotrophic, host-parasite interactions, screening of agrochemicals for fungicide, herbicide and nematicide and research into crop disease resistance has been enhanced (Mugnier, 1997).

Novel compounds can be generated by introduction of heterologous genes into hairy roots or by biotransformation. Kwon *et al.* (1997) isolated four new polyacetylene analogs from *Panax ginseng* hairy roots. These were found to significantly inhibit rat lever acyl coA cholesterol acetyltransferase. Asada *et al.* (1998) detected two novel isoprenylated flavonoids with antimicrobial activity in hairy root cultures of *Glycyrrhiza glabra*.

Production of proteins of industrial and therapeutic importance by plants is an area of intense commercial interest. A novel ribosome inactivating protein was found in cell suspension cultures derived from hairy roots of *Trichosanthes kirilowii* (Shih *et al.*, 1998). High levels of ribosome inactivating proteins may be found in hairy roots but these hairy root derived cell suspensions secreted the protein into the media. This is an advantage for product recovery. Torregrosa and Bouquet (1997) reported production of hairy roots containing coat protein of grapevine chrome mosaic virus. Although regeneration of plantlets was not achieved but, grafted plantlets with transgenic roots could be established in glass house. Hairy roots of *Lotus corniculatus* containing heterologous dihydroflavanol reductase cDNA produced higher level of condensed tannis (Bavage *et al.*, 1997).

An additional advantage of hairy root cultures is for enzymological studies. Abundant quantities of sterile, rapidly growing tissue can be generated. In hairy roots the proportion of meristematic tissue is high and the phenolic content lower than in normal plant roots, that leads to an improved enzyme activity (McLauchlan *et al.*, 1993; Walton *et al.*, 1994).

7. CONCLUSIONS AND FUTURE PROSPECTS

Agrobacterium rhizogenes-mediated genetically transformed "hairy roots" are becoming important in the recent times for the production of economically important compounds *in vitro*. In this review attempts have been made to present the potential applications of hairy root system. Emphasis has been given to the problems associated with the induction of hairy roots in the Agrobacterial infection process as well as its scaling up using bioreactors for biomass production. The review highlighted some of the promising applications of hairy roots including micropropagation (clonal propagation), germplasm conservation, production of foreign compounds (antibodies, proteins etc.) and production of genetically modified plants for useful genetic traits. Besides this, metabolic engineering of genes for key enzymes, the antisense regulation of biosynthetic pathways may also help in increasing the secondary metabolite production in hairy roots. The concomitant integration of the manipulated biosynthetic pathway with the process development in scaling up for biomass production will pave the way for commercial exploitation of hairy root system. However, some of the limitations like use of hairy roots for only root derived secondary metabolites should be addressed. Recently some reports are emerging to extend the process by using green photoautotrophic roots with this it is possible to produce metabolites with Rubisco promoters. Computer modeling and simulation approach may be extrapolated to hairy roots to evaluate product synthesis, genetic stability and structural sustainability. In our opinion now it is appropriate time to replace the cell culture system with transgenic hairy roots and develop new models of bioreactors for process development for product synthesis either secondary metabolites or foreign proteins of pharmaceutical value in human welfare or clonal propagation and genetic manipulation for commercial exploitation.

REFERENCES

Ahn JC, Hwang B, Tada H, Ishimaru K, Sasaki K and Shimomura K (1996) Polyacetylenes in hairy roots of *Platycodon grandiflorum. Phytochemistry,* **42** : 69-72.

Albiol J, Campmajo C, Casas C and Poch M (1995) Biomass estimation in plant cell cultures: A neural network approach. *Biotechnol. Prog.,* **11** : 88-92.

Allan EJ, Stuchbury T and Mordue (Luntz) AJ (1999) *Azadirachta indica* A. Juss. (Neem Tree) : *In vitro* Culture, Micropropagation and the productin of Azadiractine and other secondary metabolites. In: *Biotechnology in Agriculture and forestry, Medicinal and Aromatic Plants XI.* Vol. 43 (Ed. Bajaj YPS) Springer-Verlag Berlin pp. 11-41.

Argolo AC, Charlwood BV and Pletsch M (2000) The regulation of solasodine production by *Agrobacterium rhizogenes* transformed roots of *Solanum aviculare. Planta Medica,* **66** : 448-451.

Arroo RRJ, Develi A, Meijers H, Van de Westerolo E, Kemp AK, Croes AF and Williems GJ (1995) Effects of exogenous auxin on root morphology and secondary metabolism in *Tagetes patula* hairy root cultures. *Physiol. Plant.,* **93** : 233-240.

Asada Y, Li W and Yoshikawa T (1998) Isoprenylated flavonoids from hairy root cultures of *Glycyrrhiza glabra. Phytochemistry,* **47** : 389-392.

Asamizu T, Abiyam K and Yasuda I (1988) Anthraquinones production by hairy root culture in *Cassia obtusifolia. Yakagaku Zasshi,* **108** : 1215-1218.

Atkinson RG and Gardner RC (1991) *Agrobacterium* mediated transformation of *Pepino* and regeneration of transgenic plants. *Plant Cell Rep.,* **10** : 208-212.

Babakov AV, Bartova LM, Dridze IL, Maisuryan AN, Margulis GU, Ogaian RR, Voblikova VD and Muromtsev GS (1995) Culture of transformed horseradish roots as source of Fusicoccin- like ligands. *J. Plant Growth Reg.,* **14** : 163-167.

Baily JE (1991) Towards a science of metabolic engineering. *Science,* **251** : 1668-1674.

Bajaj YPS and Ishimaru K (1999) Genetic transformation of medicinal plants In: *Biotechnology in Agriculture and Forestry, Transgenic medicinal plants.* Vol. 45. (Ed. Bajaj YPS) Springer-Verlag Berlin, pp 1-19.

Bakkali AT, Jaziri M, Vanderheyden Y, Vanhaelen N and Homes J (1997) Lawsone accumulation in normal and transformed cultures of henna, *Lawsonia inermis. Plant Cell Tiss. Org. Cult.,* **51** : 83-87.

Bavage AD, Davies IG, Robbins MP and Morris P (1997) Expression of an *Antirrhinum* dihydroflavanolreductase gene results in changes in condensed tannin structure and accumulation in root cultures of *Lotus corniculatus* (bird's foot trefoil). *Plant Mol. Biol.,* **35** : 443-458.

Bel-Rhlid R, Chabot S, Piche Y and Chenevert T (1993) Isolation and identification of flavanoids from Ri T-DNA transformation roots (*Daucus carota*) and their significance in vesicular-arbuscular mycorrhiza. *Phytochemistry,* **35** : 381-383.

Banerjee S, Naqui AA, Mandal S, Ahuja PS (1994) Transformation of *Withania somnifera* (L) Dunal by *Agrobacterium rhizogenes* – infectivity and phytochemical studies. *Phytotherapy Res.,* **8** : 452-455.

Ben A, Bergmann, Xiaohong LM and Whetten R (1998) Suceptibility of *Paulowinia elongata* to *Agrobacterium* and production of transgenic calli and hairy roots by *in vitro* inoculation. *Plant Cell Tiss. Org. Cult.,* **55** : 45-51.

Benjamin BD, Roja G and Heble MR (1994) Alkaloid synthesis by root cultures of *Rauwolfia serpentina* transformed by *Agrobacterium rhizogenes. Phytochemistry,* **35** : 381-383.

Berlin J, Ruegenhagen C, Dietze P, Fecker LF, Goddijin OJM and Hoge JHC (1993) Increased production of seratonin by suspension and root cultures of *Peganum harmala* transformed with a tryptophan decarboxylase cDNA clone from *Catharanthus roseus. Transgenic Res.,* **2** : 336-344.

Berthomieu P and Jouanin L (1992) Transformation of rapid cycling cabbage (*Brassica oleraceae* var. *capitata*) with *Agrobacterium rhizogenes. Plant Cell Rep.,* **11** : 334-338.

Bhadra R, Vani S and Shanks JV (1993) Production of Indole alkaloids by selected bairy root lines of *Catharanthus roseus. Biotechnol. Bioeng.,* **41** : 582-592.

Bharda R and Shanks JV (1997) Transient studies of nutrient uptake, growth and indole alkaloid accumulation in heterotrophic cultures of hairy roots of *Catharanthus roseus. Biotechol. Bioeng.,* **55** : 527-534.

Bhadra R, Morgan JA and Shanks JV (1998) Transient studies of light –adapted cultures of hairy roots of *Catharanthus roseus*: growth and indole alkaloid accumulation. *Biotechnol. Bioeng.,* **60** : 670-678.

Birot AM, Boucher D, Casse-Delbaert F, Durand-Tardif M, Jouanin L, Pantot J, Robaglia C, Tepfer D, Tepfer M and Vilaine F (1987) Studies and uses of the Ri plasmids of *Agrobacterium rhizogenes. Plant Phys. Biochem.,* **25** : 1-13.

Boitel CM, Gontier E, Laberche JC, Ducrocq C and Sangvan-Norreel BS (1996) Inducer effect of Tween 20 permeabilization treatment used for release of stored alkaloids in *Datura innoxia* Mill. hairy root cultures. *Plant Cell Rep.,* **16** : 241-244.

Bolton GW, Nester EW and Gordon MP (1986) Plant phenolic compounds induce expression of the *Agrobacterium tumefaciens* loci needed for virulence. *Science,* **232** : 983-985.

Bougaud F, Bouque V and Guckert A (1999) Production of flavonoids by *Psoralea* hairy root cultures. *Plant Cell Tiss. Org. Cult.,* **56** : 96-103.

Boulter ME, Croy E, Simpson P, Shields R, Croy RRD and Shirsat AH (1990) Transformation of *Brassica napus* L. (Oilseed rape) using *Agrobacterium tumefaciens* and *Agrobacterium rhizogenes* –a comparison. *Plant Sci.,* **70** : 91-99.

Buitellar RN, Langenholl AAN, Heidstra R and Tramper J (1991) Growth and thiophene production by hairy root cultures of *Tagetus patula* in various two liquid phase bioreactors. *Enzyme Microbial. Technol.,* **13** : 487-494.

Cabrera-Ponce JC, Vegas-Garcia A and Harrera-Estrella L (1996) Regeneration of transgenic Papaya plants via somatic embryogenesis induced by *Agrobacterium rhizogenes*. *In Vitro Cell. Dev. Biol. Plant,* **32** : 86-90.

Cardarelli M, Mariotti D, Pomponi M, Spano L, Capone I and Constantino P (1987) *Agrobacterium rhizogenes* T-DNA genes capable of inducing hairy root phenotype. *Mol. Gen. Genet.,* **209** : 475-480.

Carron TR, Robbins MP and Morris P (1994) Genetic modification of condensed tannin biosynthesis in *Lotus corniculatus*: 1. Heterologous and antisense dihydroflavanol reductase down-regulate tannin accumulation in "Hairy root" cultures. *Theor. Appl. Genet.,* **87** : 1006-1015.

Chavadej S, Brisson N, Mc Neil JN and De Luca V (1994) Redirection of tryptophan leads to production of low glucosinolate canola. *Proc. Natal. Acad. Sci. USA,* **91** : 2666-2170.

Chilton MD, Tepfer DA, Peti A, David C, Casse-Delbart F and Tempe J (1982) *Agrobacterium rhizogenes* inserts T-DNA into the genome of the host plant root cells. *Nature,* **295** : 432-434.

Christen P, Roberts MF, Phillipson JD and Evants WC (1989) High yield production of tropane alkaloids by hairy root cultures of a *Datura candida* hybrid. *Plant Cell Rep.,* **8** : 75-77.

Christey MC and Sinclair BK (1992) Regeneration of transgenic Kale (*Brassica oleracea* var *acephala*). rap (*B. napus*) and turnip (*B. campestris* var *rapifera*) plants via *Agrobacterium rhizogenes*-mediated transformation. *Plant Sci.,* **87** : 161-169.

Christey MC, Sinclair BK, Braun RH and Wyke L (1997) Regeneration of transgenic vegetable Brassicas (*Brassica oleracea* and *B. campestris*) via Ri –mediated transformation. *Plant Cell Rep.,* **16** : 587-593.

Constabel CP and Towers GHN (1988) Thiarubrine accumulation in hairy root cultures of *Chaenactis douglasei. J. Plant Physiol.,* **133** : 67-72.

Couilterot E, Caron C, Trentesaux C, Chenieux JC and Audran JC (1999) *Fagara zanthoxyloides* Lam. (Rutaceae): *In vitro* culture and the production of Benzophenanthridine and Furoquinoline alanine. In: *Biotechnology in Agriculture and Forestry, Medicinal and Aromatic Plants* XI, Vol 43 [Ed. Bajaj YPS]. Spinger-Verlag Berlin pp. 43-56.

Cristen P (1999) *Centranthus* Species : *In vitro* Culture and the production of Valepotriates and other secondary metabolites. In : *Biotechnology in Agriculture and Forestry, Medicinal and Aromatic plants* XI, Vol. 43 (Ed. Bajaj YPS) Spinger Verlag Berlin pp. 42-56.

Damiani F and Aricioni S (1991) Transformation of *Medicago arborea* L. with *Agrobacterium rhizogenes* binary vector carrying the hygromycin resistance genes. *Plant Cell Rep.,* **10** : 300-303.

Damiani F, Paoloci F, Consonni G, Crea F, Tonelli C and Aricioni S (1998) A maize anthocyanin transactivator induces pigmentation in hairy roots of dicotyledonous species. *Plant Cell Rep.,* **17** : 339-344.

Davey MR, Mulligan BJ, Gartland KMA, Peel E, Sargent AW and Morgan AJ (1987) Transformation of *Solanum* and *Nicotiana* species using an Ri plasmid vector. *J. Exp. Bot.,* **38** : 1507-1516.

Davioud E, Kan C, Hamon J, Tempe J and Husson HP (1989) Production of Indole alkaloids by *in vitro* root cultures from *Catharanthus trichophyllus. Phytochemistry,* **28** : 2675-2680.

Delbeque JP, Beydon P and Chapuis L (1995) *In vitro* incorporation of radio labelled cholesterol and mevalonic acid into ecdysteron by hairy root cultures of a plant *Serratula tinctoria. Eur. J. Entomol.,* **92** : 301-307.

Delmotte FM, Delay D, Dizeau J, Guerin B and Leple JC (1991) *Phytochemistry,* **30** : 3549-3552.

Deno H, Yamagata T, Emoto T, Yoshioka T, Yamada Y and Fujita Y (1987b) Scopolamine production by root cultures of *Duboisia myoporoides* II. Establishment of a hairy root culture by infection with *Agrobacterium rhizogenes. J. Plant Physiol.,* **131** : 315-323.

Dilorio AA, Cheetham RD and Weathers PJ (1992) Growth of transformed roots in a nutrient mist bioreactor: reactor performance and evaluation. *Appl. Microbiol. Biotechnol.,* **37** : 457-462.

Diouf D, Gherbi Y, Prin Y, Franche C, Duhoux E and Bogusz D (1995) Hairy root nodulation of *Casuarina glauca*: a system for the study of symbiotic gene expression in an actinorhizal tree. *Mol. Plant Microbe Interactions,* **8** : 532-537.

Dobigny A, Tizroutine S, Gaisne C, Haicour R, Rossignol L, Ducreux G and Sihackar D (1996) Direct regeneration of transformed plants from stem fragments of potato inoculated with *Agrobacterium rhizogenes. Plant Cell Tiss. Org. Cult.,* **45** : 115-121.

Downs CG, Christey MC, Davies KM, King GA, Seelye JF, Sinclair BK and Stevenson DG (1994) Hairy roots of *Brassica napus* : II Glutamine synthase over expression alters ammonia assimilation and the response to phosphinothricin. *Plant Cell Rep.,* **14** : 41-46.

Dwivedi UV, Camplell WH, Yu J, Datta RSS, Bugos RC, Chiung VL and Podia GK (1994) Modification of lignin biosynthesis in transgenic *Nicotiana* through expression of an antisense O-methyltransferon gene from Populus. *Plant Mol. Biol.,* **26** : 61-71.

Ermayanti TM, McComb JA, O Brien PA (1994) Stimulation of synthesis and release of swainsonine from transformed roots of *Swainsona galegifolia. Phytochemistry,* **36** : 313-317.

Fecker LF, Rugenhagen C and Brien J (1993) Increased production of cadaverine and anabesine in hairy root cultures of *Nicotiana tabacum* expressing a bacterial lysine decarboxylase gene. *Plant Mol. Biol.,* **23** : 11-21.

Fei HM, Mei KF, Shen X, Ye YM, Lin ZP and Peng LH (1993) Transformation of *Gynostemma pentaphyllum* by *Agrobacterium rhizogenes*. Saponin production in hairy root cultures. *Acta Bot Sincia,* **35** : 626-631.

Flores HE and Filner P (1985) In: Neumann K, Barz W. and Reinhard E. [Eds] *Primary and secondary metabolism of plant cell cultures.* Springer, Berlin, Heidelberg, New York, pp. 174-185.

Flores HE, Hoy MW and Pikard JJ (1987b) Secondary metabolism in heterotrophic and photosynthetic root cultures of Asteraceae. *Plant Physiol.,* **83** : 36.

Flores HE, Pickard JJ and Hoy MW (1988) Production of polyacetylenes and thiophenes in heterotrophic and photosynthetic root cultures of Asteraceae. In: *Chemistry and Biology of Naturally occurring Acetylenes and related compounds* (NOARC) (Eds. Lam, J, Breheler, H, Arnason, T and Hansen, L), *Bioactive Molecules* 7, pp. 233-254.

Flores HE and Curtis WR (1992) Approaches to understanding and manipulating the biosynthetic potential of plant roots. *Ann. N.Y. Acad. Aci.,* **665** : 188.

Forde BG, Day HM, Turton JF, Shen WJ, Cullimore V and Oliver JE (1989) Two glutamine synthase genes from *Phaseolus vulgaris* L. display contrasting developmental and spatial patterns of expression in transgenic *Lotus corniculatus* plants. *The Plant Cell,* **1** : 391-401.

Fukui H, Feroj Hasan AFM, Ueoka T and Kyo M (1998) Formation and secretion of a new brown Benzoquinone by hairy root cultures of *Lithospermum erythrorhizon. Phytochemistry,* **47** : 1037-1039.

Furze JM, Rhodes MJC, Parr AJ, Robins RJ, Whitehead IM and Threlfall DR (1991) Abiotic factors elicit sesquiterpenoid phytoalexin production but not alkaloid production in transformed root cultures of *Datura stramonium. Plant Cell Rep.,* **10** : 111-114.

Gamborg OL, Murashige T, Thorpe TA and Vasil IK (1976) Plant Tissue Culture Media, *In vitro,* **7** : 473-478.

Geerling A, Hallard D, Martinez Caballeno A, Lopes Cardoso I, Vander Heijden R and Verpoorte R (1999) Alkaloid production by a *Cinchona officinalis* 'Ledgeriana' hairy root culture containing constitutive-expression constructs of tryptophan decarboxylase and strictosidine synthase cDNA from *Catharanthus roseus. Plant Cell Rep.,* **19** : 191-196.

Giovanni A, Pecchioni N, Rabaglio M and Allavena A (1997) Characterization of ornamental *Datura* plants transformed by *Agrobacterium rhizogenes. In Vitro Cell. Dev. Biol. Plant,* **33** : 101-106.

Giri A, Banerjee S, Ahuja PS and Giri CC (1997) Production of hairy roots in *Aconitum heterophyllum* Wall. using *Agrobacterium rhizogenes. In Vitro Cell. Dev. Biol. Plant,* **33** : 280-284.

Giri A and Lakshmi Narasu M (2000) Transgenic hairy roots : Recent trends and Applications. *Biotechnology Advances,* **18** : 1-22.

Giulientti AM, Parr AJ and Rhodes MJC (1993) Tropane alkaloids production in transformed root cultures of *Brugmansia candida. Planta Medica,* **59** : 428-431.

Ganicher F, Cristen P and Kaptanidis I (1995b) Production of Valepotriates by hairy root cultures of *Centranthus ruber* D.C. *Plant Cell Rep.,* **14** : 294-298.

Granicher F, Cristen P and Vuagnat P (1994) Rapid high performance liquid chromatographic quantification of Valepotriates in hairy root cultures of *Valeriana officinalis* L. var. *Sambucifolia* hairy roots. *Phytochemistry,* **38** : 103-105.

Gutierrez-Pesce P, Taylor K, Muleco R and Rugini E (1998) Somatic embryogenesis and shoot regeneration from transgenic roots of the cherry rootstock colt (*Prunus avium* × *P. pseudocerasus*) mediated by pRi 1855 T-DNA of *Agrobacterium rhizogenes. Plant Cell Rep.,* **17** : 574-580s.

Halperin SJ and Flores HE (1997) Hyoscyamine and Proline accumulation in water stressed *Hyoscyamus muticus* hairy root cultures. *In Vitro Cell Dev. Biol. Plant,* **33** : 240-244.

Hamill JD, Parr JJ, Robins RJ and Rhodes MJC (1986) Secondary product formation by cultures of *Beta vulgaris* and *Nicotiana rustica* transformed with *Agrobacterium rhizogenes. Plant Cell Rep.,* **5** : 111-114.

Hamill JD, Parr AJ, Rhodes MJC, Robins RJ and Walton NJ (1987) New routes to plant secondary products. *Bio/Technol.,* **5** : 800-804.

Hamill JD, Robins RJ and Rhodes MJC (1989) Alkaloid production by transformed root cultures of *Cinchona ledgeriana. Planta Medica,* **55** : 354-357.

Hamill JD, Robins RJ, Parr AJ, Evans PM, Furze JD and Rhodes MJC (1990) Over expressing a yeast ornithine decarboxylase gene in transgenic roots of *Nicotiana rustica* can lead to enhanced Nicotine accumulation. *Plant Mol. Biol.,* **15** : 27-38.

Han KH, Keathley DE, Davis JM and Gordon MP (1993) Regeneration of a transgenic woody legume *Robinia pseudoacacia* L., (Black locust) and morphological alternations induced by *Agrobacterium rhizogenes*-mediated transformation. *Plant Sci.,* **88** : 149-157.

Handa T, Sujimura T, Kato E, Kamada H and Takayanagi K (1995) Genetic transformation of *Eustoma grandiflorum* with *rol* genes. *Acta. Hort.,* **392** : 209-218.

Hashimoto T, Yun DJ and Yamada Y (1993) Production of tropane alkaloids in genetically engineered root cultures. *Phytochemistry,* **32** : 713-718.

Hashimoto T and Yamada Y (1994) Alkaloids biogenesis: molecular aspects. *Ann Rev. Plant Physiol Plant Mol. Biol.,* **45** : 257-285.

Hatamoto H, Boulter ME, Shirsat AH, EJ and Ellis JR (1990) Recovery of morphologically normal transgenic tobacco from hairy roots co-transformed with *Agrobacterium rhizogenes* and a binary vector plasmid. *Plant Cell Rep.,* **9** : 88-92.

Hayashi T, Totoh K, Ohnishi K, Okamura K and Asamizu T (1994) 6-Methoxy-2-benzoxazolenone in *Scoparia dulcis* and its production by cultured tissues. *Phytochemistry,* **37** : 161-164.

Hilton MG and Rhodes MJC (1990) Growth and hyoscyamine production of 'hairy root' culture of *Datura stramonium* in a modified stirred tank reactor. *Appl. Microbiol. Biotechnol.,* **33** : 132-138.

Hilton MG and Rhodes MJC (1993) Factors affecting the growth and hyoscyamine production during batch culture of transformed roots of *Datura stramonium. Planta Medica,* **59** : 340-444.

Hook I (1994) Secondary metabolites in hairy root cultures of *Leontopodium alpinum* cass (edelweiss). *Plant Cell Tiss. Org. Cult.,* **38** : 321-326.

Hooykaas PJJ, Den Dulk-Ras H and Schilperoort and Rorsch A (1988) The *Agrobacterium trumefaciens* T-DNA gene 6^b is an *onc* gene. *Plant Mol. Biol.,* **11** : 791-794.

Hoshino Y and Mii M (1998) Bialaphos stimulates shoot regeneration from hairy root of snapdragon (*Antirrhinum majus* L.) transformed by *Agrobacterium rhizogenes. Plant Cell Rep.,* **17** : 574-580.

Hu ZB and Alfermann AW (1993) Diterpenoid production in hairy root cultures of *Salvia miltiorrhiza. Phytochemistry,* **32** : 699-703.

Hua Chen Da, Chang-Jun Liu, He-Chun Ye, Guo-Feng Li, Ben-Ye Liu, Yu-Ling Meng and Xiao-ya Chen (1999) Ri –mediated transformation of *Artemisia annua* with a recombinant farnesyl diphosphate synthase gene for artemisinin production. *Plant Cell Tiss. Org. Cult.,* **57** : 157-162.

Huang Y, Diner A and Karnosky F (1991) *Agrobacterium rhizogenes* mediated genetic transformation of a conifer: *Larix decidua. In Vitro Cell. Dev. Biol. Plant,* **27** : 201-207.

Ikenaga T, Oyama T and Muranaka T (1995) Growth and steroidal saponin production in hairy root cultures of *solanum aculeatissi Plant Cell Rep.,* **14** : 413-417.

Ionkava I, Kartnig T and Alfermann W (1997) Cycloartane saponin production in hairy root cultures of *Astragalus mongholicus. Phytochemistry,* **45** : 1597-1600.

Ishimaru K, Sudo H, Salake M, Malsugama Y, Hasagewa Y, Takamoto S and Shimomura K (1990) Amarogenetin and amaroswertin and four xanthones from hairy root cultures of *Swertia japonica. Phyotchemistry,* **29** : 1563-1565.

Ishimaru K and Shimomura K (1991) Tannin production in hairy root cultures of *Geranium thunmbergii. Phytochemistry* **30** : 825-828.

Jaziri M, Yoshimatsu K, Homes J and Vanhaelen M (1988) Tropane alkaloid production by hairy root cultures of *Datura stramomium* and *Hyoscyamus niger. Phytochemistry,* **27** : 419-420.

Jaziri M, Shimomura K, Yoshimatsu K, Fauconnier KL, Marlier M, and Homes J (1995) Establishment of normal and transformed root cultures of *Artemisia annua* L. for artemisinin production. *J. Plant Physiol.,* **145** : 175-177.

Jobanovic V, Grubisic D, Giba Z, Menkovid N and Ristic M (1991) Alkaloids from hairy root cultures of *Anisodus luridus* (*Scopolia lurida* Dunal Solanaceae Tropane alkaloids). *Planta Medica,* **2** : 102.

Jones RL (1994) ASPP recommends hazardous remadiation rechnologies to DOE. *Am. Soc. Plant Physiol. News Lett.,* **21** : 12-13.

Jung G and Tepfer D (1987b) Use of genetic transformation by the Ri T-DNA of *Agrobacterium rhizogenes* to stimulate biomass and tropane alkaloid production in *Atropa belladonna* and *Calystegia sepium* roots grown *in vitro. Plant Sci.,* **50** : 145-151.

Kamo K and Blowers A (1999) Tissue specificity and expression level of gus A. under rol D., mannopine synthase and translation elongation factor 1 subunit a promoters in transgenic *Gladiolus* plants. *Plant Cell Rep.,* **18** : 809-815.

Kittipongpatana N, Hock RS and Porter JP (1998) Production of solasodine by hairy root, callus and cell suspension cultures of *Solanum aviculare Forst. Plant Cell Tis. Org. Cult.,* **52** : 133-143.

Kim CH, Lee SW and Chung IS (1994) Hairy root cultures of *Dacus carota* for anthocyanin production in fluidized–bed bioreacter. *Agric. Chem. Biotechnol.,* **37** : 237-242.

Kim S, Hoppa E and Hjortso M (1995) Hairy root growth models: Effect of different branching patterns. *Biotechnol. Prog.,* **11** : 178-186.

Kisiel W, Stojakowska A, Malarz J and Kohlmunzer S (1995) Sesquiterpene lactones in *Agrobacterium rhizogenes* transformed hairy root cultures of *Lactuca virosa. Phytochemistry,* **40** : 1139-1140.

Kisiel W and Stojakowskas A (1997) A sesquiterpene coumarin ether from transformed roots of *Tanacetum parthenium. Phytochemistry,* **46** : 5515-516.

Ko KS, Ebizuka Y, Noguchi H and Sankawa U (1995) Production of polypeptide pigments in hairy root cultures of *Cassia* plants. *Chem. Pharm.* Bull. (Tokyo), **43** : 274-278.

Kondo O, Honda H, Taya M and Kobayashi T (1989) Comparison of growth properties of carrot hairy root in various bioreactors. *Appl. Microbiol. Biotechnol.,* **32** : 291-294.

Koturba P, Macek T, Skacel F and Rumi T (1995b) Accumulation of Cadmium by hairy root culture of *Solanum nigrum* from nutrient medium. In : *Biosorption and Bioremediation,* (Eds. Mecek T, Demnerova K and Mackova M) 1-8: Prague: Abstr. Int. Symp., Merin. CSBMB.

Kumar V, Jones B and Davey MR (1991) Transformation by *Agrobacterium rhizogenes* and regeneration of transgenic shoots of the wild soybean *Glycine argyrea. Plant Cell Rep.,* **10** : 135-138.

Kutchan TM (1993) Strictosidine: from alkaloid to enzyme to gene. *Phytochemistry,* **32** : 493-506.

Kutchan TM (1995) Alkaloid biosynthesis- the basis for metabolic engineering of medicinal Plants. *Plant Cell,* **7** : 1059-1070.

Kuroyanagi M, Arakava T, Mikami Y, Yshida K, Kawahar N, Hayashi T and Ishimaru H (1998) Phytoalexins from hairy roots cultures of *Hyoscyamus albus* treated with methyl jasmonate. *J. Nat. Prod.,* **61** : 1516-1519.

Kwok KH and Doran PM (1995) Kinetic and stoichiometric analysis of hairy roots in a segmented bubble column reactor. *Biotechnol. Prog.,* **11** : 429-435.

Kwon BM, Ro SH, Kim MK, Nam JY, Jung HJ, Lee IR, Kim YK and Bok SH (1997) Polyacetylene analogs isolated from hairy roots of *Panax ginseng,* inhibit acetyl-CoA-cholesterol. *Planta Medica,* **63** : 552-553.

Lee KT, Yamakawa T, Kodama T and Shimomura K (1998) Effect of chemicals on alkaloid production by transformed roots of Belladonna. *Phytochemistry,* **49** : 2343-2347.

Lee KT, Suzuki T, Yamakawa T, Kodama T, Igarshi Y and Shimomura K (1999) Production of tropane alkaloid by transformed root cultures of *Atropa belladonna* in stirred biorectors with a stainless steel net. *Plant Cell Rep.,* **18** : 567-571.

Lidgett AJ, Dobin M, Fredericks EB, Michael A and Hamill JD (1994) Metabolic effects of over-expressing foreign ornithine decarboxylase (ODC) or arginine decarboxylase (ADC) genes in tobacco. 4th International Congr. Plant Molecular Biology. Amsterdum 1309.

Linsmaier EM and Skoog F (1965) Organic growth factor requirements of tobacco tissue cultures. *Physiol. Plant.,* **18** : 100-127.

Lodhi AH, Bongaerts RJM, Verpoorte R, Coomber SA and Charlwood BV (1996) Expression of bacterial isochorismate synthase (EC 5.4.99.6) in transgenic root cultures of *Rubia peregrina. Plant Cell Rep.,* **16** : 54-57.

Mecek T, Kotrba P, Suchova M, Skacel F, Demnerova K and Rumi T (1994) Accumulation of Cadmium by hairy root cultures of *Solanum nigrum. Biotechnol. Lett.,* **16** : 621-624.

Mecek T, Mackova M, Holubkova A, Burkhard J, Demnerove K and Pazlarova A (1995) Biodegradation of polychlorinated biophenyls of plant cells and the effect of peroxidase activity. In: *Biosorption and Bioremediation,* (Eds. Mecek T, Demnerova K and Mackova M) 1-8. Prague: Abstr. Int. Sym., Merin. CSBMB.

Mahagamasekera MGP and Doran PM (1998) Intergeneric co-culture of genetically transformed organs for the production of scopolamine. *Phytochemistry,* **47** : 17-25.

Manners JM and Way H (1989) Efficient transformation with regeneration of the tropical pasture legume *Stylosanthes humilis* using *Agrobacterium rhizogenes* and a Ti Plasmid-binary vector system. *Plant Cell Rep.,* **8** : 341-345.

Mano Y, Nabeshima S, Matsiu C and Ohkawa H (1987) Production of tropane alkaloids by hairy-root cultures of *Scopolia japonica. Agric. Bio. Chem.,* **50** : 2715-2722.

Marchant YY (1988) *Agrobacterium rhizogenes* - transformed root cultures for the study of polyacetylene metabolism and biosynthesis. In: *Chemistry and Biology of naturally occurring Acetylenes and related compounds* (NOARC) (Eds. Lam J, Breheler H, Arnason T and Hansen L), *Bioactive Molecules* 7, pp. 217-231.

Matsumoto T and Tanaka N (1991) Production of phytoecdysteroides by hairy root cultures of *Ajuga reptans* var *atropurpuera. Agric Biol. Chem.,* **55** : 10-25.

Mazarei M, Ying Z and Houtz RL (1998) Functional analysis of the Rubisco large subunit AN-methyltransferase promoter from tobacco and its regulation by light in Soybean hairy roots. *Plant Cell Rep.,* **17** : 907-912.

Mc Afee BJ, White EE, Pelcher LE and Lapp MS (1993) Root induction in Pine (*Pinus*) and Larch (*Larix*) Spp. Using *Agrobacterium rhizogenes. Plant Cell Tiss. Organ Cult.,* **34** : 53-62.

Mc Granahan GH, Leslie CA and Dandekar AM (1993) Transformation of Pecan and regeneration of trangenic plants. *Plant Cell Rep.,* **12** : 634-638.

Mc Granahan GH, Leslie CA and Uratsu SL (1988) *Agrobacterium*-mediated transformation of Walnut somatic embryos and regeneration of transgenic plants. *Bio/Technology,* **6** : 800-804.

Mc Innes E, Morgan AJ, Mulligan BJ and Davey MR (1991) Phenotypic effects of isolated pRiA4 TL-DNA *rol* genes in the presence of intact TR-DNA in trangenic plants of *Solanum dulcamara* L. *J. Exp. Bot.,* **42** : 1279-1286.

McLauchlan WR, Mc Kee RA and Evans DA (1993) The purification and immunocharacterization of N-methyl putrescine oxidase from transformed root cultures of *Nicotiana tabacum* L. cvSc 58. *Planta,* **191** : 440-445.

Merkli A, Christen P and Kapetanidis I (1997) Production of diosgenin by hairy root cultures of *Trigonella foenum-graecum* L. *Plant Cell Rep.,* **16** : 632-636.

Metzger L, Fouchault I, Glad Ch, Prost R and Tepfer D (1992) Estimation of cadmium availability using transformed roots. *Plant Soil,* **143** : 249-257.

Morgan AJ, Cox PN, Turner DA, Peel E, Davey MR, Gartland KMA and Mulligan BH (1987) Transformation of tomato using and Ri plasmid vector. *Plant Sci.,* **49** : 37-49.

Motomari Y, Shimomura K, Mori K, Kunitake H, Nakashima T, Tanaka M, Miyazaki S and Ishimaru K (1995) Polyphenol production in hairy root cultures of *Fragaria* × *ananassa. Phytochemistry,* **40** : 1425-1428.

Mugnier J, Jung G and Prioul JL (1982) Method of producing endomycorrhizian fungi with arbuscules and vesicles *in vitro*. Patent F R 82, 10768; US 4, 599, 312.

Mugnier J (1997) Mycorrhizal interactions and the effects of fungicides, nematicides and herbicides on hairy root cultures. In: *Hairy roots: Culture and Applications.* (Ed. Coran PM) Amsterdam. Harwood Publishers. Pp. 123-131.

Mukundan U and Hjortso MA (1990) Growth and thiophene accumulation by hairy root cultures of *Tagetus patula* in media of varying initial pH. *Plant Cell Rep.,* **9** : 627-630.

Murakami Y, Omoto T, Asai I, Shimomura K, Hoshihira K and Ishimaru K (1998) Rosmarinic acid and related phenolics in transformed root cultures of *Hyssopus officinalis. Plant Cell Tiss. Org. Cult.,* **53** : 75-78.

Muranaka T, Ohakawa H and Yamada Y (1992) Scopolamine release into media by *Duboisia leichhardtti* hairy root clones. *Appl. Microbiol. Biotechnol.,* **37** : 554-559.

Murashige T and Skoog F (1962) A revised medium for rapid growth and bioassays with tobacco tissue cultures. *Physiol. Plant.,* **15** : 473-497.

Nakano M, Hoshino Y and Mii M (1994) Regeneration of transgenic plants of grape vine (*Vitis vinifera* L.) via *Agrobacterium rhizogenes* mediated transformation of embryogenic calli. *J. Exp. Bot.,* **45** : 649-656.

Nakashimada Y, Uozemi N, Kobayashi T (1995) Production of plantlets for use as artificial seeds from horseradish hairy roots fragmented in a blender. *J. Ferment Bioeng.,* **79** : 458-464.

Nessler CL (1994) Metabolic engineering of plant secondary products. *Transgenic Res.,* **3** : 109-115.

Nin S, Bennici A, Roselli G, Mariotti D, Schiff S and Magherini R (1997) *Agrobacterium* mediated transformation of *Artemisia absinthum* L. (Worn Wood) and production of secondary metabolites. *Plant Cell Rep.,* **16** : 725-730.

Nishikawa K and Ishimaru K (1997) Flavonoids in root cultures of *Scutellaria baicalensis. J. Plant Physiol.,* **151** : 633-636.

Noda T, Tanaka N, Mano T, Nakeshima S, Ohkawa H and Matsui C (1987) Regeneratin of horse radish hairy roots incited by *Agrobacterium rhizogenes* infection. *Plant Cell Rep.,* **6** : 283-286.

Nguyen C, Bourgaud F, Forlot P and Guckert A (1992) Establishment of hairy root cultures of *Psoralea* species. *Plant Cell Rep.,* **11** : 424-427.

Nussbbaumer P, Kapetanidis I and Christen P (1998) Hairy roots of *Datura candida* × *D. aurea*: effect of culture composition on growth and alkaloid biosynthesis. *Plant Cell Rep.,* **17** : 405-409.

Ogasawara T, Cheba K and Tada M (1993) Production in high –yield of a naphthoquinone by a hairy root culture of *Sesamum indicum. Phytochemistry,* **33** : 1095-1098.

Ohara A, Akasaka Y, Daimon H and Mii M (2000) Plant regeneration from hairy roots induced by infection with *Agrobacterium rhizogenes* in *Crotalaria juncea* L. *Plant Cell Rep.,* **19** : 563-568.

Olszowska O, Alfermann AW and Furmanowa M (1996) Eugenol from normal and transformed root cultures of *Coluria geoides. Plant Cell Tiss. Org. Cult.,* **45** : 273-276.

Ooms G, Twell D, Bossen ME, Hoge JHC and Burrell MM (1986b) Development regulation of Ri T DNA gene expression in roots, shoots and tubers of transformed potato (*Solanum tuberosum* cv. Desiree). *Plant Mol. Biol.,* **6** : 321-330.

Oostdam A, Mol JNM and Vanderplas LHW (1993) Establishement of hairy root cultures of *Linum flavum* producing the lignan 5-methoxy podophyllotoxin. *Plant Cell Rep.,* **12** : 474-477.

Otani M, Mu M, Handa T, Kamada H and Shimada T (1993) Transformation of sweet potato (*Ipomoea batatus* (L.) Lam.) plants by *Agrobacterium rhizogenes. Plant Sci.,* **94** : 151-159.

Padmanabhan K, Cantliffe DJ, Harrell RC and Harrison J (1998) Computer vision analysis of somatic embryos of sweet potato (*Ipomoea batatas* (L.) Lam.] for assessing their ability to convert to plants. *Plant Cell Rep.,* **17** : 681-684.

Palazon J, Cusido RM, Roig C and Pinol MT (1998) Expression of the rol gene and nicotine production in transgenic hairy roots and their regenerated plants. *Plant Cell Rep.,* **17** : 384-390.

Parr AJ and Hamill JD (1987) Relationship between *Agrobacterium rhizogenes* transformed hairy roots and intact uninfected *Nicotiana* plants. *Phytochemistry,* **26** : 3241-3245.

Parr AJ, Peerless ACJ, Hamill HD, Walton NJ, Robins RJ and Rhodes MJC (1988) Alkaloid production by transformed root cultures of *Catharanthus roseus. Plant Cell Rep.,* **7** : 309-312.

Paul H, van Deelen JEM, Henken B, de Block TSM, Lange W and Krens FA (1990) Expression *in vitro* of resistance to *Heterodera schachtii* Schm. in hairy roots of an alien monotelosomic addition plant of *Beta vulgaris* transformed by *Agrobacterium rhizogenes. Euphytica,* **48** : 153-157.

Payne J, Hamill JD, RJ and Rhodes MJC (1987) Production of hyoscyamine by hairy root cultures of *Datura stramonium. Planta Medica,* **29** : 2545-2550.

Pellegrinesch A, Damon JP, Valtorta N, Paillard N and Tepfer D (1994) Improvement of ornamental characters and fragrance production in lemon scented geranium through genetic transformation by *Agrobacterium rhizogenes. Bio/Technol.,* **12** : 64-68.

Perez-Molphe-Balch E and Ochoa-Alejho N (1998) Regeneration of transgenic plants of Mexican lime from *Agrobacterium rhizogenes* transformed tissue. *Plant Cell Rep.,* **17** : 591-596.

Phelep M, Petit A and Mortin L (1991) Transformation of a nitrogen fixing tree *Allocasuriana verticillate* Lam. *Bio/Technol.,* **9** : 461-466.

Phunchindawan M, Hirata K, Sakai A and Miyamoto K (1997) Cryopreservation of encapsulated shoot primordia induced in horse radish (*Armoracia rusticana*) hairy root cultures. *Plant Cell Rep.,* **16** : 469-473.

Porter J (1991) Host range and implications of plant infection by *Agrobacterium rhizogenes. Crit. Rev. Plant Sci.,* **10** : 387-421.

Pythoud F, Sinbar VP, Nester EW and Gordon MP (1987) Increased virulence of *Agrobacterium rhizogenes* conferred by the *vir* region of pTiB0542: Applications to genetic engineering of Poplar. *Bio/Technol.,* **5** : 1323-1327.

Rao KR, Venkanna N and Lakshmi Narasu M (1998) *Agrobacterium rhizogenes* mediated transformation of *Artemisia annua. J. Sci. Indust. Res.,* **57** : 773-776.

Rech EL, Golds TJ, Husnain T, Vainstein MH, Jones B, Hammat N, Mulligan BJ and Davey MR (1989) Expression of a chimaeric Kanamycin resistance gene introduced into the wild soybean *Glycine canescens* using a co-integrate Ri plasmid vector. *Plant Cell Rep.,* **8** : 33-36.

Redenbaugh K and Walkder K (1990) Role of artificial seeds in alfalfa breeding. In: *Plant tissue culture: Applications and Limitations* (Ed. Bhojwani SS) Elsevier Amsterdam pp-102-135.

Rhodes MJC, Robins RJ, Hamill JD, Parr AJ and Walton NJ (1987a) Secondary product formation using *Agrobacterium rhizogenes* transformed hairy root cultures. *T.C.A. Newsletter,* **53** : 2-15.

Rhodes MJC, Parr JJ, Ceielutt A and Aird ELH (1994) Influence of exogenous hormone on the growth and secondary metabolite formation in transformed root cutures. *Plant Cell Tiss. Org. Cult.,* **38** : 143-151.

Rijhwani SK and Shansks JV (1998) Effect of elicitor dosage and exposure time on biosynthesis of indole alkaloids by *Catharanthus roseus* hairy root cultures. *Biotechnol Prog.,* **14** : 442.

Rijhwani SK, Ho CH and Shanks JV (999) *In vitro* 31 P. and multilabel 13 C. N.M.R. measurements for evaluation of plants metabolic pathways. *Metab. Eng.,* **1** : 12-25.

Rodriguez-Mediola MA, Stafford A, Cresswell R and Aria-Castro C (1991) Bioreactors for growth of plant roots. *Enzyme Microbe. Technol.,* **13** : 697-702.

Roy MC (1989) Plant growth response to *Agrobacterium rhizogenes* M. App. Sci. Thesis, Lincoln College, University of Canterbury, Newzealand.

Rugini E and Moriotti D (1991) *Agrobacterium rhizogenes* T-DNA genes and rooting in woody species. *Acta. Hort.,* **300** : 301-308.

Satio K, Yoshimatsu K and Murakoshi T (1990) Genetic transformation of fox-glove (*Digitalis purpurea*) by chimeric foreign genes and production of cardioactive glycosides. *Plant Cell Rep.,* **9** : 121-124.

Satio K, Yamazaki M, Kawaguchi A and Murakoshi I (1991) Metabolism of solanaceous alkaloids in transgenic plant teratomas integrated with genetically engineered genes. *Tetrahedron,* **47** : 5955-5968.

Satio K, Yamazaki M, Anzai H, Yoneyama K and Murakoshi I (1992) Transgenic herbicide-resistant *Atropa belladonna* using an Ri plasmid vector and inheritance of the transgenic trait. *Plant Cell Rep.,* **11** : 219-224.

Sandra I, Pitta-Alvarez and Guelietti AM (1999) Influence of chitosan, acetic acid and citric acid on growth and tropane alkaloid production in transformed roots of *Brugmansia candida.* Effect of medium pH and growth phase. *Plant Cell Tiss. Org. Cult.,* **59** : 31-38.

Santos PM, Figueiredo AC, Olivera MM, Barroso JG, Pedro LG, Deans SG, Younus AKM and Scheffer JJC (1998) Essential oils from hairy root cultures and from fruits and roots of *Pimpinella anisum. Phytochemistry,* **48** : 455-460.

Sasaki K, Udagava A, Ishimaru H, Hayashi Y, Alfermann AW, Nakanishi F and Shimomura K (1998) High forskolin production in hairy roots of *Coleus forskohlii. Plant Cell Rep.,* **17** : 457-459.

Sato K, Yamazaki T, Okuyama E, Yoshihira K and Shimomura K (1991) Anthraquinone production by transformed root cultures of *Rubia tictorum*: Influence of phytohormones and sucrose concentration. *Phytochemistry,* **30** : 2977-2978.

Sauerwein M and Shimomura K (1991) Indole alkaloids in hairy root cultures of *Amnosia elliptica. Phytochemistry,* **30** : 3277-3280.

Sauerwein M, Yamazaki T and Shimomura K (1991a) Hernandulcin in hairy root cultures of *Lippia dulcis. Plant Cell Rep.,* **9** : 579-581.

Sauerwein M, Wink M and Shimomura K (1992) Influence of light and phytohormones on alkaloid production in transformed root cultures of *Hyoscymus albus. J. Plant Physiol.,* **140** : 147-152.

Savary BJ and Flores HE (1994) Biosynthesis of defense related proteins in transformed root cultures of *Tricosanthes kirilowii* Maxim. var. *japonicum* (Kitam). *Plant Physiol.,* **106** : 1195-1204.

Schnoor JL, Licht LA, Mc Cutcheon SC, Wolfe NL and Carreira LH (1995) Phytoremediation of organic and nutrient contaminants. *Env. Sci. Technol.,* **29** : 318 A-326 A.

Sevon N, Drager B, Hiltunen R and Oksman-Caldentey KM (1997) Characterization of transgenic plants derived from hairy roots of *Hyoscyamus muticus. Plant Cell Rep.,* **16** : 605-611.

Shahin EA, Sukhainda K, Simpson RB and Spivey R (1986) Transformation of cultivated tomato by a binary vector in *Agrobacterium rhizogenes*: transgenic plants with normal phenotypes harbor binary vector T DNA, but no Ri-Plasmid T-DNA. *Theor. Appl. Genet.,* **72** : 770-777.

Shanks JV, Bhadra R, Morgan J, Rijhwani S and Vani S (1998) Quantification of metabolites in the indole alkaloid pathway of *Catharanthus roseus:* Implications for metabolic engineering. *Biotech. Bioeng.,* **58** : 333-338.

Shanks J and Morgan J (1999) Plant "hairy root" culture. *Curr. Opinion Biotechnol.,* **10** : 151-155.

Shnaks JV, Rijhwani SK, Morgan J, Vani S, Bhadra R and Ho CH (1999) Quantification of metabolic fluxes for metabolic engineering of plant products In : *Plant Cell and Tissue Culture for Food Ingredient Production.* (Eds. Fu TJ, Singh G and Curtis WR) Kluwer Academic/Plenum Publishing, New York. Pp 45-60.

Sharp JM and Doran PM (1990) Characteristics of growth and tropane alkaloid synthesis in *Atropa belladonna* roots transformed by *Agrobacterium rhizogenes J. Biotechnol.,* **16** : 171-186.

Shen WH, Petit A, Guern J and Tempe J (1998) Hairy roots are more sensitive to auxin than normal roots. *Proc. Natl. Acad. Sci. USA,* **35** : 3417-3421.

Shih NJR, Mc Donald KA, Dandekar AM, Girbes T, Iglesias R and Jackman AP (1998) A novel type-1 ribosome inactivating protein isolated from the supernatant of transformed suspension cultures of *Trichosanthes kiriowii. Plant Cell Rep.,* **17** : 531-537.

Shin WS and Hjortso MA (1998) A tissue embedding technique for measuring the structure to hairy root mats of *Tagetes erecta. Korean J. Chem Eng.,* **15** : 150-156.

Shimomura K, Sudo H, Saga H and Kamada H (1991) Shikonin production and secretion by hairy root cultures of *Lithospermum erythrorhizon. Plant Cell Rep.,* **10** : 282-285.

Shin DI, Podila GK, Huang Y and Karnosky DF (1994) Transgenic Larch expressing genes for herbicide and insect resistance. *Can J. For. Res.,* **24** : 2059-2067.

Signs MW and Flores HE (1990) The biosynthetic potential of plant roots. *Bio-Essays,* **12** : 282-285.

Sim SJ, Chang HN, Liu JR and Jung KH (1994) Production and secretion of Indole alkaloids in hairy root cultures of *Catharanthus roseus:* Effects of *insitu* adsorption, fungal elicitation and permeabilization. *J. Ferment. Bioeng.,* **78** : 229-234.

Smith TC, Weathers PJ and Cheetam RD. Effects of gibbrellic acid on hairy root cultures of *Artemisia annua:* Growth and artemisinin production. *In vitro Cell Dev. Biol.,* **33** : 75-79.

Sommer S, Siebert M, Bechthold AA and Heide L (1998) Specific induction of secondary product formation in transgenic plant cell cultures using an inducible promoter. *Plant Cell Rep.,* **17** : 891-896.

Srinivasan V, Pestchanker L, Moser S, Hirasuna TJ, Taticek RA Shuler MK (1995) Taxol production in bioreactor: Kinetics of biomass accumulation, nutrient uptake and taxol production by cell suspension of *Taxus baccata. Biotechnol. Bioengg.,* **47** : 666-676.

Stachel SE, Messens E, Van Montagu M and Zambryski P (1985) Identification of signal molecules produced by wounded plant cells that activate T-DNA transfer in *Agrobacterium tumefaciens. Nature,* **318** : 624-629.

Stachel SE, Timmerman B and Zambryski PC (1986) Generation of single-stranded T-DNA molecules during the initial stages of T-DNA transfer from *Agrobacterium tumefaciens* to plant cells. *Nature,* **322** : 706-712.

Stiller J, Nasinec V, Svoboda S, Nemcova B and Machackova T (1992) Effects of agrobacterial oncogenes in kidney vetch (*Anthyllis vulneraria* L.). *Plant Cell Rep.,* **11** : 363-367.

Stougaard J, Abildsten D and Marcher KA (1987) The *Agrobacterium rhizogenes* pRi TL – DNAsegment as a gene vector system for transformation of plants. *Mol. Gen. Genet.,* **207** : 251-255.

Stummer BE, Smith SE and Langridge (1995) Genetic transformation of *Verticordia grandis* (Myrtaceae) using wild-type *Agrobacterium rhizogenes* and binary *Agrobacterium* vectors. *Plant Sci.,* **111** : 51-62.

Subroto AM, Kwok KH, Hamill JD and Doran PM (1996) Co-culture of genetically transformed roots and shoots for Synthesis, Translocation and Biotransformation of secondary metabolites. *Biotech. Bioeng.,* **49** : 481-494.

Tada H, Shimomura K and Ishimaru K (1995b) Polyacetylenes in *Platycodon grandiflorum* hairy roots and campanulaceous plants. *J. Plant Physiol.,* **145** : 7-10.

Tada H, Nakashima T, Kuntake H, Mori K, Tanaka M and Ishimaru K (1996) Polyacetylenes in hairy root cultures of *Campanula medium.* L. *J. Plant Physiol.,* **147** : 617-619.

Tanaka N, Takao M and Matsumoto T (1994) *Agrobacterium rhizogenes* mediated transformation and regeneration of *Vinca minor* L. *Plant Tiss. Cult. Lett.,* **11** : 191-198.

Taya M, Yoyama A, Kondo O and Kobayashi T (1989a) Growth characteristics of plant hairy roots and their cultures in bioreactors. *J. Chem. Eng. Japan,* **22** : 84-89.

Taya M, Yoyama A, Nomura R, Kondo O, Matsui C and Kobayashi T (1989b) Production of peroxidase with horse radish hairy root cells in a two step culture system. *J. Ferment. Bioeng.,* **67** : 31-34.

Taya M, Mine K, Kinoka M, Tone S and Ichi T (1992) Production and release of pigments by cultures of transformed hairy roots of red beet. *J. Ferment. Bioeng.,* **73** : 31-36.

Teoh KH, Weathers PJ, Cheetham RD and Walcerz DB (1996) Cryopreservation of transformed (hairy) roots of *Artemisia annua. Cryobiology,* **33** : 106-117.

Tepfer D (1989) Ti T-DNA from *Agrobacterium rhizogenes*: a source of genes having applications in rhizosphere biology and plant development, ecology and evolution. In: Kosuge T., Nester E.W. [Eds] *Plant – microbe interations, molecular and genetic perspectives.* Vol. 3, Mc Graw Hill; New York, pp 294-342.

Tescione LD, Ramakrishan D and Curtis WR (1997) The role of liquid mixing and gas–phase dispersion in a submerged, sparged root reactor. *Enzyme Microb. Tech.,* **20** : 207-213.

Thomas MR, Rose RJ and Nolan KE (1992) Genetic transformation of *Medicago truncatula* using *Agrobacterium* with genetically modified Ri and disarmed Ti plasmid. *Plant Cell Rep.,* **11** : 113-117.

Thompson DV, Melchers LS, Iyer KB, Schilperoort RA and Hooykaas PJJ (1988) Analysis of the complete nucleotide sequence of the *Agrobacterium tumefaciens* vir B operon. *Nucleic Acids Res.,* **16** : 4621-4636.

Toivonen L (1993) Utilization of hairy root cultures for production of secondary metabolites. *Biotechnol. Progr.,* **9** : 12-20.

Toivonen L, Ojala M and Kauppinen V (1990) Indole alkaloid production by hairy root cultures of *Catharanthus roseous*: growth kinetics and fermentations, *Biotechnol. Lett.,* **12** : 519-524.

Torregrosa L and Bouquet A (1997) *Agrobacterium rhizogenes* and *Agrobacterium tumefaciens* co-transformation to obtain grapevine hairy roots producing the coat protein of grapevine chrome mosaic nepovirus. *Plant Cell Tiss. Org. Cult.,* **49** : 53-62.

Trotin F, Moumou Y and Vasseur J (1993) Flavonol production by Fagopyrum esculentum hairy roots and normal root cultures. *Phytochemistry,* **33** : 929-931.

Trulson AJ, Simpson RB and Shahin EA (1986) Transformation of cucumber (*Cucumis sativus* L.) plants with *Agrobacterium rhizogenes. Theor. Appl. Genet.,* **73** : 11-15.

Trypsteen M, Van-Lijsekettens M, Van Severen R and Van Montagu M (1991) *Agrobacterium rhizogenes* mediated transformation of *Echinacea purpurea. Plant Cell Rep.,* **10** : 85-89.

Uozumi N, Asano Y and Kobayashi T (1994) Micropropagation of horse radish hairy roots by means of adventitious root primordia. *Plant Cell Tiss. Org. Cult.,* **36** : 183-188.

Uozumi N, Ohtake Y, Nakashimada Y, Morikawa Y, Tanaka N and Kobayashi T (1996) Efficient regeneration from a GUS transformed Ajuga hairy roots. *J. Ferment. Bioeng.,* **81** : 374-378.

Uozumi N and Kobayashi T (1997) Arificial seed production through hairy root regeneration. In: *Hairy roots: Culture and Applications.* [Ed. Doran PM] Amsterdam. Harwood Publishers. Pp. 113-121.

Van Altvorst AC, Bino RJ, Van Dijk AJ, Lamers AMJ, Lindhout WH, Mark FV and Dons JJM (1992) Effects of the introduction of *Agrobacterium rhizogenes* rol genes on tomato plant and flower development. *Plant Sci.,* **83** : 77-85.

Vanhala L, Hiltunen R and Oksman-Caldentey KM (1995) Virulence of different *Agrobacterium* strains on hairy root formation of *Hyoscyamus muticus. Plant Cell Rep.,* **14** : 236-240.

Veluthambi K, Ream W and Gelvin SB (1988) Virulence genes, borders and overdrive generate single T-DNA molecules from the A6 Ti plasmid of *Agrobacterium tumefaciens. J. Bacteriol.,* **170** : 1523-1532.

Visser RGF, Hesseling-Meinders A, Jacobsen E, Nijdam H, Witholt B and Feenstra WJ (1989a) Expression and inheritance of inserted markes in binary vectors carrying *Agrobacterium rhizogenes* transformed potato (*Solanum tuberosum* L.) *Theor. Appl. Genet.,* **78** : 708-714.

Vasser RGF, Jacobsen E, Witholt and Feenstra WJ (1989b) Efficient transformation of potato *Solanum tuberosum* L. using a binary vector in *Agrobacterium rhizogenes. Theor. Appl. Genet.,* **78** : 594-600.

Walton NJ, Peerless ACJ, Robins RJ, Rhodes MJC, Boswell HD and Robins DJ (1994) Purification and properties of Putrescine N-methyltransferase from transformed roots of *Datura stramonium* L. *Planta,* **193** : 9-15.

Washida D, Shimomura K, Nakajima Y, Takido M and Kitanaka S (1998) Ginsenosides in hairy roots of a *Panax hybrid. Phytochemistry,* **49** : 2331-2335.

Weathers PJ, Cheetham RD, Follanskie E and Teoh K (1994) Artemisinin production by transformed roots of *Artemisia annua. Biotechnol. Lett.,* **16** : 1281-1286.

Weathers PJ, Hemmavanh DD, Walcerz DB, Cheetham RD and Smith TC (1997) Interactive effects of nitrate and phosphate salts, sucrose, and inoculam culture age on growth and sesquiterpene production in *Artemisia annua* hairy root cultures. *In Vitro Cell. Dev. Biol. Plant.,* **33** : 306-312.

Whitney PJ (1992) Novel Bioreactcors for the growth roots transformed by *Agrobacterium rhizogenes. Enzyme Microbe. Technol.,* **14** : 13-17.

Williams RD and Ellis BE (1993) Alkaloids from *Agrobacterium rhizogenes* transformed *Papaver somniferum* cultures. *Phytochemistry,* **32** : 719-723.

Wongsamuth R and Doran PM (1997) Hairy roots as an expression system for production of antibodies. In: *Hairy roots culture and application.* [Ed. Doran PM] Harwood academic publishers, pp. 89-97.

Wyslouzil BE, Whipple M, Chatterjee C, Walcerz DB, Wathers PJ and Hart DP (1997) Mist deposition onto hairy root cultures: Aerosol modeling and experiments. *Biotechnol. Prog.,* **13** : 185-194.

Wyslouzil BE, Waterbury RG and Weathers PJ (2000) The growth of single roots of *Artemisia annua* in nutrient mist reactors. *Biotechnol. Bioeng.,* **70** : 143-150.

Yang DC and Choi YE (2000) Production of transgenic plants via *Agrobacterium rhizogenes* mediated transformation of *Panax ginseng. Pant Cell Rep.,* **19** : 491-496.

Yamanaka M, Ishibashi K, Shimomura K and Ishimaru K (1996) Polyacetylene glucosides in hairy root cultures of *Lobelia cardinalis. Phytochemistry,* **41** : 183-185.

Yonemitsu H, Shimomura K, Satake M, Mochida S, Tanaka M, Endo T and Kaji A (1990) Lobeline production by hairy root cultures of *Lobelia inflata* L. *Plant Cell Rep.,* **9** : 307-310.

Yoshikawa T and Furuya T (1987) Saponin production by cultures of *Panax ginseng* transformed with *Agrobacterium rhizogenes. Plant Cell Rep.,* **6** : 449-453.

Yoshimatsu K and Shimomura K (1992) Transformation of opium poppy (*Papaver somniferum* L.) with *Agrobacterium rhizogenes* MAFF 03-01724. *Plant Cell Rep.,* **11** : 132-136.

Yoshimatsu K, Yamaguchi H and Shimomura K (1996) Traits of *Panax ginseng* hairy roots after cold storage and cryopreservation. *Plant Cell Rep.,* **15** : 555-560.

Yu S, Kwok KH and Doran PM (1996) Effect of sucrose, exogenous product concentration and other culture conditions on growth and steroidal alkaloid production by *Solanum aviculare* hairy roots. *Enzyme Microbiol Technol.,* **18** : 238-243.

Yun DJ, Hashimoto T and Yamada Y (1992) Metabolic engineering of medicinal plants: transgenic *Atropa belladonna* with an improved alkaloid composition. *Proc. Natl. Acad. Sci. USA,* **89** : 11799-11803.

Zambryski P (1988) Basic processes underlying *Agrobacterium* –mediated DNA transfer to palnt cells. *Ann. Rev. Genet.,* **22** : 1-30.

Zambryski P, Tempe and Schell J (1989) Transfer and function of T-DNA genes from *Agrobacterium* Ti and Plasmids in Plants. *Cell,* **56** : 193-201.

Zhan XC, Jones DD and Kerr A (1988) Regeneration of flax plants transformed by *Agrobacterium rhizogenes. Plant Mol.,* **11** : 551-560.

Zhou Y, Hirotani M, Yshikava T and Furuya T (1997) Flavonoids and phenylethanoids from hairy root cultures of *Scutellaria bacicalensis. Phytochemistry,* **44** : 83-87.

Chapter 3

PARTICLE BOMBARDMENT MEDIATED TRANSFORMATION OF COTTON

Dennis E McCabe[1], Zhongjin Lu[2] and Maliyakal E John★

8777 Air Port-Road, Middleton, WI 53562, USA, (DEM), Environmental Research Laboratory, University of Arizona, 2601 Air Port Drive, Tucson, AZ 85706, USA (ZL) and, 3622 Rolling Hill Drive, Middleton, WI 53562, USA (MEJ)

Summary

Agrobacterium-mediated transformation of cotton is a very reliable method, but lacks broad cultivar independence necessary for rapid commercialization. Many of the elite commercial cotton varieties are recalcitrant to regeneration and therefore not directly amenable to Agrobacterium-mediated transformation. Moreover presence of a rigid cell wall causes special problems for the introduction of DNA into plant cells by biological or chemical methods. Therefore several physical methods, such as particle bombardment, sonication, microinjection, laser beam application and silicon whisker bombardment, have been developed to over come this problem. Using the above methods, DNA can be delivered to meristems and recipient cells and allowed to develop into transgenic plants. We have used particle bombardment to produce elite commercial transgenic cotton varieties. Several private companies as well as universities have also developed particle bombardment devices based on electric discharge or compressed gas/air to drive DNA coated microparticles into meristems. Transgenic events are followed by expression of marker genes and or selection regime based on antibiotic or herbicides. Stable integration of input DNA occur in

---

[1]Former Monsanto Company scientist

★Corresponding author : E-mail : maliyakaljohn@hotmail.com

[2]Current address : Seaphire International, 4500 N. 32nd St., Suite 100, Phoenix, AZ 85018 USA

about 0.2% of bombarded cotton meristems. These transformation events are stable through several generations, tested and maintain the transgene expression. Improving the transformation efficiency is the most important challenge facing the particle bombardment technique.

Keywords : Cotton, GUS, marker genes, meristem, particle bombardment, transformation, transgenic

Abbreviations : Beta-glucuronidase (GUS); Green fluorescent protein (GFP), Chloramphenicol acetyltransferase (CAT)

1. INTRODUCTION

Plant transformation is a process by which an identifiable DNA segment is inserted into a plant genome. The introduced DNA is integrated into the chromosome, which may or may not be transcriptionally active, and is transmitted to the sexual offspring. Thus, transformation enables us to modify the genetic makeup of a plant, alter gene expression, and enhance agronomic or product traits. Plant transformation is especially challenging due to the presence of rigid cell walls that act as barriers to DNA molecules. Hence, methodologies have been developed to remove the cell wall. Once the cell wall is removed, the resulting protoplast with the plasmalemma allows access to DNA. A few cells take up the new DNA. In order to obtain a transgenic plant, the cells that have taken up the DNA must be isolated and grown into a whole plant. However, not all cells are totipotent. Thus, transformation procedures that depend on regeneration of protoplasts through tissue culture are limited to certain cultivars. To overcome this shortcoming, direct DNA introduction into growing meristems has been developed.

Transformation requires a reporter gene (marker gene) or a selection process that distinguishes the cells that took up DNA from those that did not. Several marker genes have been identified and corresponding assays developed for plant transformation work. We will briefly summarize marker genes and general transformation methods used in plant genetic engineering. A detailed description of particle bombardment, a premier method for cotton transformation follows.

2. MARKER AND SELECTION GENES USEFUL IN COTTON TRANSFORMATION

Reporter genes are essential for the identification of transformation events in cotton or other plants. High sensitivity, qualitative and quantitative detection assays, lack of background activities, tolerance to N or C terminus fusion and relative ease of assay are important factors in selecting reporter systems in plants. The most widely used reporter gene in plants is the *uidA* gene of *Escherichia coli*. The *uidA* gene encodes a β-glucuronidase (GUS) enzyme which is a stable hydrolase that catalyzes the cleavage of a variety of β-glucuronides (Stoeber, 1991). The GUS gene has been cloned and sequenced (Jefferson *et al.*, 1986). Its monomeric molecular weight is 68 kDa and it tolerates many detergents, various ionic conditions and temperatures, and has no cofactor requirements. It is inhibited by heavy metal ions and EDTA is recommended in assay buffers. It is active at any physiological pH with the optimum between 5.2 and 8.0. The gene can be used for translational fusion as it tolerates large amino-terminal additions (Jefferson *et al.*, 1987; Guivarc'h *et al.*,· 1996). Its commercially available substrates include: p-nitrophenyl glucuronide (PNPG), 4-methyl umbelliferyl glucuronide (MUG), 5-bromo-4-chloro-3-indolyl glucuronide (X-GLUC), naphthol AS-B1 glucuronide (NAG) and resorufin glucuronide (ReG). Detection of β-glucuronidase activity can be determined by histochemical or fluorometric assays. Fluorogenic substrate MUG, is specific for β-glucuronidase and has an excellent signal-to-noise ratio.

Another marker gene useful for transformation is the green fluorescent protein (GFP). GFPs are a class of proteins responsible for the green light emitted by many marine invertebrates (Leffel *et al.*, 1997). They are highly fluorescent and are activated by an energy transfer process via luciferase or a Ca^{2+} activated photoprotein. GFP does not require exogenous substrates or cofactors and therefore is useful in localizing gene expression in transformed tissues. GFP was cloned from the bioluminescent jellyfish *Aequorea victoria* (Prasher *et al.*, 1992; Chalfie *et al.*, 1994). A number of mutant forms of GFPs that exhibit rapid formation of the chromophore and higher excitation peaks at 470-490 nm have been isolated. Thus, these mutants have higher detection sensitivity (Heim *et al.*, 1995). However, GFP in stable transformed plants show only faint fluorescence, probably due to its low expression caused by primary structural features, such as AT content or cryptic intron sequences (Haseloff, 1995)

Other reporter or selection genes useful in cotton transformation include: chloramphenicol acetyltransferase (CAT) from transposon Tn9 of *E. coli* (Gorman, 1982), aminoglycoside 3-phosphotransferase II (APH: Bevan *et al.*, 1983), and fire fly luciferase (*luc*: Ow *et al.*, 1986).

Detection of CAT activity can be accomplished by the radiolabelled substrate [^{14}C]-acetylcoenzyme A or [^{14}C]-chloramphenicol. The CAT enzyme inactivates chloramphenicol by acetylation to produce acetylated chloramphenicol [1-acetyl chloramphenicol (1-AC0, 3-acetyl chloramphenicol (3-AC) and 1,3-diacetyl chloramphenicol (1,3-AC)]. The radiolabelled products are separated by thin layer chromatography and detected by autoradiography (Umbeck *et al.*, 1987).

Aminoglycoside 3-phosphotransferase, also known as neomycin phosphotransferase II, inactivates antibiotics, neomycin, kanamycin, geneticin (G418) and paromomycin through phosphorylation. The antibiotic kanamycin is a trisaccharide composed of one deoxystreptamine and two glucosamine units. It impairs the protein synthesis in plant mitochondria and chloroplasts resulting in chlorosis. Both radioactive and nonradioactive assays are available for NPTII (Curtis *et al.*, 1995). Kanamycin has been successfully used as a selection agent for transgenic cotton (Umbeck *et al.*, 1987). Glufosinate is a nonselective herbicide that may be useful in cotton transformation. *Streptomyces* spp. produce bialaphos, a natural analog of glufosinate. A gene conferring resistance to bialaphos (*bar*) encoding phosphinothricin acetyltransferase (PAT) was isolated from *S. viridochromogenes*. The *bar* gene was introduced into several cotton varieties and shown to confer resistance to the herbicide (Keller *et al.*, 1997). Other herbicides that may have selection potential include 2,4-D (2,4-dichlorophenoxy acetic acid) and glyphosphate. A list of potential marker and selection genes is given in Table 1.

3. GENERAL METHODS OF TRANSFORMATIONS

DNA can be introduced into regenerable cells by biological, physical or chemical means (Potrykus, 1991). *Agrobacterium* infection (Barton *et al.*, 1983; Fraley, 1986), electroporation (Shimamoto, 1989; D'Halluin, 1992), polyethylene glycol (Golovkin *et al.*, 1993), microinjection (Morikawa, 1994; Gong, 1988), particle bombardment (Klein *et al.*, 1987; Christou, 1992), laser microbeam (Guo *et al.*, 1995), and silicon whisker bombardment (Frame *et al.*, 1994; Wang *et al.*, 1995) are some of the techniques developed. Recent attempts to transform cotton (*Gossypium hirsutum* L; Texas cultivar CUBQHRPIS) via *Agrobacterium* infection of shoot apex may be applicable for other genotypes (Zapata *et al.*, 1999). Protoplast transformation can generate many independent transformants at a time, provided the cells can be regenerated into fertile plants. Thus, this technology can be efficient and economical. The

Table 1. Reporter/selectable marker genes for plant transformation

Enzyme (gene)	Selection Agent	Reference
β-glucuronidase (*uidA* or *GUS*)	—	Jefferson *et al.*, 1986
		Guivarch *et al.*, 1996
Luciferase (*luc*)	—	Ow *et al.*, 1986
Green fluorescent protein (*gfp*)	—	Leffel *et al.*, 1997
		Prasher *et al.*, 1992
		Chalfie *et al.*, 1994
β-galactosidase (*lacZ*)	—	Matsumoto *et al.*, 1988
Chloramphenicol acetyltransferase	Chloramphenicol	Gorman *et al.*, 1982
Neomycin phosphotransferase II (*nptII*)	G418/kanamycin	Bevan *et al.*, 1983
Hygromycin phosphotransferase (*hpt*)	Hygromycin	Elzen *et al.*, 1985
Streptomycin phosphotransferase (*spt*)	Streptomycin	Jones *et al.*, 1987
Amino glycoside-3-adenyltransferase (*aadA*)	Spectinomycin	Svab *et al.*, 1990
Phosphinothricin acetyltransferase (*bar*)	Phosphinothricin★	Block *et al.*, 1987
Dihydropteroate synthase (*sul*)	Sulfonamides★	Guerineau *et al.*, 1990
5-enolpyruvylshikimate-3-phosphate synthase (*epsps*)	Glyphosate★	Della-Cioppa *et al.*, 1987
Dihydrodipicolinate synthase (*dhps*)	S-aminoethyl L-cysteine	Perl *et al.*, 1993
Aspartate kinase (*ak*)	Lysine and threonine	Perl *et al.*, 1993
2,4-D monooxygenase	2,4-dichlorophenoxyacetic acid★	Lyon *et al.*, 1993
Acetohydroxy acid synthase (*ahas*)	Sulfonylureas,★ Imidazolinones★	Miki *et al.*, 1990
Acetolactate synthase (*als*)	Primisulfuron★	Harms *et al.*, 1991

★Herbicide resistance

difficulties in regenerating some major poaceous crops have been overcome recently (D'Halluin *et al.*, 1992; Golovkin *et al.*, 1993; Lazzeri and Shewry, 1993). The utility of protoplast transformation in cotton is limited by the ability to regenerate plants. Finer and McMullen used particle bombardment of embryogenic tissue from Coker (*G. hirsutum* L. cv Coker 310) to recover transgenic plants (Finer and McMullen, 1990).

4. GENE TRANSFER TO MERISTEMS: PARTICLE BOMBARDMENT

Rapid commercialization of genetically engineered cotton requires a reliable method of introducing genes directly into elite varieties of germplasms. Thus development of cotton transformation methods have been targeted toward faster turn around time, cultivar independence, insertion of multiple genes and high transformation efficiency. While no one method meets all of these criteria, particle bombardment (gene gun; biolistics), one method for physically transferring DNA through cell walls, holds promise in many respects. Klein and colleagues demonstrated that tungsten micro particles coated with DNA can be accelerated through onion (*Allium cepa*) cell walls, using a modified gunpowder device (Klein *et al.*, 1987). The epidermal cells receiving the DNA express the introduced gene. Many chemically inert metals, such as gold, tungsten, palladium, rhodium and platinum, can be used as projectiles. The original gene gun device was substantially refined by McCabe and colleagues (Accell® Agracetus, Madison, WI: McCabe *et al.*, 1988). They introduced genes into a number of recalcitrant crops using the Accell device (Christou *et al.*, 1992; McCabe *et al.*, 1988; McCabe and Martinell, 1993). Since then, a number of other gene gun devices have been developed to deliver microparticles to cells. The metal particle carrying DNA is lodged in the cytoplasm or nucleus by the propulsion. It is not clear whether only DNA deposited in the nucleus is utilized or nuclear transportation of DNA from the cytoplasm occurs. Transcription and translation of transgenes occur within hours after bombardment. This expression, termed transient expression, can be detected by means of a reporter gene included in the input plasmid. Several thousand copies of each plasmid are resident on each microparticle. It appears that many of these copies contribute to transient expression; however, only a few copies are eventually integrated into the plant genome (Cooley *et al.*, 1995). Particle bombardment transformation of elite cotton cultivars has been very successful (McCabe and Martinell, 1993; John, 1996; John and Keller, 1996).

4.1. Instrumentation

A number of particle bombardment devices are used in delivering DNA to plant cells. The major difference among them is the mechanics of producing the force to accelerate the microparticles. Commercially available PDS-1000He (BioRad, Hercules, CA) uses pressurized helium (450-2200 psi) as the acceleration power. A vacuum is applied in the apparatus to aid in the acceleration of the particles. A newer version, Helos Gene Gun (BioRad), is a hand-held, low-pressure helium (100-600 psi) gun that eliminates the need for a vacuum. Helium was also used as the accelerating force by Takeuchi *et al.* (1992). These authors describe a simple and inexpensive helium driven device to accelerate microparticles. Morikawa *et al.* (1994) describes a pneumatic device using compressed air as the acceleration force. The device is equipped with pumps to compress air (up to 260 kg/cm^2) into an air chamber. DNA-coated gold particles are smeared over a polyethylene projectile and accelerated toward a stopper by compressed air. A commercially available air gun is also used as a pressure pulse generator (Sautter, 1994).

In the *Accell* device, shock waves are generated by exploding a droplet of water through an electrical discharge. This shock wave propels the microparticles (McCabe *et al.*, 1992). The gold particles (1 to 3 μm in diameter) are coated with DNA and accelerated toward plant tissues. The extent of particle penetration to specific cell layers (L1, L2, and L3) and damage to the tissue can be controlled, since the electric discharge gun allows fine-tuning. A detailed description of the electric apparatus appears in US patent # 5,120,657 (McCabe *et al.*, 1992). The accelerator consists of a spark chamber (containing electrodes) connected to an offset reverberation chamber over which the carrier sheet (bearing the microparticles) is placed. The carrier sheet is a flat plastic coated metalized Mylar cut to fit over the top of the reverberation chamber. Located above the carrier sheet is a 100 mesh stainless steel retainer screen which restrains the carrier sheet from impacting the biological target during bombardment. Plant tissue adhering to agar or other gels in a petri dish is inverted over the retainer screen in the path of the flight of the particles. The entire assembly is flooded with helium and can partially evacuated to assist in increasing the acceleration of the particles.

The gap between the electrodes is adjusted to 1-1.5 mm and bridged by a 10 μl droplet of distilled water prior to discharge. During the operation an adjustable high voltage source is used to charge a 2 Microfarad capacitor to the desired voltage (between 4 and 25 Kilovolts).

After disconnecting the voltage source from the capacitor, the capacitor is discharged through the electrodes. The electric discharge causes an electric arc to jump between the two electrodes, instantly vaporizing the droplet of water. The shock wave created enters the reverberation chamber where it echoes from the walls until the many overlapping wave fronts finally hit the carrier sheet and propel it toward the retainer screen. When the carrier sheet impacts, the gold particles are propelled forward and become embedded in the plant tissue.

4.2. Protocol

A number of protocols and procedures have been reported for particle bombardment mediated cotton transformation (Finer and McMullan, 1990; McCabe and Martinell, 1993; Chlan *et al.*, 1995). The procedure reported by McCabe and Martinell has been reproducible and is summarized below (McCabe and Martinell, 1993; McCabe *et al.*, 1992) The procedure uses an *Accell®* gene gun.

DNA preparation: Mix 10 mg of amorphous crystalline gold (~1 mm; Degussa, South Plainfield, NJ) with 100 μl of 0.1 M spermidine (free base) and 1 to 20 μg of DNA by vortexing. Precipitate the DNA by the addition of 100 μl of 2.5 M $CaCl_2$ and leave at room temperature for ten minutes. Discard the supernatant by brief centrifugation in a microfuge. Resuspend the microparticles in 20 mls of 100% ethanol by gentle sonication. Coat the carrier sheet (18 × 18 mm squares of ½ mil metabolized Mylar; Dupont 50 MMC) at a calculated rate of 0.05 mg per square centimetre. Drain the excess ethanol and dry the carrier sheets.

4.3. Seed axes preparation

Surface sterilize cotton seeds in 50% Chlorox bleach for three minutes and rinse with distilled water before immersing in an antibiotic solution of 200 mg/l; carbenicillin, 125 mg/l cefotaxime; 30 mg/l Bravo WP and 30 mg/l Benlate fungicides for three hours in the dark at room temperature. Incubate the seeds a second time for another three hours in fresh antibiotic solution and drain the solution before allowing the seeds to germinate in the dark at 15°C for additional 12 hours. Select the seeds with one to four millimetres of exposed radicles and remove the seed axes under sterile conditions. Wash the excised seed axes repeatedly in distilled water before subjecting to microdissection to remove the embryonic leaves. Wash the microdissected seed axes in distilled water and incubate them in standard OR ccb medium made with fresh benzylaminopurine (BAP), but no NAA for 12 hours at 15°C in the dark (Barwale *et al.*, 1986).

Plate the seed axes vertically (10 to 25 each) on 12% xanthan gum in petri dishes and subject them to bombardment. After bombardment seed axes are left in the dark to recover and then plate them on fresh OR ccb medium with BAP but no NAA for up to two days. Care is taken to keep the meristem from contacting any other medium.

4.4. Screening for transformation events

The inclusion of a marker gene, such as β-glucuronidase (GUS), allows the screening of the bombarded cotton seed axes for transformation events (Jefferson *et al.*, 1987). After three or more weeks, the portions of leaves from the seedlings are tested for GUS expression. Histochemical staining of a piece of each leaf will allow mapping the nodes or axillary buds subtending the transformed leaves. Nontransformed tissue is pruned to allow growth of the transformed buds. The selective pruning is continued until all the leaves of the growing plant are GUS positive. Various plant tissues arise from three physiological cell layers (L1, L2, and L3) of the meristem (Sussex, 1989). It is assumed that the dicot epidermal layer is derived from L1 layer in the meristem. When the L1 layer is severely damaged, as may happen to some meristems during bombardment, the L2 layer is likely to be converted to L1. Another possibility is the conversion of L2 to L1 by high cytokinin levels in the regeneration medium. The L2 and L3 layers form the vascular system, the middle mesophyll layer in the central region of the leaf blade and most importantly, the germline. The intensity of the bombardment or the depth of bead penetration into the embryonic axes influences the likelihood of germline transformation events. Selective pruning and forcing of the axillary buds cause a few transformed cells from L1, L2 or L3 to gain ascendancy and populate an entire tissue layer of its origin (McCabe and Martinell, 1993). Thus, two types of transformants may be generated: epidermal and germline. In epidermal transformants, only the epidermal cell layer (L1) contain the transgene and therefore they do not pass the transgene into their progenies. If L2 or L3 layer is transformed (germline) the transgenes will be passed on to its progeny. Those plants are referred to as germline transformants. McCabe and colleagues devised a screen based on the expression of

GUS in tissue layers to distinguish epidermal or germline transformants very early in plant growth (McCabe *et al.*, 1988). This protocol was first developed for soybeans. It was found that when GUS was present in the vascular system of a leaf, the majority of the progeny was transgenic, indicating germline transformation. This system was then adapted to cotton, where cross sections of either leaf or petiole were stained to identify specific tissues expressing GUS gene and the status of the transformant as epidermal or germline was inferred (McCabe and Martinell, 1993). Thus, if one or more cotton vascular bundles express

Table 2. Timeline for transgenic cotton development by particle bombardment.

	Stage	# Plants	# Months from Start	Comments
1)	Bombardment	10,000 explants	0	—
2)	1st screen (GUS positive)	133 (1.33%)	1-1.5	- GUS expression
3)	2nd screen (GUS positive)	27 (0.27%)	3.0	- Insect/herbicide resistance - Test feasible on leaf biomass - Germline identification (leaf)
4)	Mature R_0 plant	25 (0.25%)	6.0	- Insect/herbicide test - Fiber quality test - Germline confirmation (pollen)
5)	Mature R_1 plant	8 (0.08%)	11.0	- Fiber quality test - Insect/herbicide test - Field test

Transformation frequency is based on the number of seed axes bombarded and total number of transformants recovered expressing GUS. Assume bombardment of 10,000 DP 50 seed axes (explants) with *GUS* plasmid. The first screen in 1 to 1. 5 months resulted in the identification of 133 putative transformants, the remaining ones were discarded. The second screen reduced this number to 27 plants. In 6 months a total of 25 transformants (including 8 germlines) were identified.

Table 3. Particle bombardment mediated cotton transformation frequencies.

Experiment	Explants Bombarded	Survived (%)	Total Transgenics (%)	Epidermal Frequencies	Germline Frequencies
#1	13,292	6,475 (49%)	33 (0.25%)	24 (0.18%)	9 (0.07%)
#2	10,913	6,057 (56%)	45 (0.41%)	36 (0.33%)	9 (0.08%)
#3	12,278	5,368 (44%)	17 (0.14%)	10 (0.08%)	7 (0.06%)
#4	11,378	4,892 (43%)	27 (0.24%)	19 (0.17%)	8 (0.07%)
#5	12,930	6,888 (53%)	43 (0.33%)	25 (0.19%)	18 (0.14%)
#6	8,799	3,812 (43%)	19 (0.22%)	14 (0.16%)	5 (0.06%)
#7	13,159	7,105 (54%)	41 (0.31%)	31 (0.24%)	10 (0.08%)
#8	12,587	6,420 (51%)	35 (0.28%)	27 (0.21%)	8 (0.06%)
#9	5,013	2,305 (46%)	12 (0.24%)	5 (0.1%)	7 (0.14%)
#10	4,993	2,447 (49%)	10 (0.2%)	6 (0.12%)	4 (0.08%)
Total	105,342	51,769 (49%)	282 (0.27%)	197 (0.19%)	85 (0.08%)

Ten independent experiments involving DP50 variety and various plasmid constructs are shown. All plasmids contained a 35S-GUS gene. Percent survival and transformation frequencies are based on the number of seed axes bombarded. In each experiment one or several plasmids were bombarded into seed axes. Data show the variation in transformation efficiencies from experiment to experiment. Percent survival was determined 6 to 8 weeks after bombardment.

Table 4. Characteristics of germline and epidermal cotton transformants

Germline	Epidermal
Only L3 layer transformed	Only L1/L2 (epidermis) layer transformed
Pollen is GUS positive	Pollen is GUS negative
Transgene transmitted through seeds (R_1)	Seeds are non-transgenic
Propagation through seeds	Vegetative propagation (R_0)
Transformation frequency is 0.09% (4/5000)	Transformation frequency is 0.19% (9/5000)
Homozygotes [R_1] in 11-12 months	Fiber tests in 6 months
Commercial development	Gene screen

GUS, the shoot is assumed to be a germline transformant. If only epidermal or cortical layers stained for GUS, the transformants are considered to be epidermal transformants. Confirmation of the epidermal or germline transformation event is done at the anthesis stage by staining pollen for GUS. Pollen were incubated at 37°C for four hours, then scored as GUS positive (germline) or negative (epidermal). The progeny

Table 5. Inherited transgene activity In transformed cotton acetoacetyl CoA reductase activity mmol/min/mg.

Plant Number	R_1	R_2	R_3	R_4
6888-72	0.0	0.0	0.0	0.0
6888-75	-	0.055	0.066	NT
6888-76	-	0.153	0.063	NT
6888-79	-	0.169	0.160	0.141

Acetoacetyl CoA reductase activity in the progenies of primary transformant #68887. DP50 was transformed with *Alcaligenes eutrophus* gene acetoacetyl CoA reductase gene linked to cotton promoter E6 (John and Keller 1996)· A primary transformant, #68887, was identified through GUS screen and the fibers tested for acetoacetyl CoA reductase activity. Enzymatic activities shown are in mmol/min/mg. The seeds of #6888-7 were germinated and fibers from four plants tested for enzymatic activity (R_1). One of the plants # 68887-2 was negative for reductase activity, although it was positive for GUS. All four plants were then followed through another three more generations (R_2, R_3 and R_4). NT = not done.

Table 6. Cotton Varieties Transformed with Particle Bombardment.

Cultivar	Introduced Genes	EP	GL	Reference
Delta & Pine Land 50	Carrot extensin	43	4	John and Crow, 1992
	Flavobacterium parathon hydrolase (*opd*)	0	2	Miller *et al.*, 1993
	Agrobacterium hormone genes (*iaam*, *iaaH* and *ipt*)	27	5	John, 1994
	Sheep keratins	102	41	John, unpublished
	Catharanthus tryptophan decarboxylase (*tdc*)	10	2	John, 1999
	Peroxidase genes from cotton, *Arabidopsis,* tomato and tobacco	29	18	John and Keller, 1996
	Alcaligene β-ketothiolase, acetoacetyl CoA reductase and PHA synthase	208	79	John and Keller, 1996
	Streptomyces phosphinothricin acetyl transferase (*bar*)	30	13	Keller *et al.*, 1997
Delta & Pine Land 90	Cotton, *Arabidopsis,* tomato and tobacco peroxidase	10	4	John, 1999
	Agrobacterium hormone genes (*iaam, iaaH,* and *ipt*)	21	3	John, 1994
	Streptomyces phosphinothricin acetyl transferase (*bar*)	2	-	Keller *et al.*, 1997
Suregrow 501 & 125	*Catharanthus* tryptophan decarboxylase (*tdc*)	15	1	John unpublished
	Agrobacterium hormone genes (*ipt*)	8	4	Miller *et al.*, 1993
	Alcaligene β-ketothiolase, acetoacetyl CoA reductase and PHA synthase	29	14	John and Keller, 1996
Coker 312	*Streptomyces* phosphinothricin acetyl transferase (*bar*)	16	7	Keller *et al.*, 1997
	Cotton, *Arabidopsis*, tomato, and tobacco peroxidase	3	-	John, 1999
	Pseudomonas indigo gene and *E. coli* tryptophanase	11	10	John unpublished
	Streptomyces tryosinase	127	183	John unpublished
El Dorado	*Streptomyces* phosphinothricin acetyl transferase (*bar*)	2	-	Keller *et al.*, 1997

Table 6. Continued

Cultivar	Introduced Genes	EP	GL	Reference
Pima S-6	Cotton peroxidase	4	4	John, 1999
	Streptomyces phosphinothricin acetyl transferase (*bar*)	14	5	Keller *et al.*, 1997
Sea Island	*E. coli* β-glucuronidase (GUS)	-	-	Finer and McMullen, 1990

Cotton cultivars were transformed with particle bombardment. The protocol used for transformation is described in the text. The genes were linked to either cauliflower mosaic virus 35S promoter or to the cotton promoters, E6 or FbL2A (John 1996, John and Keller 1996). Genes, *opd, iaaM, iaaH, ipt*, various peroxidases, tyrosinase, β-ketothiolase, indigo, acetoacetyl CoA reductase, PHA synthase and tryptophanase were cloned by PCR using published sequence information. Gene *tdc* was a gift of Dr. V. De Luca, Plant Biotechnology Institute, Saskatoon, Canada. The *bar* gene was cloned at Cetus Corp. Emeriville, CA. Abbreviations, EP and GL denote epidermal and germline events respectively.

Table 7. Comparison of cotton transformation methods.

Characteristics	*Agrobacterium*	Protoplast Gene Transfer	Particle Bombardment
Cycle time			
Agronomic traits	2-6 months	2-6 months	3-4 months
Product traits	10-14 months	10-14 months	6 months
Cultivar	Dependent (Coker/Acala)	Dependent (Coker)	Independent (Upland/ barbadense)
Tissue culture/ Regeneration/ Somoclonal variations	Yes	Yes	No
Antibiotic Selection	Yes	Yes	No
R_1 homozygote	14-18 months	14-18 months	11-12 months
Transgenic plant quality	Variable	Variable	Consistent (90 to 95% excellent)
Multigene transfer	Limited	Not limited	Not limited
Efficiency	High	(Limited data)	Low
Commercialization	7 - 10 years	(No data)	5 - 6 years*

*Based on estimation (not actual data).
Cycle time refers to the earliest time or time period, after initiation of the experiment, when results can be assessed. Agronomic traits include insect, or herbicide resistance. Product traits indicate fiber traits, such as strength, length and micronaire or other chemical and physical parameters, or fiber yield. R_1 homozygote can be identified through pollen staining for marker gene activity. Transgenic plant quality refers to morphological aspects of the plant. Many of the transformants regenerated from tissue culture may be sterile. Also some of these plants show deficiencies in growth and yield.

of the germline events (R_1) are tested after seed germination for GUS activity and the segregation pattern is established. The R_1 plants can be tested at anthesis for homozygosity by pollen staining. A timeline required for cotton transformation is summarized in Table 2.

4.5. Transgenic cotton

The transformation frequencies for cotton by particle bombardment is low. Only 0.1 to 0.2% of bombarded meristems are recovered as transgenic plants (Table 3). Not all factors that affect efficiency have been determined. With a given set of bombardment parameters, different cultivars differ in their transformation efficiencies. Thus, fine tuning the conditions is required for each cultivar. Normally the epidermal transformation frequencies are two to three times greater than that of germline transformation (Table 3). The epidermal transformants are useful for testing the effects of various genes. For example, a herbicide or an insect resistant gene can be inserted into cotton and the epidermal transformants tested for corresponding resistance in leaf biomass. Similarly, the effects of fiber modification genes can be tested in the fibers of epidermal transformants (Table 4), since fibers are epidermal in origin (John and Crow, 1992). The germline transformants are screened for high expression of the transgene and for suitable agronomic traits and subjected to many field trials and selection before marketing. Backcrossing of transgenic lines to improve crop traits also appears to be important.

The heritability and stability of the transgene in the transformants are confirmed by Southern and enzyme analyses from several generations (McCabe and Martinell, 1993; Keller *et al.*, 1997). The transgene is present in subsequent generations as part of the genomic DNA suggesting integration into cotton genome. We tested several generations of a transgenic cotton containing a bacterial gene for acetoacetyl CoA reductase involved in the production of polyhydroxybutyrate. As shown in Table 5, R_2, R_3 and R_4 generations of the primary transformant expressed the enzyme. Similar results were also obtained from other transformants expressing the marker gene GUS. We have introduced several genes into various cotton cultivars, increasing the confidence level that this methodology is widely applicable (Table 6).

Particle bombardment requires fairly sophisticated instrumentation and is labor intensive. However, as shown in Table 7, there are many positive attributes to this technology for generating transformed cotton in a timely fashion. Further technological refinements, which would increase transformation frequency, would broaden its applications to routine cotton transformation in academic laboratories. Development of an antibiotic or herbicide selection regimen after particle bombardment may be useful in this regard (Chlan *et al.*, 1995).

REFERENCES

Barton KA, Binns AN, Matzke, AJM and Chilton MD (1983) Regeneration of intact tobacco plants containing full-length copies of genetically engineered T-DNA and transmission of T-DNA to R_1 progeny. *Cell,* **32** : 1033-1043.

Barwale UB, Kerns HR and Widholm JM (1986) Plant regeneration from callus cultures of several soybean genotypes via embryogenesis and organogenesis. *Planta,* **167** : 473-481.

Bevan MW, Flavell RB and Chilton MD (1983) A chimaeric antibiotic resistance gene as a selectable marker for plant cell transformation. *Nature,* **304** : 184-187.

Chalfie M, Tu Y, Euskirchen G, Ward WW and Prasher DC (1994) Green fluorescent protein as a marker for gene expression. *Science,* **263** : 802-805.

Chlan CA, Lin J, Cary JW and Cleveland TE (1995) A procedure for biolistic transformation and regeneration of transgenic cotton from meristematic tissue. *Plant Mol. Biol. Rep.,* **13** : 31-37.

Christou P, Ford TL and Kofron JM (1992) The development of a variety-independent gene-transfer method for rice. *Trends Biotechnol.,* **10** : 239-246.

Cooley J, Ford T and Christou P (1995) Molecular and genetic characterization of elite transgenic rice plants produced by electric-discharge particle acceleration. *Theor. Appl. Genet.,* **90** : 97-104.

Curtis IS, Power JB and Davey MR (1995) NPTII assays for measuring gene expression and enzyme activity in transgenic plants. *Methods Mol. Biol.,* **49** : 149-159.

De Block M, Botterman J, Vandewiele M, Dockx J, Thoen C, Gossele V, Movva NR, Thompson C, Vanmontagu M and Leemans J (1987) Engineering herbicide resistance in plants by expression of a detoxifying enzyme. *EMBO J.,* **6** : 2513-2518.

Della-Cioppa G, Bauer SC, Taylor ML, Rochester DE, Klein BK, Shah DM, Fraley RT, and Kishore GM (1987) Targeting of herbicide-resistant enzyme from *Escherichia coli* to chloroplasts of higher plants. *Bio-Technology,* **5** : 579-584.

D'Halluin K, Bonne E, Bossut M, DeBeuckleleer M and Leemans J (1992) Transgenic maize plants by tissue electroporation. *Plant Cell,* **4** : 1495-1505.

Elzen PJM van den, Townsend J, Lee KY and Bedbrook JR (1985) A chimaeric hygromycin resistance gene as a selectable marker in plant cells. *Plant Mol. Biol.,* **5** : 299-302.

Finer JJ and McMullen MD (1990) Transformation of cotton (*Gossypium hirsutum L.*) via particle bombardment. *Plant Cell Rep.* **8** : 586-589.

Fraley RT, Rogers SG, and Horsch RB (1996) Genetic transformation in higher plants. *Crit. Rev. Plant Sci.,* **4** : 1-46.

Frame BR, Drayton PR, Bagnall SV, Lewnau CJ, Bullock WP, Wilson HM, Dunwell JM, Thompson JA, and Wang K (1994) Production of fertile transgenic maize plants by silicon carbide whisker-mediated transformation. *Plant J.,* **6** : 941-948.

Golovkin MV, Abraham M, Morocz S, Bottka S, Feher A, and Dudits D (1993) Production of transgenic embryogenic plants by direct DNA uptake into maize protoplasts. *Plant Sci.,* **90** : 41-52.

Gong Z, Shen W, Zhou G, Huang, J, and Qian S (1988) Introducing exogenous DNA into plants after pollination DNA traverse pathway of pollen tube to reach embryonic sac. *Sci. Sin. Ser B,* **31** : 1080-1084.

Gorman CM, Moffat LF and Howard BH (1982) Recombinant genomes that express chloramphenicol acetyltransferase in mammalian cells. *Mol. Cell Biol.,* **2** : 1044-1051.

Guerineau F, Brooks L, Meadows J, Lucy A, Robinson C and Mullineaux P (1990) Sulfonamide resistance gene for plant transformation. *Plant Mol. Biol.*, **15** : 127-136.

Guivarch A, Caissard JC, Azmi A, Elmayan T, Chriqui D and Tepfer M (1996) In situ detection of Expression of gus reporter gene in transgenic plants: Ten years of blue genes. *Transgenic Res.*, **5** : 281-288.

Guo Y, Liang H and Berns MW (1995) Laser-mediated gene transfer in rice. *Physiol. Plant*, **93** : 19-24.

Harms CT, DiMaio JJ, Jayne SM, Middlesteadt LA, Negrotto DV, Thompson-Taylor H, and Montoya AL (1991) Primisulfuron herbicide-resistant tobacco plants mutant selection *in vitro* by Adventitious shoot formation from cultured leaf discs. *Plant Sci.*, **79** : 77-85

Haseloff J and Amos B (1995) GFP in plants. *Trends Genet.*, **11** : 328-329.

Heim R, Cubitt AB and Tsien RY (1995) Improved green fluorescence. *Nature*, **373** : 663-664.

Jefferson RA, Burgess SM and Hirsh D (1986) β-glucuronidase from *Escherichia coli* as a gene fusion marker. *Proc. Natl. Acad Sci. USA*, **83** : 8447-8451.

Jefferson RA, Kavanagh TA and Bevan MW (1987) GUS fusions: β-glucuronidase as a sensitive and Versatile gene fusion marker in higher plants. *EMBO J.*, **6** : 3901-3907.

John ME and Crow LJ (1992) Gene expression in cotton (*Gossypium hirsutum L.*) fiber: Cloning of the mRNAs. *Proc. Natl. Acad. Sci. USA*, **89** : 5769-5773.

John ME (1994) Progress in genetic engineering of cotton for fiber modifications. In: *Challenging the Future: Proceedings of the World Cotton Research Conference.* (Eds. Constable GA, Forrester NW) Melbourne: CSIRO pp 292-296.

John ME (1996) Structural characterization of genes corresponding to cotton fiber mRNA, E6: reduced E6 protein in transgenic plants by antisense gene. *Plant Mol. Biol.*, **30** : 297-306.

John ME and Keller G (1996) Metabolic pathway engineering in cotton: Biosynthesis of polyhydroxybutyrate in fiber cells. *Proc. Natl. Acad. Sci. USA*, **93** : 12768-12773.

John ME (1999) Transgenic cotton plants producing heterologous peroxidase. US patent # 5, 869, 720.

Jones JDG, Svab Z, Harper EC, Hurwitz CD and Maliga P (1987) A dominant nuclear streptomycin resistance marker for plant cell transformation. *Mol. Gen. Genet.*, **210** : 86-91.

Klein TM, Wolf ED, Wu R and Sanford JC (1987) High-velocity microprojectiles for delivering nucleic acids into living cells. *Nature*, **327** : 70-73.

Keller G, Spatola L, McCabe DE, Martinell B, Swain W and John ME (1997) Transgenic cotton resistant to herbicide bialaphos. *Transgenic Res.*, **6** : 385-392.

Lazzeri PA and Shewry PR (1993) Biotechnology of cereals. *Biotechnol. Genet. Eng. Rev.*, **11** : 79-146.

Leffel SM, Mabon SA and Stewart CN (1997) Applications of green fluorescent protein in plants. *BioTechniques*, **23** : 912-918.

Lyon BR, CousinsYL, Llewellyn DJ and Dennis ES (1993) Cotton plants transformed with a bacterial degradation gene are protected from accidental spray drift damage by the herbicide 2, 4-dichlorophenoxyacetic acid. *Transgenic Res.*, **2** : 162-169.

Matsumoto S, Takebe I and Machida Y (1988) *Escherichia coli lacZ* gene as a biochemical and histochemical marker in plant cells. *Gene,* **66** : 19-29.

McCabe DE, Swain WF, Martinell BJ and Christou P (1988) Stable transformation of soybean (*Glycine max*) by particle acceleration. *BioTechnology,* **6** : 923-926.

McCabe DE, Martinell BJ and Glaser DA (1992) Apparatus for genetic transformation. US patent # 5, 120, 657.

McCabe DE and Martinell BJ (1993) Transformation of elite cotton cultivars via particle bombardment of meristems. *BioTechnology,* **11** : 596-598.

Miki BL, Labbe H, Hattori J, Ouellet, T, Gabard J, Sunohara, G, Charest, PJ and Iyer VN (1990) Transformation of *Brassica napus* canola cultivars with *Arabidopsis thaliana* acetohydroxy acid synthase genes and analysis of herbicide resistance. *Theor. Appl. Genet.,* **80** : 449-458.

Miller MJ, Franklin N, Keller G and John ME (1993)Trangenic fiber for bioremediation: Expression and characterization of parathion hydrolase in transgenic cotton fiber and application to parathion pesticide degradation for water decontamination. *In Vitro Cell Dev. Biol. Animal,* 29A : 85A.

Morikawa H and Yamada Y (1985) Capillary microinjection into protoplasts and intranuclear localization of injected materials. *Plant Cell Physiol.,* **26** : 229-236.

Morikawa H, Chiba K, Irifune K, Iida A, Seki M, Nishihara M, Tanaka T, Yamashita T and Asakura N (1994) Transformation of plant cells and tissues by pneumatic particle gun. In : *Particle bombardment technology for Gene Transfer.* (Eds. Yang NY, Christou P) Oxford University Press, New York, pp 52-59.

Ow DW, Wood KV, DeLuca M, de Wet JR., Helinski DR and Howell SH (1986) Transient and stable expression of the firefly luciferase gene in plant cells and transgenic plants. *Science,* **234** : 856-859.

Perl A, Galili S, Shaul O, Ben-Tzvi I and Galili G (1993) Bacterial dihydrodipicolinate synthase and desensitized aspartate kinase: Two novel selectable markers for plant transformation. *Bio-Technology,* **11** : 715-718.

Potrykus I (1991) Gene transfer to plants assessment of published approaches and results. *Annu. Rev. Plant Physiol. Plant Mol. Biol.,* **42** : 205-225.

Prasher DC, Eckenrode VK, Ward WW, Prendergast FG and Cormier MJ (1992) Primary structure of the *Aequorea victoria* green-fluorescent protein. *Gene,* **111** : 229-233.

Sautter C (1994) Particle bombardment microtargeting device for shooting to meristems. In: *Particle bombardment technology for gene transfer.* (Eds. Yang NY, Christou P) Oxford University Press, New York, pp 60-69.

Shimamoto K, Terada R, Izawa T and Fujimoto H (1989) Fertile transgenic rice plants regenerated From transformed protoplasts. *Nature,* **338** : 274-276.

Stoeber F (1991) Stude sur le glucurondiase et le glucornide permease chez E. coli. These de Docteur des Science Paris.

Sussex IM (1989) Developmental programming of the shoot meristem. *Cell,* **56** : 225-230.

Svab Z, Harper EC, Jones JDG and Maliga P (1990) Aminoglycoside-3 - adenyltransferase confers resistance to spectinomycin and streptomycin in *Nicotiana tabacum. Plant Mol. Biol.,* **14** : 197-205.

Takeuchi Y, Dotson M and Keen NT (1992) Plant transformation: a simple particle bombardment Device based on flowing helium. *Plant Mol. Biol.,* **18** : 835-839.

Umbeck P, Johnson G, Barton K and Swain W (1987) Genetically transformed cotton *Gossypium hirsutum* L plants. *BioTechnology,* **5** : 263-266.

Wang K, Drayton P, Frame B, Dunwell J and Thompson J (1995) Whisker-mediated plant transformation: An alternative technology. *In Vitro Cell Dev. Biol. Plant,* **31** : 101-104.

Zapata C, Park SH, El-Zik KM and Smith RH (1999) Transformation of a Texas cotton cultivar by using Agrobacterium and shoot apex. *Theor. Appl. Genet.,* **98** : 252-256.

Chapter 4

TRANSGENIC BARLEY - FROM THE LABORATORY TO THE FIELD

A Ritala[1★], AM Nuutila[1], M Salmenkallio-Marttila[1], K Aspegren[2], R Aikasalo[3], U Kurten[1,4], J Tammisola[1,5], TH Teeri[2], L Mannonen[1,6] and V Kauppinen[1]

[1]*VTT Biotechnology, P.O. Box 1500, FIN-02044-VTT, Finland*
[2]*Institute of Biotechnology, P.O. Box 56, FIN-00014 University of Helsinki, Finland*
[3]*Boreal Plant Breeding Ltd., Myllytie 10, FIN-31600 Jokioinen, Finland*
[4]*Present address : Leo Pharma Oy, Rajatorpantie 41C, FIN-01640 Vantaa, Finland*
[5]*Present address : Ministry of Agriculture and Forestry, P.O. Box 30, FIN-00023 Government, Helsinki, Finland*
[6]*Present address: National Food Administration, P.O. Box 5, FIN-00531 Helsinki, Finland*

Summary

Production of transgenic barley was carried out in this project from the laboratory to the field. Cell and tissue culture and gene transfer techniques were developed for the successful transformation of barley. Here, we present a case study summarizing this research and describe how complementation of the malt enzyme spectrum was obtained using genetic engineering. The applicability of the α-amylase promoter and the gene coding for endo-1,4-β-glucanase was demonstrated to lead to amounts of thermotolerant β-glucanase in transgenic barley cell lines which were sufficient to reduce wort viscosity during mashing. Consequently, the malting quality of two barley cultivars, Kymppi and Golden Promise, was modified to better meet the requirements of the brewing process. In order to prepare for commercial cultivation of transgenic barley, a risk assessment to measure transgene flow in field conditions through pollen dispersal was also performed.

Keywords : Anther culture, barley, embryo culture, electroporation, field trial, gene transfer, *Hordeum vulgare* L., marker

★Corresponding author : E-mail : anneli.ritala@vtt.fi

gene, microspore culture, particle bombardment, risk assessment, transgenic, thermotolerant endo-1,4-β-glucanase, transgene flow.

Abbreviations : *nptII*, gene coding for neomycin phosphotransferase II (NPTII); *egl1*, gene coding for thermotolerant endo-1,4-β-glucanase (EGI).

1. INTRODUCTION

Since the early 1980s, the development of plant genetic engineering has given rise to a number of practical applications ranging from improved food quality to production of vaccines and biodegradable plastics in plants (e.g. Mahon *et al.,* 1998; Conway and Toenniessen, 1999; DellaPenna, 1999; Hanley *et al*., 2000). Molecular breeding can be used for precise and efficient improvement of traditional plant breeding characteristics, but also provides the possibility of introducing essentially new traits to plants, thus offering great promise for the generation of value-added new plant varieties for the production of high-quality raw materials. In this respect, plant varieties producing special oils, starches and fibres, as well as pharmaceutical compounds, medicinal proteins and edible vaccines are either under development or have been bred using modern biotechnological methods (e.g. Lassner, 1997; Mahon *et al.,* 1998; Heyer *et al*., 1999; Poirier, 1999).

The processing qualities of cereals can be modified by altering the structural grain constituents or the enzyme activities that mobilize storage reserves of the seeds. β-Glucan is the main constituent of the barley endosperm cell wall. During malting, the β-glucan polymer is hydrolysed by β-glucanase. A decrease in soluble β-glucans in wort improves the filtration of beer. Native barley β-glucanase is susceptible to irreversible thermoinactivation at temperatures above 55°C. This results in suboptimal degradation of cell wall β-glucans during mashing. Complementation of the barley genome with a thermotolerant β-glucanase gene would ensure sufficient β-glucanase activity during mashing. In order to be able to produce improved malting barley by genetic engineering, several alterations must be made. A schematic presentation of the whole project aiming towards complementation of the malt enzyme spectrum is presented in Figure 1.

For successful transformation barley cell and tissue culture are needed.

Therefore the development of gene transfer techniques and tissue culture systems have progressed in parallel (Figure 1). In the case of barley, the first reports cover procedures for cell and tissue culture, isolation and regeneration of protoplasts and both transient and stable transformation of tissue cultures (e.g. Lührs and Lörz, 1987; Kartha *et al.*, 1989; Mendel *et al.*, 1989; Yan *et al.*, 1990; Jähne *et al.*, 1991; Lazzeri *et al.*, 1991; Lee *at al.*, 1991; Ritala *et al.*, 1993; Harwood *et al.*, 1995).

2. CELL AND TISSUE CULTURE OF BARLEY

Tissue culture is an essential component of plant biotechnology. Potentially, embryogenic cells are good material for gene transfers,

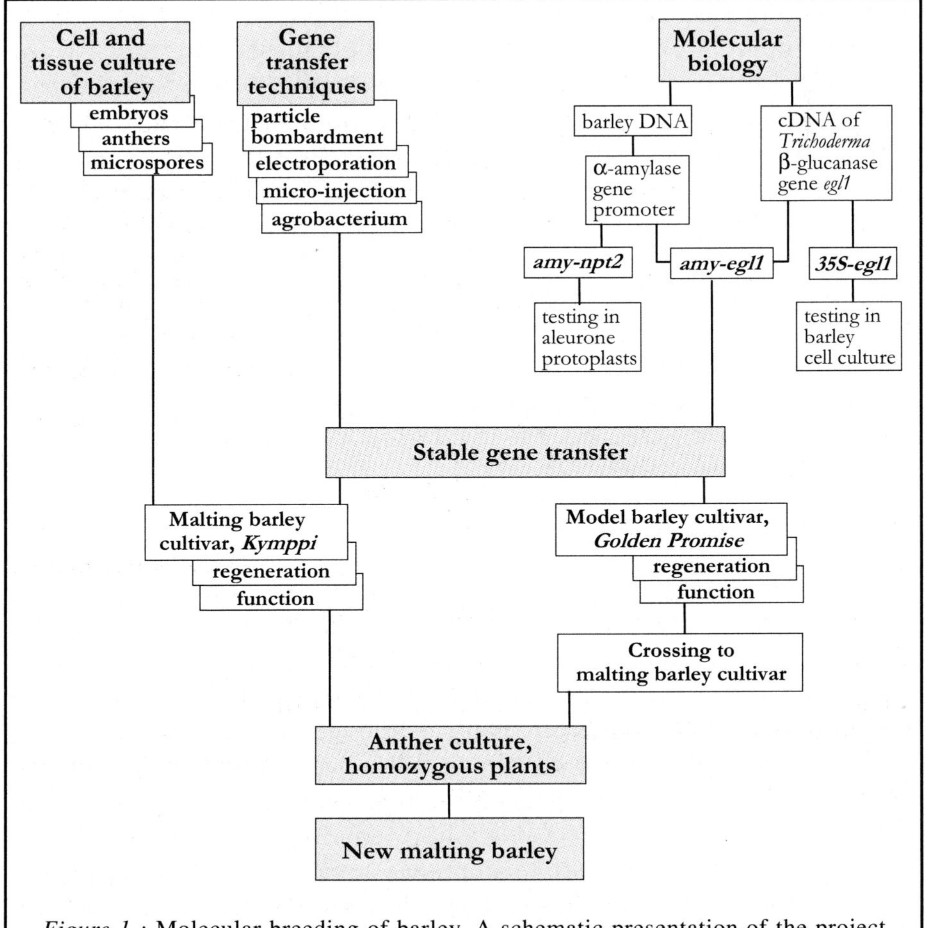

Figure 1 : Molecular breeding of barley. A schematic presentation of the project aiming towards improved malting quality of barley.

because chimerism of a transgenic plant can be avoided. However, care must be taken to ensure that plants derived from tissue cultures are true-to-type. It is important to realize that uncontrolled instability occurs when plant cells are cultured *in vitro*. This source of variability, somaclonal variation, is at least partly due to increased genetic variability (e.g. Evans, 1989; Karp, 1993).

Plant regeneration through tissue culture can be accomplished via culture of zygotic embryos, organogenesis or somatic embryogenesis. Since the first descriptions of somatic embryogenesis by Steward *et al.,*, (1958) and Reinert (1958), it has been recognized as an important pathway for the regeneration of plants from cell culture systems (Zimmerman, 1993). Theoretically all healthy, undamaged cells are totipotent; although in practice high-frequency production of embryogenic cells from diverse genotypes or from explants other than those obtained from spatially and temporally discrete tissues has met with considerable difficulties (Carman, 1990). For *Daucus carota* and some other plant species almost any part of the plant taken at any time of development will produce somatic embryos in culture: excised embryos (Smith and Krikorian, 1990), hypocotyls (Wake *et al.,* 1992), secondary phloem (Janick *et al.,* 1989), apical meristems (Kamada *et al.,* 1989), petiole segments (Ammirato, 1983) and taproots (Steward *et al.,* 1958). For many species, however, only certain regions of the plant at a certain developmental stage may respond to treatment. This is the case with many monocotyledonous plants. For example in barley, somatic embryogenesis has mainly been induced from juvenile tissues, *e.g.* immmature embryos and inflorescences (Karp and Lazzeri, 1992).

2.1. Embryo culture of barley

Immature embryos and polyembryogenic cell mass of barley (*Hordeum vulgare* L.) have often been used as target material for gene transfer by particle bombardment (Wan and Lemaux, 1994; Koprek *et al.,* 1996). The transformation efficiency depends greatly on the ability of the target material to regenerate into green plants. Although acceptable regeneration frequencies have been achieved, albinism has been a problem. The choice of culture medium affects regeneration efficiency and the proportion of green plants obtained. Many studies have clearly demonstrated that the nitrogen balance of the culture medium has a major effect on the somatic embryogenesis of several species. Especially the nitrate:ammonium ratio and the addition of reduced nitrogen in the form of amino acids affect somatic embryogenesis (Reinert *et al.,* 1967; Stuart and Strickland, 1984;

Meijer and Brown, 1987; Grimes and Hodges, 1990; Shetty and Asano, 1991). Furthermore, the nitrogen requirements vary between different stages of embryogenesis. This might be due to differences in the endogenous amino acid metabolism during morphogenetic processes such as embryogenesis and germination.

We were able to enhance the regeneration efficiency of polyembryogenic cultures of barley cv. Kymppi (Figure 2) by optimizing the nitrogen composition of the polyembryogenesis and regeneration media. Furthermore, we were able to reduce the percentage of albino regenerants from 8.0% to 2.3% (Nuutila *et al.,* 2000). Our results suggest that the nitrogen composition of the polyembryogenesis medium is not as crucial as the composition of the regeneration medium. This result is consistent with those of Mordhorst and Lörz (1993), who found that when the nitrogen composition of media used for barley anther culture was altered the plant regeneration was strongly affected, whereas the other culture stages were only moderately influenced. The polyembryogenesis stage clearly benefited from the addition of organic nitrogen. Addition of organic nitrogen to embryogenesis medium in the

Figure 2 : Embryo culture of barley: A) isolated immature embryo on induction medium, B) induced polyembryogenic calli, C) somatic embryos on regeneration medium, D) seed-set in regenerated fertile barley plants.

form of different amino acids has been shown to enhance differentiation and somatic embryogenesis in barley (Rengel and Jelaska, 1986; Olsen, 1987; Muyuan *et al.*, 1990). In the regeneration medium the amount of organic nitrogen and the ratio of organic to inorganic nitrogen were the factors with the greatest effect on regeneration. In the optimized medium the amount of organic nitrogen (glutamine) was low (0.7 mM) and the ratio of inorganic:organic nitrogen was 58:1. However, if organic nitrogen was totally omitted from the medium, the regeneration frequency decreased (Nuutila *et al.*, 2000). These results are consistent with those obtained by Olsen (1987) and Mordhorst and Lörz (1993).

We were also able to improve the regeneration by modifying the copper content of the medium (Nuutila *et al.*, 2000). Elevated levels of copper have been shown to improve regeneration in barley (Dahleen, 1995). Furthermore, our observations confirmed the earlier report of Dahleen (1995) that the optimal concentration of copper for barley cultures is genotype-dependent (Nuutila *et al.*, 2000).

2.2. Anther and microspore culture of barley

Cultures of anthers or isolated microspores have been used increasingly in plant breeding in recent years. Androgenic haploids have been produced in several genera, but most effort has been directed to economically important plants such as cereals and vegetables (Foroughi-Wehr and Wenzel, 1989). For example several barley cultivars have been released as doubled haploids (Pickering and Devaux, 1992). The main advantage of using the anther and microspore culture techniques for plant production is the rapid and complete homozygosity of the offspring. This makes phenotype selection for qualitative and particularly for quantitative inherited characters easier, and breeding more efficient. Much work has been carried out to develop reliable and effective anther and microspore culture systems. The effects of various factors, e.g. genotype, developmental stage and growth conditions of the donor plant, pretreatment, medium components and culture conditions, on the culture response have been determined.

The physiological state of the donor plants, which varies with season and culture conditions, affects the anther and microspore culture response. Critical environmental factors include light intensity, photoperiod, temperature and nutrition. To ensure good and even quality of the donor plants, they are usually grown in controlled environmental chambers. We followed the regeneration capacity of isolated barley microspores

(*Hordeum vulgare* L. cv. Kymppi) for one year (Ritala *et al.*, 2001a). In general, the regeneration varied from batch to batch. Although the plant material was produced in an artificial growth chamber, strong seasonal variation was observed (Figure 3). No explanation for this could be found. The number of regenerated plants was significantly lower if the donor barley was potted in soil during winter (from November to February), whereas better results were obtained with material grown during spring, summer and autumn. A similar observation of the variability of regeneration frequency with external season was earlier reported by Jähne and co-workers (1994, personal communication)

It has been shown that microspores need a signal to be switched from the gametophytic to the sporophytic developmental pathway. Cold-shock and osmotic stress pretreatments have been applied to a variety of plant species (for a review see Jähne and Lörz, 1995). In barley a common pretreatment has been at + 4°C for four weeks in the dark (Mejza *et al.,* 1993). The optimal duration of the cold pretreatment period has also been shown to be genotype-dependent (Powell, 1988). As a result of cold storage, increased numbers of microspores are arrested

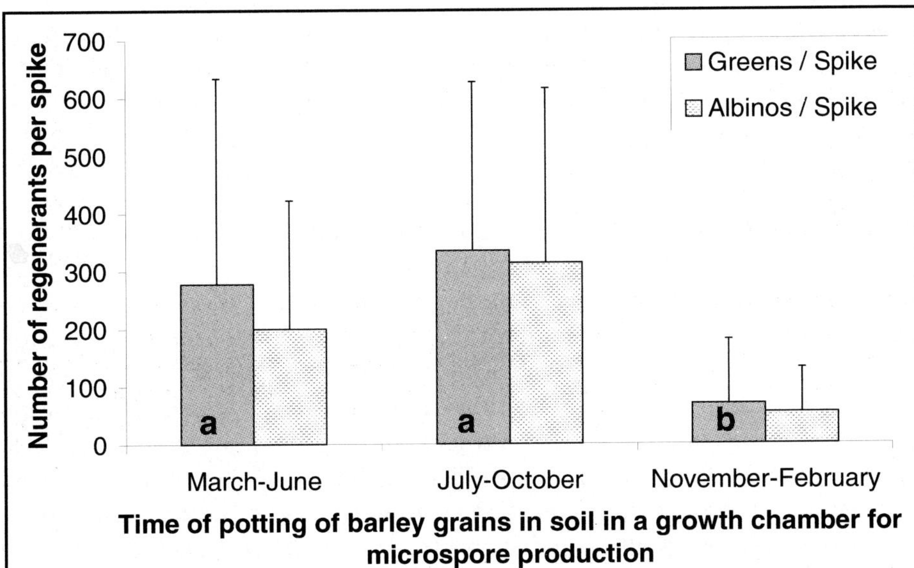

Figure 3 : Strong seasonal variation in regeneration capacity of isolated microspores (*Hordeum vulgare* L. cv. Kymppi) was observed. The regeneration capacity was followed for one year. The different letters (a, b) denote a statistically significant difference ($p < 0.001$) at the 95 % confidence level.

during the first mitosis, as starch production is blocked (Foroughi-Wehr and Wenzel, 1989). The cold treatment also retards degradation of the tapetum, allowing a higher proportion of microspores to change their developmental pattern from gametophytic to sporophytic. Cold treatment may also provide a non-specific shock, resulting in the establishment of endogenous cellular conditions which favour the desired development. The optimal length of the cold pretreatment for the Finnish barley cultivar Kymppi proved to be four weeks at + 4°C in the dark (Ritala *et al.*, 2001a).

Several culture systems and media compositions have been used for anther and microspore culture, e.g. the modified MS medium (Foroughi-Wehr *et al.*, 1976) and the N6 medium (Chu *et al.*, 1975). The concentration and type of nitrogen in the medium, especially supplementation with amino acids, is important for green plant regeneration (Olsen, 1987; Mordhorst and Lörz, 1993). Sucrose is the most commonly used carbon source, but maltose was observed to promote culture response (Hunter, 1987) and somatic embryogenesis (Orshinsky *et al.*, 1990). Furthermore, the replacement of sucrose by maltose stabilized the osmotic pressure of the medium. The effects of maltose in the isolation and growth media were tested on the regeneration of green plants from isolated barley microspores (cv. Kymppi). A twofold increase in number of regenerated green plants was obtained by increasing the maltose concentration of the microspore isolation medium from 0.175M to 0.25M and of the culture medium from 0.175M to 0.325M (Salmenkallio-Marttila *et al.*, 1995b).

The sub-culturing regime had a marked effect on the regeneration of green plants from isolated barley microspores (cv. Kymppi). First the microspores were cultured in liquid medium for four weeks, after which the microspore mass was collected and plated on solid regeneration medium. Tsukahara and Hirosawa (1992) reported a tenfold increase in plantlet regeneration by dehydration of rice calli. The same dehydration treatment was tested with four weeks old microspore mass. We also wanted to make the regeneration procedure faster and easier and therefore tried plating the microspore mass after three weeks of culture by pipetting the suspension as small aliquots on the solid regeneration medium. The dehydration treatment improved regeneration by 40%. The change in subculture regime had an even greater effect, and an almost tenfold increase in the yield of green plants was obtained by shortening the suspension culture time of the developing proembryo mass (Salmenkallio-Marttila *et al.*, 1995b).

Careful choice of the above-mentioned microspore culture steps resulted in an average regeneration frequency of 300 green plants per starting material spike from the Finnish barley cultivar Kymppi (Figure 4, Salmenkallio-Marttila *et al.,* 1995b; Ritala *et al.,* 2001a). In order to evaluate the fertility of the microspore culture-derived plants, 76 plants were transferred to soil mix after rooting without any colchicine treatment. Of these plants 59 % were fertile, which is in accordance with the reported spontaneus doubling frequency of barley, varying from 50 % to over 90 % (Pickering and Devaux, 1992; Gudu *et al.,* 1993). Thus the tissue culture did not appear to cause any reduced fertlity and the use of colchicine treatment would probably have further improved the fertility.

3. GENE TRANSFER TO BARLEY

The first transgenic barley plants were produced by particle bombardment (Jähne *et al.,* 1994; Ritala *et al.,* 1994; Wan and Lemaux, 1994; Hagio *et al.,* 1995). Since then, other techniques have also resulted in transgenic barley plants (Funatsuki *et al.,* 1995; Salmenkallio-Marttila

Figure 4 : Microspore culture of barley: A) isolated barley microspores cultured for one week, B) embryogenic microspore culture after five weeks of culture, C) plantlets regenerating after seven weeks of culture, D) seed-set in a regenerated fertile barley plant.

et al., 1995a; Matthews *et al.*, 1997; Tingay *et al.*, 1997; Zhang *et al.*, 1999; Nobre *et al.*, 2000). Although, Koprek *et al.* (1996) reported successful transformation of several different barley varieties, the most reproducible results have been obtained by gene transfer by the bombardment of embryos of the variety Golden Promise (Wan and Lemaux, 1994) The majority of gene transfers aiming at commercial applications have been carried out using this method (e.g. Jensen *et al.*, 1996; Nuutila *et al.*, 1999; Horvath *et al.*, 2000).

Recently, barley has also been shown to be transformable by *Agrobacterium*. Tingay and co-workers (1997) reported a very efficient system resulting in transformation frequencies as high as 4 % of the treated Golden Promise embryos. It is very promising that the same group repeated the experiments with another variety, Schooner (Matthews *et al.*, 1997) and that another group was able to adapt the original transformation system with the variety Golden Promise (Qureshi *et al.*, 1997).

3.1. Production of transgenic plants by electroporation

Three- to four week old cultures of isolated barley microspores (*Hordeum vulgare* L. cv. Kymppi) were used for protoplast isolation (Salmenkallio-Marttila and Kauppinen, 1995). Protoplasts failed to divide without the support of nurse cells. As the nurse cultures used were highly regenerable, culture well inserts were used to ensure the separation of protoplast cultures from the nurse cells. The importance of nurse cultures for the regeneration of barley protoplasts has also been demonstrated in other studies (Jähne *et al.*, 1991; Funatsuki *et al.*, 1992).

Two different media were tested for protoplast culture: modified N6 medium (Chu *et al.*, 1975) derived from the medium used for microspore culture, and L1 medium used for barley protoplast culture by Lazzeri *et al.* (1991). Sustained protoplast divisions were obtained when L1 medium was used. In the protoplast culture and regeneration procedure the same factors have importance as in the microspore culturing method, e.g. the maltose concentration of the culturing and regeneration media, the growth regulators used, the amount and nature of the nitrogen source(s) and the time of cultivation in the different media. Maltose is the preferable carbon source for the culture and regeneration of barley protoplasts. For the division of protoplasts the growth regulator 2,4-D is necessary. The addition of coconut water to the L1 protoplast medium is also important for the division and growth of the protoplasts (Salmenkallio-Marttila and Kauppinen, 1995).

For regeneration of plants the essential factors are sequential lowering of the sugar concentration of the medium and addition of the plant growth regulator 6-benzyl-aminopurine, which promotes differentiation of the proembryos. Several hundred green plants were regenerated from the cultures. Plantlets regenerated from protoplasts were potted in soil within 4-5 months of collecting the spikes for microspore culture (Salmenkallio-Marttila and Kauppinen, 1995).

Barley protoplasts isolated from cultured microspores (cv. Kymppi) were electroporated and cultured to produce transgenic plants (Salmenkallio-Marttila *et al.*, 1995a). Many variables affect the efficiency of gene transfer through electroporation, including the field strength, duration and number of electrical pulses, DNA concentration and form and buffer composition (Nagata, 1989; Van Wert and Saunders, 1992). However, one variable which is very difficult to control is variation between the protoplast preparations. This factor causes considerable variation in the level of transformation efficiency between experiments that have been performed with different batches of protoplasts.

The time schedule for the production of transgenic barley plants by electroporation is shown in Figure 5. In total it took seven months from isolation and transformation of protoplasts to obtain seeds from transgenic plants. The results of enzyme activity assays and Southern blot hybridization analyses of transformed plants and progeny support the conclusion that copies of the transgene gene can be incorporated into the barley genome using electroporation and that the transferred gene is inherited and expressed by the progeny (Salmenkallio-Marttila *et al.*, 1995a). Electroporation is a good method for the production of transgenic barley plants. With our transformation protocol no selectable marker genes were used. The transformed protoplasts were regenerated without any selection and the plants regenerated were screened for the transferred gene. The use of selectable marker genes and selective agents (e.g. herbicides, antibiotics) might give rise to problems concerning consumer acceptance of the products and therefore transformation methods which allow production of transgenic plants without any additional selectable markers are of particular interest.

3.2. Production of transgenic plants by particle bombardment

Zygotic embryos can be germinated *in vitro* very easily. The use of

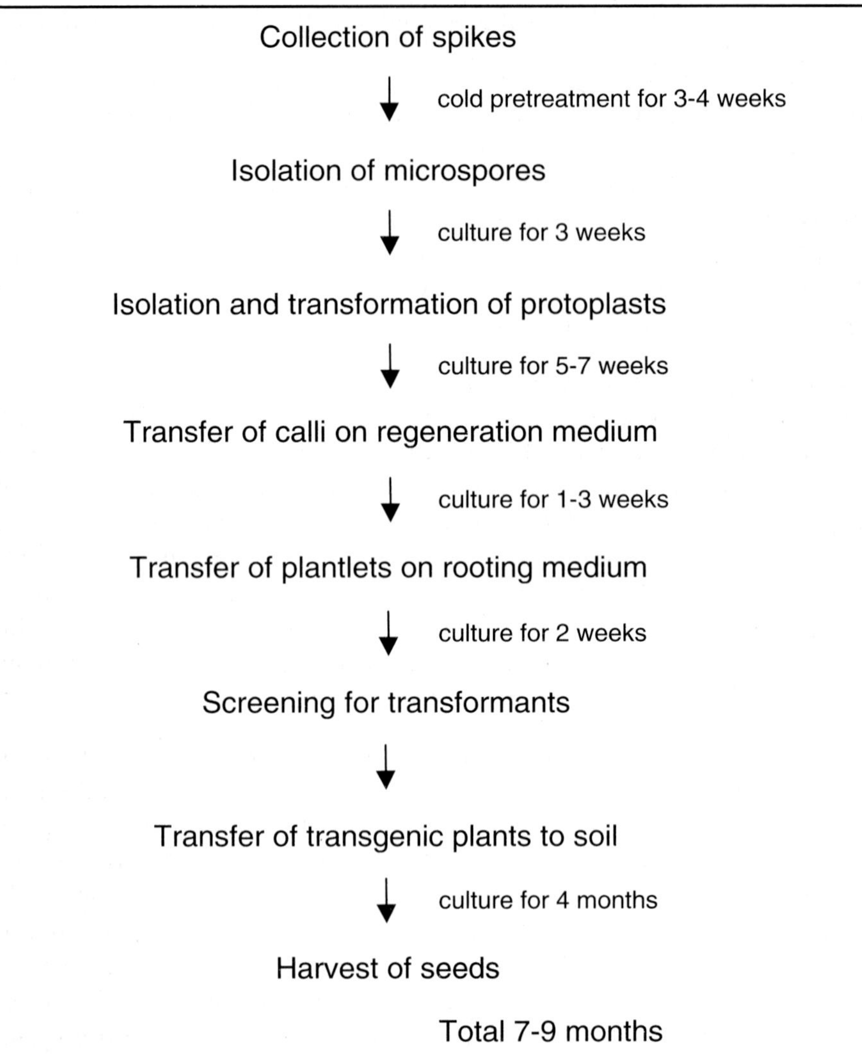

Figure 5 : Time schedule for the production of transgenic barley plants from protoplasts isolated from cultured microspores.

the meristem of an embryo as a target for gene transfer provides a relatively fast and simple way of producing transgenic plants. A grass embryo has several separate meristematic regions and for example in a barley grain, depending on its size and stage of development, even 16 separately mutatable meristems can exist (Jacobsen, 1966). When the meristem of an embryo is used as a target, the main problems are in targeting of the gene transfer to the desired regions and in chimerism of the obtained transgenic plants. However, if transgenes are passed to the next generation, these problems are solved. The first transgenic, fertile barley (*Hordeum vulgare* L. cv. Kymppi) was obtained by particle bombardment of the embryonic axis side of immature embryos (Ritala *et al.*, 1994). Successful production of transgenic plants through transformation of meristematic regions of embryos has also been reported for soybean (McCabe *et al.*, 1988; Christou *et al.*, 1990) and maize (Lowe *et al.*, 1995).

We followed the inheritance and stability of the transgene (*nptII*) of the first transgenic barley line (Ritala *et al.*, 1994) through six generations (Table 1). PCR-analysis of the T_2 generation revealed that 79 % of the plants carried the *nptII* gene. This was consistent with the expected Mendelian ratio and there was no statistically significant difference. Of these transgenic plants, 82 % expressed the transferred gene, indicating that in some of the plants the foreign gene had been silenced (Ritala *et al.*, 1995). Inheritance of the transferred gene was confirmed in the T_5 generation by Southern blot hybridization. The Southern blot hybridization pattern was passed unchanged to the T_5 plants, which revealed that no rearrangements had occurred. The overall appearance of the transgenic plants was normal when compared to non-transgenic barley plants.

Table 1. Observations of the stability of the transferred *nptII* gene through six generations of transgenic barley plants

Generation	Number of plants analyzed	Number of plants showing amplification of the *nptII* gene (PCR)	Number of plants expressing NPTII	Number of plants showing showing hybridization of the *nptII* probe (Number of analyzed plants)
T_0	1	1	1	1 (1)
T_1	1	1	1	1 (1)
T_2	22	22	21	ND
T_3	26	22	26	ND
T_4	25	ND	19	ND
T_5	30	ND	22	8 (8)

ND : Not done

The most reproducible transformation method of barley has been the bombardment of immature embryos of the variety Golden Promise (Wan and Lemaux, 1994). The time schedule for the production of transgenic barley plants by bombardment of immature embryos is shown in Figure 6. In total it took seven months from the isolation and bombardment of immature embryos to obtain seeds from transgenic plants. Our own results (Nuutila *et al.*, 1999) indicate very strongly that this transformation system is cultivar dependent. With the Finnish malting barley cultivar Kymppi the bombardment of 13 000 embryos with the method of Wan and Lemaux (1994) resulted in only one transgenic line. However, the bombardment of Golden Promise embryos with the same gene constructs resulted rather easily in transgenic plants. From 250 bombarded Golden Promise embryos, 38 transgenic plants were produced. We have estimated an approximately twenty-fold difference in the regeneration frequency of these varieties in favour of Golden Promise, but this does not explain the almost one thousand-fold difference in transformation frequency (unpublished results and Nuutila *et al.,* 1999).

4. CASE : PRODUCTION OF TRANSGENIC BARLEY WITH IMPROVED MALTING QUALITY

In malting barley the main interest has been in the enhanced ability of the hydrolytic enzymes to degrade the b-glucans of the grain in at the elevated temperatures prevailing during mashing. Transformation of thermotolerant bacterial (Olsen *et al.*, 1991) and fungal (Mannonen *et al.,* 1993) genes or of engineered native barley genes (Fincher, 1994) have been chosen as approaches for obtaining greater thermostability of b-glucanase.

4.1. Testing of the α-amylase promoter in aleurone protoplasts

Our aim was to express the thermotolerant fungal endo-1,4-β-glucanase during germination of the barley seed, thus providing active enzyme for mashing. The regulation of gene expression by plant growth regulators is well documented. In germinating barley grains, gibberellins synthesized by the embryo stimulate the production of hydrolytic enzymes, in particular α-amylase, in the aleurone layer. In order to study tissue-specific regulation of α-amylase promoters, we developed a method using isolated aleurone protoplast. By this method we confirmed correct regulation and expression of the gene of interest under the regulation of an α-amylase promoter (Salmenkallio *et al.,* 1990).

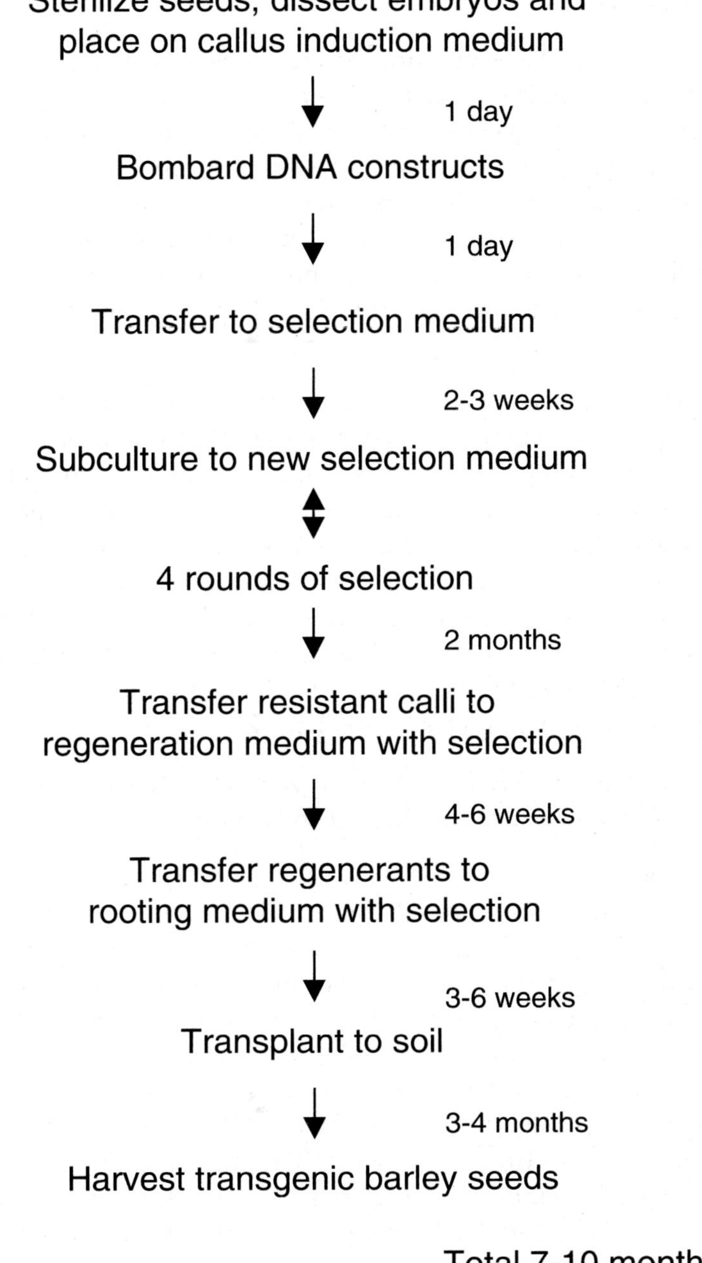

Figure 6 : Time schedule for the production of transgenic barley plants from bombarded immature embryos.

4.2. Testing of the fungal β-glucanase gene in barley cell culture

In order to test the expression of the fused fungal β-glucanase gene (*egl1*) we introduced it into suspension cultured cells of barley (Aspegren *et al.*, 1995). The fungal secretion signal was demonstrated to function in the transgenic barley cells. Quantitation of heterologous protein (EGI) in the culture medium of the transgenic barley cell line showed that the transgenic culture secreted EGI to the medium at a level of 0.5 mg/L, which is 0.4 % of the extracellular protein. The observed production level is below the amounts reported for the production of heterologous EGI by yeast (Penttilä *et al.,* 1987).

The EGI produced by barley cells was shown to be thermotolerant (Table 2), and the potential of heterologous EGI produced by barley to improve the mashing process was studied. We carried out mashing experiments with additions of native and heterologous EGI and compared the results with controls performed without enzyme additions. A significant reduction of soluble ß-glucan in the culture broth was observed when EGI of *Trichoderma reesei* or heterologous EGI from transgenic barley cells was used in the mashing (Figure 7). The filterabilities of broths from both EGI-treated mashing experiments were improved, as were the extract yields (Aspegren *et al.,* 1995).

4.3. Transgenic barley seeds expressing thermotolerant endo-1,4-β-glucanase during germination

In the next stage, we introduced a gene (*egl1*) encoding for a thermotolerant fungal endo-1,4-β-glucanase (EGI) to barley cultivars Golden Promise and Kymppi (Nuutila *et al.*, 1999). The *egl1* was expressed during germination under the control of the α-amylase promoter, thus providing a thermotolerant EGI enzyme that is active under high temperature mashing conditions (Table 2). The fungal EGI consists of a catalytic domain, a linker region and a cellulose binding domain (Penttilä *et al.,* 1986; Srisodsuk, 1994). This native EGI has been found to undergo proteolytic cleavage. However, the smaller unit, which contains the active core (catalytic domain) of the enzyme, remains enzymatically active (Srisodsuk, 1994). Our results suggest that the heterologous EGI from transgenic barley seeds contains only the active core of the enzyme and has no cellulose binding domain. The activity and thermostability of the heterologous EGI were not lost as a result of these changes (Nuutila *et al.,* 1999).

Table 2. Comparison of heterologous endo-(1,4)-β-glucanase I (EGI) and the native *Trichoderma reesei* enzyme (adapted from Nuutila *et al.*, 1999).

	EGI produced by transgenic seeds (Golden Promise and Kymppi)	EGI produced by transgenic barley cell culture	Active core of EGI produced by *T. reesei*	EGI produced by *T. reesei*
pI	4.10	4.10	4.30	4.65, 4.50
Molecular weight	50 kDa	52 kDa	48 kDa	55 kDa
β-glucanase activity	+	+	+	+
Thermo-tolerance	Active after 2 h at 65°C	Active after 2 h at 65°C	Active after 2 h at 65°C	Active after 2 h at 65°C
Binding to crystalline cellulose[a]	-	-	-	+

[a]Presence of cellulose-binding domain

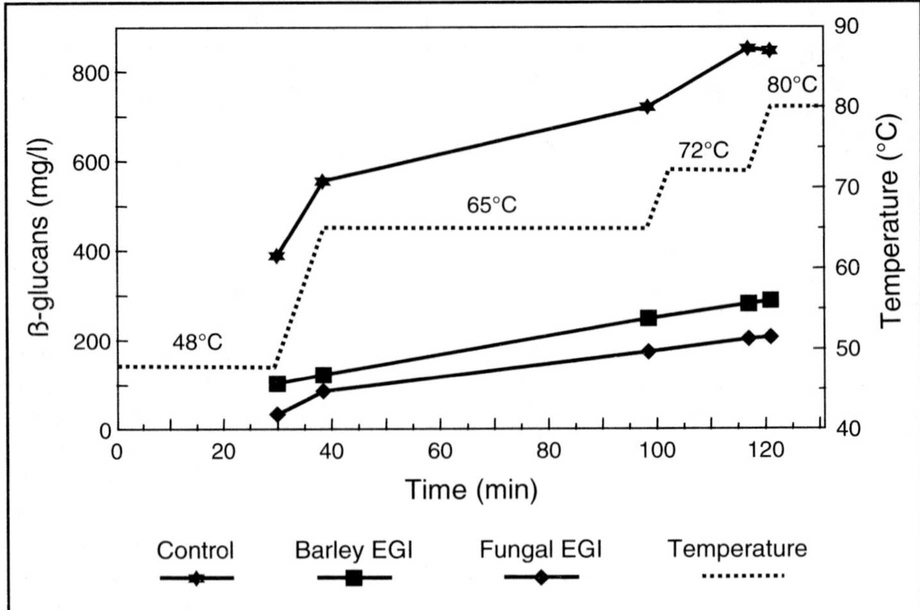

Figure 7 : Accumulation of soluble ß-glucans during mashing. Mashings performed without enzyme additions, with addition of heterologous EGI produced by barley cells and with EGI of *Trichoderma reesei* (modified from Aspegren *et al.* 1995)

We were able to complement the enzyme spectrum of two barley cultivars, Golden Promise and Kymppi, with a thermotolerant EGI of fungal origin. The amount of thermotolerant β-glucanase produced by the seeds (approx. 0.025% soluble seed protein) was shown to be sufficient to reduce wort viscosity by decreasing the soluble β-glucan content. This should lead to improved filterability of the wort and a better extract yield (Nuutila *et al.,* 1999).

5. FIELD TRIALS AND RISK ASSESSMENT IN CULTIVATION OF TRANSGENIC BARLEY

In order to demonstrate that transgenic plants are true-to-type, field trials are needed. It is important to realize that uncontrolled instability occurs when plant cells are cultured and transformed *in vitro* (Choi *et al.*, 2000). Furthermore, the transformation event cannot be fully controlled e.g. in respect of the integration locus and copy number. In general, the transformation strategy includes the production of a high number of transgenic plants and an efficient screening protocol for the desired phenotype and transgene expression.

Molecular breeding is becoming a useful tool in the improvement of plants and plant-based raw materials. Varieties with value-added traits are developed and different production lines must be kept separate. For good management practices, knowledge of relevant gene flow parameters is required. When varieties with novel traits are being developed, safety questions should also be considered. The impacts of different genetically modified plants should always be considered on a case-by-case basis according to the actual trait and plant in question. Gene flow of transgenes does not differ from that of other genes and common population genetic principles apply in all cases. Essential factors that deserve study in the cultivation of novel varieties are gene flow between fields, survival of seeds in the field after cultivation, occurrence of established natural populations, and potential exchange of genetic material with weeds or wild species through hybridization.

5.1. Agronomic properties of transgenic barley plants

The field performance of transgenic barley lines was observed by breeders at Boreal Plant Breeding Ltd. (Jokioinen, Finland) during several summers. The growth, flowering and ripening rhythms were identical between transgenic and non-transgenic plants. Differences in plant height and grain yield were observed in the case of the first transgenic line produced through bombardment of immature embryos (Ritala *et al.*, 1994). The mean height and grain yield were higher in this particular transgenic line when compared to non-transgenic controls. This was shown to be caused by the *in vitro* phase during the production of transgenic plants. The same phenomenon has been observed by breeders with anther culture-derived barley plants (Reino Aikasalo and Eero Nissilä, Boreal Plant Breeding Ltd., Jokioinen, Finland, personal communication).

5.2. Measuring the transgene flow in the cultivation of transgenic barley

The pollen-mediated dispersal of transgenes in the cultivation of transgenic barley was also studied in field trials (Ritala *et al.*, 2001b). A transgenic barley line carrying a marker gene coding for neomycin phosphotransferase II (*nptII*) was used as a pollen donor (Salmenkallio-Marttila *et al.,* 1995a). For maximum resolution a cytoplasmically male-sterile barley line was utilized as recipient and the flow of *nptII* transgene was monitored at distances of 1, 2, 3, 6, 12, 25, 50 and 100 meters from the donor plots of 225 m^2 and 2000 m^2. Male-fertile plots at a distance of one meter were included to measure the transgene flow in normal barley. The experimental set-up is shown in Figure 8.

The number of seeds obtained from male-sterile heads diminished rapidly with distance and only a few seeds were found at distances of 50 and 100 meters (Figure 9). The molecular biological analysis revealed that all the seeds obtained from male-sterile heads at a distance of one meter were transgenic, as anticipated. However, only 3 % of the distant seeds (50 m) actually carried the transgene, whereas most of them resulted from fertilization with non-transgenic background pollen. This background pollen was mainly due to pollen leakage in some male-sterile

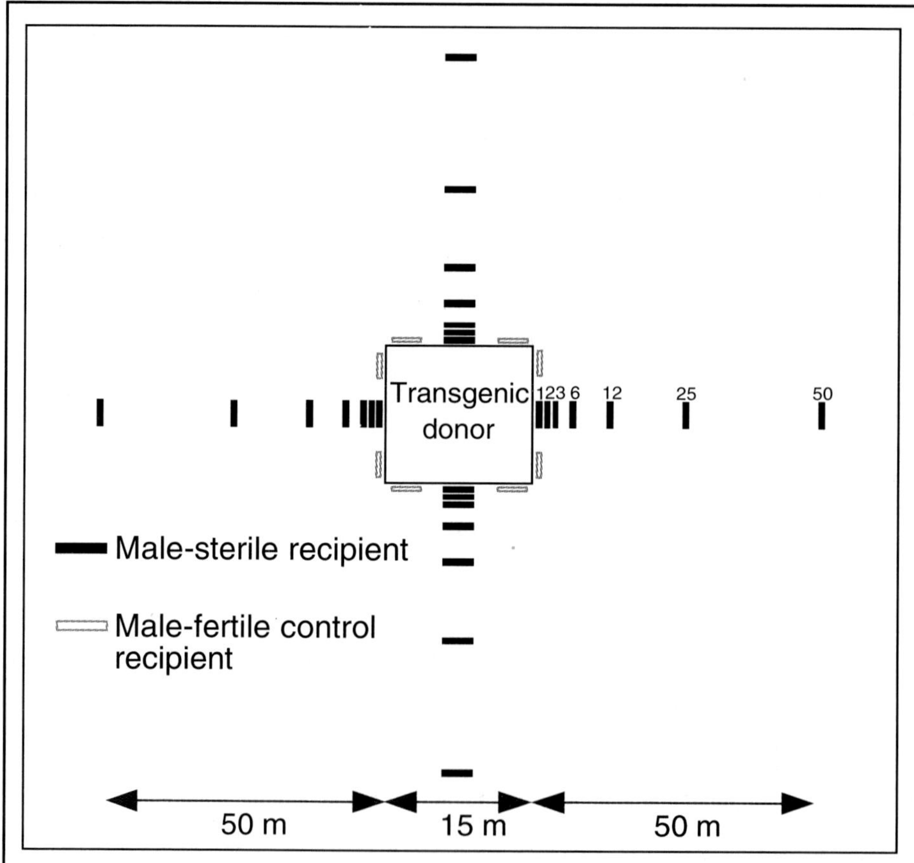

Figure 8 : Experimental set-up of the field trials at Boreal Plant Breeding, Jokioinen, Finland, in 1996 and 1997. In the cultivation scale experiment (2000 m^2) the distance of 100 m for the male-sterile recipient plots was included (modified from Ritala *et al.*, 2001b).

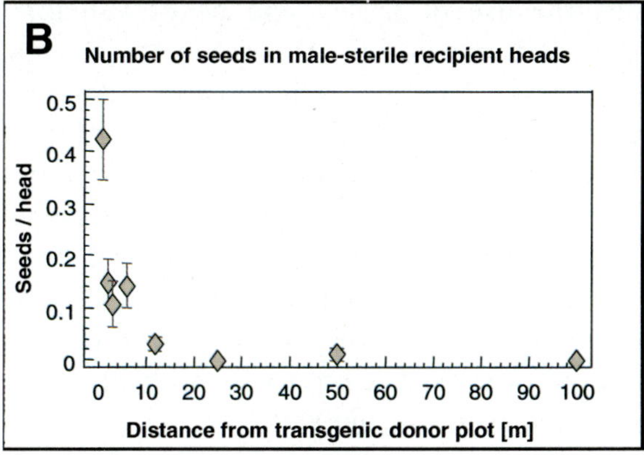

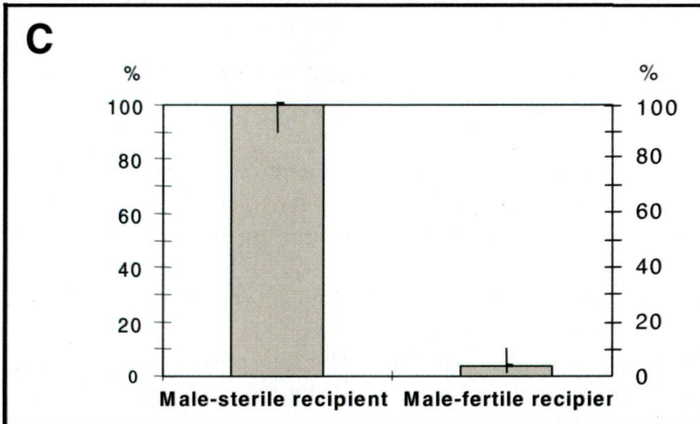

Figure 9 : Measuring gene flow through pollen dispersal in the cultivation of transgenic barley: A) transgenic donor plot, B) seed-set in recipient male-sterile barley plots in the dominant wind direction, C) the cross-fertilization frequencies of male-sterile and male-fertile recipient plots at a distance of 1 m from the donor area.

heads. In normal male-fertile barley the cross-fertilization frequency varied from 0 % to 7 % at a distance of one meter, depending on weather conditions on the heading day (Figure 9). We conclude that, due to competing self-produced and non-transgenic background pollen, the possibility of cross-pollination between a transgenic barley field and an adjacent field cultivated with normal barley is very low (Ritala *et al.*, 2001b).

6. CONCLUSIONS AND FUTURE PROSPECTS

The global cultivation area of transgenic crop plants is growing rapidly, the main species being maize, soybean, cotton, canola and potato (McDougall, 2000a, 2000b). It is obvious that gene transfer techniques can also broaden the use of barley and be of valuable assistance in barley breeding. As a precaution, the safety of the use of the new varieties must be assessed before their release. Although the risks and benefits depend mainly on the actual trait under consideration, knowledge of transgene flow parameters is also needed for determining good cultivation practices. Suitably designed risk management procedures implemented in farming will make it possible to transfer the benefits of genetic engineering into the production of high quality plant-based raw materials. With the present pace of research on barley it seems obvious that within a few years even this crop will probably be included in the list of transgenic plants on the market.

REFERENCES

Ammirato PV (1983) The regulation of somatic embryo development in plant cell cultures: suspension culture techniques and hormone requirements. *Bio/ Technology*, **1** : 68-74.

Aspegren K, Mannonen L, Ritala A, Puupponen-Pimiä R, Kurtén U, Salmenkallio-Marttila M, Kauppinen V and Teeri TH (1995) Secretion of a heat-stable fungal b-glucanase from transgenic, suspension cultured barley cells. *Mol. Breed.*, **1** : 91-95.

Carman JG (1990) Embryogenic cells in plant tissue cultures: Occurence and behavior. *In Vitro Cell. Dev. Biol.*, **26** : 746-753.

Choi HW, Lemaux PG and Cho MJ (2000) Increased chromosomal variation in transgenic versus nontransgenic barley (*Hordeum vulgare* L.) plants. *Crop Sci.*, **40** : 524-533.

Christou P, McCabe DE, Martinelli BJ and Swain WF (1990) Soybean genetic engineering - commercial production of transgenic plants. *Trends Biotechnol.*, **8** : 145-151.

Chu CC, Wang CC, Sun CS, Hsü C, Yin KC, Chu CY and Bi FY (1975) Establishment of an efficient medium for anther culture of rice through comparative experiments on the nitrogen sources. *Sci. Sin.*, **18** : 659-668.

Conway G and Toenniessen G (1999) Feeding the world in the twenty-first century. *Nature*, **402** : C55-C58.

Dahleen L (1995) Improved plant regeneration from barley callus cultures by increased copper levels. *Plant Cell Tiss. Org. Cult.*, **43** : 267-269.

DellaPenna D (1999) Nutritional genomics. Manipulating plant micronutrients to improve human health. *Science*, **285** : 372-375.

Evans DA (1989) Somaclonal variation - genetic basis and breeding applications. *Trends Genet.*, **5** : 46-50.

Fincher GB (1994) Potential for the improvement of malting quality of barley by genetic engineering. In: *Improvement of Cereal Quality by Genetic Engineering* (Eds Henry RJ and Ronalds RA), Plenum, New York and London pp. 135-138.

Foroughi-Wehr B, Mix G, Gaul H and Wilson HM (1976) Plant production from cultured anthers of *Hordeum vulgare* L. *Z. Pflanzenzücht.*, **77** : 198-204.

Foroughi-Wehr B and Wenzel G (1989) Androgenic haploid production. *IAPTC Newslett.*, **58** : 11-18.

Funatsuki H, Kuroda H, Kihara M, Lazzeri PA, Müller E, Lörz H and Kishinami I (1995) Fertile transgenic barley generated by direct gene transfer to protoplasts. *Theor. Appl. Genet.*, **91** : 707-712.

Funatsuki H, Lörz H and Lazzeri PA (1992) Use of feeder cells to improve barley protoplast culture and regeneration. *Plant Sci.*, **85** : 179-187.

Grimes HD and Hodges TH (1990) The inorganic NO_3:NH_4 ratio influences plant regeneration and auxin sensitivity in primary callus derived from immature embryos of indica rice (*Oryza sativa* L.). *J. Plant Physiol.*, **136** : 362-367.

Gudu S, Procunier JD, Ziauddin A and Kasha KJ (1993) Anther culture derived homozygous lines in *Hordeum bulbosum*. *Plant Breed.*, **110** : 109-115.

Hagio T, Hirabayashi T, Machii H and Tomotsune H (1995) Production of fertile transgenic barley (*Hordeum vulgare* L.) plant using hygromycin-resistance marker. *Plant Cell Rep.*, **14** : 329-334.

Hanley Z, Slabas T and Elborough KM (2000) The use of plant biotechnology for the production of biodegradabale plastics. *Trends Plant Sci.*, **5** : 45-46.

Harwood WA, Bean SJ, Chen DF, Mullineaux PM and Snape JW (1995) Transformation studies in *Hordeum vulgare* using a highly regenerable microspore system. *Euphytica*, **85** : 113-118.

Heyer AG, Lloyd JR and Kossmann J (1999) Production of modified polymeric carbohydrates. *Curr. Opin. Biotechnol.*, **10** : 169-174.

Horvath H, Huang J, Wong O, Kohl E, Okita T, Kannangara G and von Wettstein D (2000) The production of recombinant proteins in transgenic barley grains. *Proc. Natl. Acad. Sci. USA*, **97** : 1914-1919.

Hunter CP (1987) Plant generation method. *European patent application*, No. 87200773.7

Jacobsen P (1966) Demarcation of mutant-carrying regions in barley plants after ethylmethane-sulfonate seed treatment. *Rad. Bot.*, **6** : 313-328.

Janick J, Kitto SL and Kim YH (1989) Production of sythetic seed by desiccation and encapsulation. *In Vitro Cell. Dev. Biol.*, **25** : 1167-1172.

Jensen LG, Olsen O, Kops O, Wolf N, Thomsen KK and von Wettstein D (1996) Transgenic barley expressing a protein-engineered, thermostable (1,3-1,4)-â-glucanase during germination. *Proc. Natl. Acad. Sci. USA*, **93** : 3487-3491.

Jähne A, Becker D, Brettschneider R and Lörz H (1994) Regeneration of transgenic, microspore-derived, fertile barley. *Theor. Appl. Genet.*, **89** : 525-533.

Jähne A, Lazzeri PA and Lörz H (1991) Regeneration of fertile plants from protoplasts derived from embryogenic cell suspensions of barley (*Hordeum vulgare* L.). *Plant Cell Rep.*, **10** : 1 - 6.

Jähne A and Lörz H (1995) Cereal microspore culture. *Plant Sci.*, **109** : 1-12.

Kamada H, Kobayashi K, Kiyosue T and Harada H (1989) Stress induced somatic embryogenesis in carrot and its application to synthetic seed production. *In Vitro Cell. Dev. Biol.*, **25** : 1163-1166.

Karp A (1993) Are your plants normal? - Genetic instability in regenerated and transgenic plants. *Agro-Food-Industry Hi-Tech*, **3** : 7-12.

Karp A and Lazzeri PA (1992) Regeneration, stability and transformation of barley. In: *Barley: Genetics, Biochemistry, Molecular Biology and Biotechnology*. (Ed. Shewry PR) Wallingford, C.A.B. International. pp. 549 - 572.

Kartha KK, Chibbar RN, Georges F, Leung N, Caswell K, Kendall E and Qureshi J (1989) Transient expression of chloramphenicol acetyltransferase (CAT) gene in barley cell cultures and immature embryos through microprojectile bombardment. *Plant Cell Rep.*, **8** : 429-432.

Koprek T, Haensch R, Nerlich A, Mendel RR and Schulze J (1996) Fertile transgenic barley of different cultivars obtained by adjustment of bombardment conditions to tissue response. *Plant Sci.*, **119** : 79-91.

Lassner M (1997) Transgenic oilseed crops: a transition from basic research to product development. *Lipid Technol.*, **9** : 5-9.

Lazzeri PA, Brettschneider R, Lührs R and Lörz H (1991) Stable transformation of barley via PEG-induced direct DNA uptake into protoplasts. *Theor. Appl. Genet.*, **81** : 437-444.

Lee BT, Murdoch K, Topping J, Jones MGK and Kreis M (1991) Transient expression of foreign genes introduced into barley endosperm protoplasts by PEG-mediated transfer or into intact endosperm tissue by microprojectile bombardment. *Plant Sci.*, **78** : 237-246.

Lowe K, Bowen B, Hoerster G, Ross M, Bond D, Pierce D and Gordon-Kamm B (1995) Germline transformation of maize following manipulation of chimeric shoot meristems. *Bio/Technol.*, **13** : 677-682.

Lührs R and Lörz H (1987) Plant regeneration from embryogenic cultures of spring- and winter-type barley (*Hordeum vulgare* L.) varieties. *Theor. Appl. Genet.*, **75** : 16-25.

Mahon BP, Morre A, Johnson PA and Kingston HG (1998) Approaches to new vaccines. *Crit. Rev. Biotechnol.*, **18** : 257-282.

Mannonen L, Kurtén U, Ritala A, Salmenkallio M, Hannus R, Aspegren K, Teeri TH and Kauppinen V (1993) Biotechnology for the improvement of malting barley. *Proc. 24th Congr. Eur. Brew. Conv.*, Oslo pp. 85-93

Matthews P, Brettell R, Feig S, Gubler F, Jacobsen J, Thornton S, Tigay S, Wang MB and Waterhouse P (1997) Transformation of barley - a progress report. *Proc. 8th Australian Barley Tech. Symb.*, Queensland, 2:8.6-2:8.8.

McCabe DE, Swain WF, Martinell BJ and Christou P (1988) Stable transformation of soybean (*Glycine max*) by particle acceleration. *Bio/Technology*, **6** : 923-926.

McDougall P (2000a) Biotechnology in crops protection and production. Part I - Commercial developments. *Agro Food Ind. Hi-Tech.*, 32-37.

McDougall P (2000b) Biotechnology in crops protection and production. Part II - Research and development. *Agro Food Ind. Hi-Tech.*, 10-15.

Meijer WEG and Brown DCW (1987) Role of exogenous reduced nitrogen and sucrose in rapid high frequency somatic embryogenesis in *Medicago sativa*. *Plant Cell Tiss. Org. Cult.*, **10** : 11-19.

Mejza SJ, Morgant V, DiBona DE and Wong JR (1993) Plant regeneration from isolated microspores of *Triticum aestivum*. *Plant Cell Rep.*, **12** : 149-153.

Mendel RR, Müller B, Schulze J, Kolesnikov V and Zelenin A (1989). Delivery of foreign genes to intact barley cells by high-velocity microprojectiles. *Theor. Appl. Genet.*, **78** : 31-34.

Mordhorst AP and Lörz H (1993) Embryogenesis and development of isolated barley (*Hordeum vulgare* L.) microspores are influenced by the amount and composition of nitrogen sources in culture media. *J. Plant Physiol.*, **142** : 485-492.

Muyuan Z, Abing X, Miaobao Y, Chunnong H, Zhilong Y, Linji Y and Jianjun Y (1990) Effects of amino acids on callus differentiation in barley anther culture. *Plant Cell Tiss. Org. Cult.*, **22** : 201-204.

Nagata T (1989) Cell biological aspects of gene delivery into plant protoplasts by electroporation. *Int. Rev. Cytol.*, **116** : 229-255.

Nobre J, Davey MR, Lazzeri PA and Cannell ME (2000) Transformation of barley scutellum protoplsts: regeneration of fertile transgenic plants. *Plant Cell Rep.*, **19** : 1000-1005.

Nuutila AM, Hämäläinen J and Mannonen L (2000) Optimization of media nitrogen and copper concentrations for regeneration of green plants from polyembryogenic cultures of barley (*Hordeum vulgare* L.). *Plant Sci.*, **151** : 85-92.

Nuutila AM, Ritala A, Skadsen R, Mannonen L and Kauppinen V (1999) Expression of thermotolerant endo-(1,4)-â-glucanase in transgenic barley seeds during germination. *Plant Mol. Biol.*, **41** : 777-783.

Olsen FL (1987) Induction of microspore embryogenesis in cultured anthers of *Hordeum vulgare*. The effects of ammonium nitrate, glutamine and asparagine as nitrogen sources. *Carlsberg Res. Commun.*, **52** : 393-404.

Olsen O, Borriss R, Simon O and Thomsen KK (1991). Hybrid *Bacillus* (1-3,1-4)-b-glucanases: engineering thermostable enzymes by construction of hybrid genes. *Mol. Gen. Genet.*, **225** : 177-185.

Orshinsky BR, McGregor LJ, Johnson GIE, Hucl P and Kartha KK (1990) Improved embryoid induction and green shoot regeneration from wheat anthers cultured in medium with maltose. *Plant Cell Rep.*, **9** : 365-369.

Penttilä M, Lehtovaara P, Nevalainen H, Bhikhajhai R and Knowles J (1986) Homology between cellulase genes of *Trichoderma reesei*: complete nucleotide sequence of the endoglucanase I gene. *Gene,* **45** : 253-263.

Penttilä ME, Suihko ML, Lehtinen U, Nikkola M, Knowles JKC (1987) Construction of brewer's yeast secreting fungal endo-ß-glucanase. *Curr. Genet.*, **12** : 413-420.

Pickering RA and Devaux P (1992) Haploid production: approaches and use in plant breeding. In: *Barley: genetics, biochemistry, molecular biology and biotechnology* (Ed. Shewry PR), Wallingford: C.A.B. International pp. 519-547.

Poirier Y (1999) Production of new polymeric compounds in plants. *Curr. Opin. Biotechnol.*, **10** : 181-185.

Powell W (1988) The influence of genotype and temperature pre-treatment on anther culture response in barley (*Hordeum vulgare* L.). *Plant Cell Tiss. Org. Cult.*, **12** : 291-297.

Qureshi JA, Sigh RR, Basri Z, Stewart RR, Burton RA, Kollmorgen JF and Fincher GB (1997) Strategies for genetic transformation of elite Australian barley varieties. *Proc. 8th Australian Barley Tech. Symb.* Queensland, **2** : 8.9.

Reinert J (1958) Morphogenese und ihre Kontrolle an Gewebekulturen aus Carotten. *Naturwiss.,* **45** : 344-345.

Reinert J, Tazawa M and Semenoff S (1967) Nitrogen compounds as factors of embryogenesis *in vitro*. *Nature*, **216** : 1215-1216.

Rengel Z and Jelaska S (1986) The effect of L-proline on somatic embryogenesis in long-term callus culture of *Hordeum vulgare*. *Acta Bot. Croat.*, **45** : 71-75.

Ritala A, Aikasalo R, Aspegren K, Salmenkallio-Marttila M, Åkerman S, Mannonen L, Kurtén U, Puupponen-Pimiä R, Teeri TH and Kauppinen V (1995) Transgenic barley by particle bombardment. Inheritance of the transferred gene and characteristics of transgenic barley plants. *Euphytica*, **85** : 81-88.

Ritala A, Aspegren K, Kurtén U, Salmenkallio-Marttila M, Mannonen L, Hannus R, Kauppinen V, Teeri TH and TM Enari (1994) Fertile transgenic barley by particle bombardment of immature embryos. *Plant Mol. Biol.*, **24** : 317-325.

Ritala A, Mannonen L, Aspegren K, Salmenkallio-Marttila M, Kurtén U, Hannus R, Mendez Lozano J, Teeri TH and Kauppinen V (1993) Stable transformation of barley tissue culture by particle bombardment. *Plant Cell Rep.*, **12** : 435-440.

Ritala A, Mannonen L and Oksman-Caldentey KM (2001a) Factors affecting the regeneration capacity of isolated barley microspores (*Hordeum vulgare* L.). Submitted to *Plant Cell Rep.*

Ritala A, Nuutila AM, Aikasalo R, Kauppinen V and Tammisola J (2001b) Measuring gene flow in the cultivation of transgenic barley. Submitted to *Crop Sci.*

Salmenkallio M, Hannus R, Teeri TH and Kauppinen V (1990) Regulation of a-amylase promoter by gibberellic acid and abscisic acid in barley protoplasts transformed by electroporation. *Plant Cell Rep.*, **9** : 352-355.

Salmenkallio-Marttila M, Aspegren K, Åkerman S, Kurtén U, Mannonen L, Ritala A, Teeri TH and Kauppinen V (1995a) Transgenic barley (*Hordeum vulgare* L.) by electroporation of protoplasts. *Plant Cell Rep.*, **15** : 301-304.

Salmenkallio-Marrtila M and Kauppinen V (1995b) Efficient regeneration of fertile plants from protoplasts isolated from microspore cultures of barley (*Hordeum vulgare* L.) *Plant Cell Rep.*, **14** : 253-256.

Salmenkallio-Marttila M, Kurtén U and Kauppinen V (1995b) Culture conditions for efficient induction of green plants from isolated microspores of barley. *Plant Cell Tiss. Org. Cult.*, **43** : 79-81.

Shetty K and Asano Y (1991) The influence of organic nitrogen sources on the induction of embryogenic callus in *Argotis alba* L. *J. Plant Physiol.*, **139** : 82-85.

Smith DL and Krikorian AD (1990) Somatic proembryo production from excised, wounded zygotic carrot embryos on hormone-free medium: evaluation of the effects of pH, ethylene and activated charcoal. *Plant Cell Rep.*, **9** : 34-37.

Srisodsuk M (1994) Mode of action of *Trichoderma reesei* cellobiohydrolase I on crystalline cellulose. VTT Publications 188. VTT, Espoo.

Steward FC, Mapes MO and Mears K (1958) Growth and organized development of cultured cells. II. Organization of cultures grown from freely suspended cells. *Am. J. Bot.*, **45** : 704-708.

Stuart DA and Strickland SG (1984) Somatic embryogenesis from cell cultures of Medicago sativa L. I The role of amino acid additions to the regeneration medium. *Plant Sci. Lett.*, **34** : 165-174.

Tingay S, McElroy D, Kalla R, Fieg S, Wang M, Thornton S and Brettell R (1997) *Agrobacterium tumefaciens*-mediated barley transformation. *Plant J.*, **11** : 1369-1376.

Tsukahara M and Hirosawa T (1992) Simple dehydration treatment promotes plantlet regeneration of rice (*Oryza sativa* L.) callus. *Plant Cell Rep.*, **11** : 550-553.

Van Wert SL and Saunders JA (1992) elctrofusion and electroporation of plants. *Plant Physiol.*, **99** : 365-367.

Wake H, Akasaka A, Umetsu H, Ozeki Y, Shimomura K and Matsunaga T (1992) Enhanced germination of artificial seeds by marine cyanobacterial extract. *Appl. Microbiol. Biotechnol.*, **36** : 684-688.

Wan Y and Lemaux PG (1994) Generation of large number of independently transformed fertile barley plants. *Plant Physiol.*, **104** : 37-48.

Yan Q, Zhang X, Shi J and Li J (1990). Green plant regeneration from protoplasts of barley (*Hordeum vulgare* L.). *Kexue Tongbao*, **35** : 1581-1583.

Zhang S, Cho MJ, Koprek T, Yun R, Bregitzer P and Lemaux PG (1999) Genetic transformation of commercial cultivars of oat (*Avena sativa* L.) and barley (*Hordeum vulgare* L.) using *in vitro* shoot meristematic cultures derived from germinated seedlings. *Plant Cell Rep.*, **18** : 959-966.

Zimmerman JL (1993) Somatic embryogenesis: A model for early development in higher plants. *Plant Cell*, **5** : 1411-1423.

Chapter 5

GENETIC TRANSFORMATION OF SUNFLOWER (*HELIANTHUS ANNUUS* L.)

Günther Hahne★

Institut de Biologie Moléculaire des Plantes, CNRS et Université Louis Pasteur, 12, Rue du Général Zimmer, F-67084 Strasbourg, France

Summary

Sunflower is one of the major crops worldwide cultivated for the production of edible oil, although it has reached this importance only within the last 20 years. Interest in biotechnological approaches and genetic engineering is therefore quite recent and investigations into this subject are less numerous than for other major crop species. Many of the protocols that are successful with species belonging to other genera and families have proven unsuccessful with sunflower and related Helianthus species. Indirect regeneration systems are difficult to master, generally of low efficiency, and callus or cell suspension cultures loose their regeneration potential rapidly. In consequence, most published transformation protocols are based on direct regeneration systems that require delicate fine tuning of the regeneration, transformation and selection conditions in order to reach practically useful transformation efficiencies. At the present (published) state of the art, transformation of sunflower remains delicate and labor-intensive. Nevertheless, transformation of sunflower has become a routine procedure in a number of laboratories as demonstrated by the number of field trials for which permission has been granted.

Keywords : *Agrobacterium tumefaciens,* biotechnology, genetic engineering, grafting, particle bombardment

---

★Corresponding author : E-mail : gunther.hahne@ibmp-ulp.u-strasbg.fr

1. INTRODUCTION

The beginnings of man's use of sunflower are lost in history. American Indians have used sunflower seeds and oil for food and cosmetic purposes already in pre-Columbian times but today, we know but little about cultivation practices and their importance for the diet at that time. In spite of this long history and although sunflower has been figuring among the four most important oil crops worldwide for the past decades, it must be considered as a recent crop to be cultivated on such scale. Sunflower has gained worldwide importance no earlier than the early seventies of the twentieth century (Bonjean, 1993). Curiously, for a species originally endemic to North America, the genetic improvement necessary for large-scale economic development has originated in Eastern Europe and spread from there to the other major production regions including back to the Americas.

The impact of plant breeding on the economy of a plant species can easily be tracked and illustrated for the example of sunflower. It was the development of high-oil varieties by Russian plant breeders in the nineteenth century (Diepenbrock, 1987) that laid the foundation for the cultivation of sunflower on industrial scale. However, the real trigger that promoted this species among the most important producers of edible oil was the introduction of highly performing F_1 hybrids based on the cytoplasmic male sterility (cms) system developed by Leclercq (Leclercq, 1969). Today, sunflower growers almost exclusively use hybrid seed because of its superior properties (yield, uniformity, disease resistance etc.). Major limitations to a further increase of the market share of sunflower include imperfect resistance or tolerance to a number of diseases and abiotic stresses, comparatively low overall yield (sunflower is mainly grown under limited input conditions), and changing market requirements. Conventional breeding has addressed many of these issues in the past and will continue to do so in the future. However, the impact of genetic improvement can be expected to soar with the exploitation of genetic engineering approaches that are hoped to confer novel traits to this crop, which would remain inaccessible otherwise. The complex breeding scheme involved in the production of F_1 hybrid seed material requires the creation and constant maintenance of suitable restorer- and maintainer lines (Bonjean, 1993; Schuster, 1993). Strategies for the introduction of foreign genes, in particular transgenes, into a high-yielding sunflower variety will have to take this circumstance into account.

1.1. The crop

The species *Helianthus annuus* originates from the Western plains of North America, a warm and rather dry habitat (Service, 1995). As can be expected, considering its origin, sunflower is rather efficient in extracting water from the soil. In deep soil, its vast root system can exploit a large soil volume. Root penetration and water use capacity may exceed a depth of 2 m (Service, 1995). Sunflower is therefore considered a drought tolerant crop although maximal yields are obtained under conditions with adequate water supply. Similar considerations hold true for fertilizer and pesticide use. A high yield level obviously depends on optimal nutrient supply and pest management. However, while production performance of sunflower under low-input conditions or high-stress soils (drought, salinity) is certainly limited, it compares favorably with other commonly grown crops.

The most common uses of sunflower include production of edible oil from oilseed types, as well as confectionery sunflower for the home and birdfeed market. The cakes remaining after oil extraction are used for animal feed, and on a more limited scale, sunflower is also used as a fodder plant, either green or after silage. Sunflower is also appreciated for horticultural and ornamental purposes where the most important traits are size, flower color and long flowering period.

In modern varieties, triglycerides may contribute to over 50-60% of the dry weight of the kernels, corresponding to >80% of the embryo's dry weight. Sunflower oil is rich in unsaturated fatty acids, particularly in linoleic acid (typically around 65%) and oleic acid (typically around 20%) (Bonjean, 1993). This composition is highly depending on the genetic background. Thus, oleic acid content may exceed 80% in high oleic sunflower varieties.

However, for a given variety, the oil composition is highly variable under environmental influence. Temperature is a decisive factor in this respect. A variety yielding 60-70% linoleic acid under temperate growth conditions may produce oil with less than 30% linoleic acid when cultivated in Mediterranean or subtropical climate (Bonjean, 1993). Alongside with characters influencing the overall grain yield, oil yield and oil composition are breeding goals of prime importance. In view of a market that increasingly requires special quality oils with defined composition, mastering the oil quality and the stability of its composition become of particular importance.

The most important causes for yield losses are fungal pathogens including *Sclerotinia sclerotiorum* (White Rot), *Botrytis cinerea* (Grey Mold), *Plasmopara halstedii* (Downy Mildew), *Phomopsis helianthi* (formerly *Diaporthe*; Stem Canker), *Alternaria helianthi, Phoma oleracea* var. *helianthi-tuberosi* (Phoma Black Stem), and *Puccinia helianthi* (Sunflower Rust) (Schuster, 1993). These fungal pathogens have a high impact on the grain yield and oil quality and thus figure among the most important concerns of sunflower growers and breeders alike. Their control requires, for the most part, specific chemical treatments, and some are to date without effective remedy (*e.g., S. sclerotiorum*). Resistance or tolerance to a number of pathogens is controlled by genes that have been introgressed from compatible wild *Helianthus* species by interspecific hybridization but the problem of fungal diseases is far from being solved. In contrast, bacterial and viral diseases are of no particular concern.

Sunflower cultivated in warm regions of the former USSR and regions bordering the Mediterranean Sea are subject to great yield losses due to a parasitic plant, *Orobanche cumana* Wallr. (broomrape). This parasite is extremely difficult to control due to its being a plant, thus requiring herbicides for chemical control that are bound to affect sunflower as well. Broomrape completes a large part of its growth cycle under ground and thus remains inaccessible to most commonly used chemical or mechanical control measures (Sackston, 1992).

In Southern and Western Europe, insects are not perceived as a problem whereas elsewhere, a large number of insect pests may cause serious yield losses (Rogers, 1992). Losses to insects occur during seed storage, the pre-establishment phase (planting to seedling) or on the post-establishment level (young plant to maturity). Adequate crop rotation schemes as well as judiciously chosen planting dates may be necessary to complement the traditional use of chemical pest management schemes. Pest tolerant cultivars are available to a limited extent. Insect resistance represents one target where genetic engineering can be expected to have an impact on crop management schemes in the near future.

1.2. The plant

The cultivated sunflower (*Helianthus annuus* L.) belongs to the genus *Helianthus* which is comprised of 67 annual or perennial species (Heiser, 1978; Heiser, 1976). Classification within this genus is complex and subject to debate — notably due to frequent interspecific hybridization events that occurred naturally in the recent past. Only two species of this genus, *i.e. H. annuus* (the cultivated sunflower) and *H. tuberosus* (Jerusalem artichoke) are cultivated on a commercial scale for human consumption; others are appreciated for their ornamental value. A number of species tend to be noxious weeds in their region of origin (Heiser *et al.*, 1969).

Sunflower seeds are large achenes (6-10 mm). For practical considerations, they consist of the kernel (the embryo) and the hull (the pericarp). In oilseed types, the hull is black, thin and adheres to the kernel while relatively thick black-and-white striped hulls that can be easily removed characterize non-oilseed varieties. The latter seeds are generally larger than the former ones, which have a more favorable kernel-to-hull ratio. The seedlings with large, fleshy cotyledons develop into tall (1.5-3 m) plants that may consist of a single stem with a single terminal inflorescence, or become more or less heavily branched plants bearing one terminal inflorescence on each branch. Branched genotypes flower over a long period and are therefore often used for ornamental purposes as well as pollinators (male parent lines) in the hybrid production scheme, while the female parents resemble the production lines in that they are unbranched for more synchronous seed production. Floral induction, *i.e.* the transition from a vegetative apical meristem to a floral meristem, occurs early in development, only a few weeks after germination, long before the flower bud becomes visible. Anthesis occurs 8-12 weeks after germination. Floral induction may in some varieties be influenced by day-length or other environmental cues but in general, rather depends on internal signals related to the developmental stage (Steeves *et al.*, 1969; Steeves and Sussex, 1989).

As is characteristic for the *Tubuliflorae* of the *Asteraceae* family, the inflorescence is composed of two types of flowers, *i.e.* sterile ray flowers with fused petals and fertile disk flowers that are incomplete and lack petals. These flowers are arranged in a spiral pattern responding to the Fibonacci numbers, yielding a radial overall symmetry. Depending on the variety, disk and ray flowers may be colored to different degrees by anthocyanins, resulting in more or less dark disk- or ray flowers. The disk flowers open progressively over a period of 5-10 days, depending on the size of the inflorescence, following a centripetal pattern. 12-15 days after pollination, the embryo is fully formed and enters the seed-filling stage. The time to full maturity is approx. 2-3 months. Immature embryos can be extracted and cultured *in vitro* as early as a few days after pollination. This possibility is routinely exploited by many plant breeders in order to shorten the generation time. Sunflower anatomy and development are described in greater detail in the publication edited by Carter (Carter, 1978).

1.3. Transgenic solutions

Genetic engineering can be expected to help in the solution of a number of problems with which sunflower breeders are confronted. In fact, the list of possible contributions of genetic engineering to the breeding of superior sunflower varieties matches well the list of general breeding objectives. The most important issues concern product quality, increased yield potential, and reduction of yield losses.

In many crop species, the transgenic solutions that have been proposed to date to commercial growers for solving existing problems concern the introduction of genes that confer resistance to insect predators, viral diseases, and particular herbicides. Historically, they were the first traits to become available to genetic engineering because they rely on comparatively simple genetic concepts and constructs. Many of these, however, are of little relevance to sunflower improvement since the issues they address are not problematic for sunflower: viral diseases are no major concern, and insect pests represent a comparatively minor nuisance that can be controlled efficiently by conventional means. Herbicide resistance may provide a certain comfort to sunflower growers and may become part of a strategy to combat broomrape but at present, does not appear to enjoy high priority among the breeding objectives. Control of volunteer sunflower, already a problematic issue, may in fact become more difficult with the integration of herbicide resistance traits into the majority of sunflower varieties.

In contrast, fungal diseases cause major yield losses in sunflower cultures. Genetic tolerance and chemical treatment do exist for certain diseases but do not cover the whole range of pathogens. No truly efficient treatment or tolerance are available for pathogens such as *S. sclerotiorum*. Unfortunately, reliable and efficient transgenic solutions to limit the impact of fungal pathogens are not yet available although promising results are being obtained (Datta *et al.*, 1999; Marchant *et al.*, 1998; Tabei *et al.*, 1998). As a crop, sunflower would certainly benefit from advances in this field.

The demand for oil with particular fatty acid composition is steadily increasing, both for food and non-food uses. In view of the high dependence on environmental conditions of the fatty acid composition of sunflower oil, a stabilizing contribution by metabolic engineering would appear desirable. Similarly, the creation of varieties with particular fatty acid profiles (*e.g.*, high oleic, mid-oleic or highly unsaturated oil varieties) might become easier and more straightforward when metabolic engineering approaches will have been integrated in the overall strategy. Additional characters addressing product quality include improved amino acid balance of the green matter and pressing cakes since the lack of essential amino acids is the most limiting factor for their use as animal feed. The introduction of totally novel products, hitherto not produced by sunflower, can also be envisaged by genetic engineering. Examples include "molecular farming" where products of pharmaceutical activity are produced in plants (Borisjuk *et al.*, 1999; Fischer *et al.*, 1999; Franken *et al.*, 1997; Gruber and Theisen, 2000) or the production of novel compounds such as poly-hydroxybutyric-acid based plastics (Dunwell, 1999). Sunflower would appear to be a suitable host for the latter example since it requires a large production of acetyl-CoA, a requirement satisfied by oil-producing plants.

Measures to increase the general yield potential of sunflower, *e.g.* by optimizing efficiency of photosynthesis, water use, or modification of developmental patterns and general plant architecture would certainly be interesting in view of increased competitiveness of this crop. However, such genetic engineering projects require further advances in our fundamental understanding of the involved processes before they can be envisaged on a practically feasible scale.

Any application of genetic engineering strategies to sunflower improvement requires mastery of the cell biological techniques necessary for the production of transgenic plants. Although large scale plant transformation has become routine for many major crop plants as well as laboratory models, in some cases even without the necessity for *in vitro* culture phases (Bechtold *et al.*, 1994; Hiei *et al.*, 1994), the application of established protocols to sunflower has met with remarkable difficulties. As a consequence, the production of genetically engineered sunflowers is lagging behind its economic importance. It is the objective of the present review to summarize the approaches that have been evaluated for their relative meirts for plant regeneration from various tissues of *H. annuus,* to review the gene transfer techniques that have been used, and to consider the practical impact genetic engineering has had to date on the crop plants, sunflower.

2. *IN VITRO* CULTURE-POSSIBILITIES AND LIMITATIONS

The possibility to regenerate fertile plants from isolated tissues and even isolated cells has been demonstrated in many plant species, including sunflower. In principle, suitable explants can be unicellular (protoplasts isolated from a variety of tissues) or multi-cellular, *i.e.* excised organs or fractions thereof (Fig. 1). Under the influence of suitable culture conditions and hormonal stimuli, cell divisions can be induced *de novo* in such explants, leading to the formation of undifferentiated growth (callus or cell suspension cultures) or directly to differentiated structures, *i.e.* adventitious meristems or somatic embryos. Where a more or less prolonged callus phase exists, such differentiated structures can usually

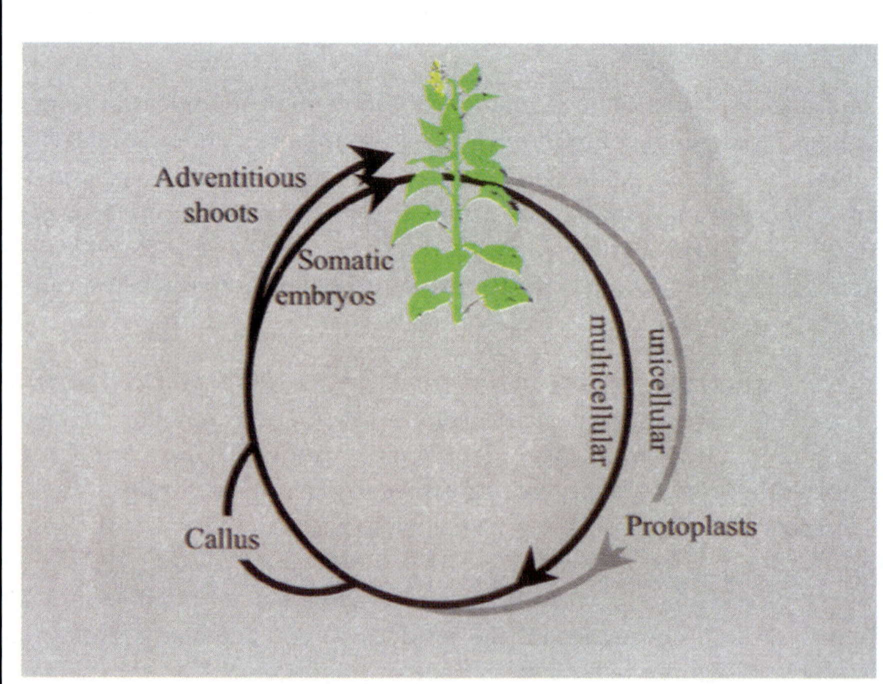

Figure 1 : Summary and schematic representation of the approaches in *in vitro* culture that have been tested in view of sunflower transformation.

be induced from callus or cell suspension cultures. Whether the regeneration of whole plants proceeds via the embryogenic route with the direct formation of bipolar structures from proembryonic masses, or via the formation of adventitious meristems leading to the development of shoots from which roots can be induced at a later step; and whether these morphogenic masses originate from a single cell or are of multicellular origin, depends on a multitude of factors. These factors include (i) the plant species under consideration, (ii) the cultured organ or cell type, (iii) the experimental conditions. In most cases, one or the other developmental pathway will predominate or be exclusively accessible for a given combination of factors. While the mode of regeneration may be of practical importance for the choice of an adequate transformation technique, it is of little consequence for the prediction of the quality of the regenerated or transformed plants.

2.1. Sunflower tissue culture - a short overview

In the case of sunflower, the vast majority of reports on *in vitro* culture of *Helianthus annuus* is recent. Studies on this species can be found in the early literature on plant tissue culture but they are rare (De Ropp, 1946; Henderson *et al.*, 1952; Hildebrandt *et al.*, 1946; Kandler, 1952). Protocols and culture schemes that are successful for plant regeneration in other species have often proven difficult or impossible to adapt to sunflower. One of the reasons for this difficulty is that the regeneration potential from callus cultures or cell suspension cultures is often lost after only short culture periods. Almost all described cases where plants were regenerated from sunflower callus cultures originated from immature embryos as donor material and were obtained within a very short time span after the induction of the callus (Espinasse and Lay, 1989; Espinasse *et al.*, 1989; Wilcox McCann *et al.*, 1988; Witrzens *et al.*, 1988). Occasionally, plants could be regenerated from callus induced on other tissues or from long-term callus cultures (Greco *et al.*, 1984; Lupi *et al.*, 1987; Paterson and Everett, 1985). However, no published callus-based regeneration protocols for sunflower have ever been developed to the efficiency necessary to form the basis of a practically useful transformation protocol.

For sunflower, regeneration efficiencies appear to be higher in direct regeneration systems. In such systems, adventitious meristems or proembryogenic masses are induced without an intervening callus phase. Both caulogenic and embryogenic protocols have been established for several tissues. Somatic embryos (Finer, 1987; Freyssinet and Freyssinet,

1988; Jeannin and Hahne, 1991) and adventitious shoots (Jeannin and Hahne, 1991) have been induced on immature cultured sunflower embryos. Large numbers of adventitious shoots can be regenerated from cotyledons of mature seeds, seedlings, and primary leaves of young seedlings (Ceriani *et al.*, 1992; Chraibi B *et al.*, 1992a; Chraibi B *et al.*, 1991; Knittel *et al.*, 1991; Konov *et al.*, 1998; Nataraja and Ganapathi, 1989; Paterson and Everett, 1985; Power, 1987).

Induction of either kind of regeneration response from more mature tissues such as expanded leaves, petioles, stem segments or roots is rather delicate to master and, judging from the published evidence, is limited to rare cases and very particular conditions if it occurs at all. A limitation that is common to almost all regeneration systems described so far for *H. annuus,* with very few exceptions, is a strong genetic determination of the morphogenic potential. Unlike the situation in species such as tomato where a monogenic mode of inheritance has been described for the regeneration potential from protoplasts (Koornneef *et al.*, 1993), the genetics of the morphogenic capacity is much more complex in sunflower and appears to be controlled by several QTLs (Bolandi *et al.*, 2000; Deglene *et al.*, 1997; Flores Berrios *et al.*, 2000a; Flores Berrios *et al.*, 2000b; Flores Berrios *et al.*, 2000c; Sarrafi *et al.*, 1996b). Nevertheless, breeding for superior regeneration capacity is possible within certain limits (Burrus *et al.*, 1991; Everett *et al.*, 1987; Paterson and Everett, 1985).

After this short overview of approaches to the regeneration of sunflower plants from *in vitro* cultured tissues, we shall examine those regeneration systems in more detail that have been evaluated in view of the production of transgenic plants.

2.2. Regeneration from protoplasts

Isolation of protoplasts has been shown to be possible from a variety of tissues. The yields that can be obtained are comparatively high, and callus formation from such protoplasts is possible at high frequency (Bohorova *et al.*, 1986; Fischer *et al.*, 1996; Guilley and Hahne, 1989; Lenee and Chupeau, 1986; Moyne *et al.*, 1988). The regeneration of fertile plants from such calli is, however, extremely difficult and proceeds at very low frequency; published values, based on the number of plated protoplasts, are usually well below the percent range, often by several orders of magnitude (Fischer *et al.*, 1992; Henn *et al.*, 1995; Krasnyanski and Menczel, 1993; Schmitz and Schnabl, 1989; Trabace *et al.*, 1995;

Wingender *et al.*, 1996). Numerous investigations have dealt with this difficulty and were conducted in search for isolation and culture conditions more favorable to subsequent plants regeneration (Barth *et al.*, 1993; Chanabé *et al.*, 1989; Chibbar *et al.*, 1987; Dupuis *et al.*, 1990; Lenée and Chupeau, 1989; von Keller *et al.*, 1997). Unfortunately, all these studies have remained unfruitful in terms of a reliable and efficient protocol for the regeneration of plants from isolated sunflower protoplasts. They did, however, demonstrate that this difficulty is likely to be an inherent property of this species, and they shed light on a peculiar behavior of sunflower protoplasts. Depending on the culture conditions, sunflower protoplasts develop either into the characteristic loose or compact callus-like colonies known from protoplasts culture systems of most other species, or they may form very compact structures which have been termed "pro-embryoids" for reasons of their suggestive appearance (Chanabé *et al.*, 1989; Dupuis *et al.*, 1990; Moyne *et al.*, 1988). Such "pro-embryoids" have never developed beyond a very early stage, and plants have never been regenerated from cultures yielding these structures. Their embryonic nature has been contested on grounds of histological data (Fischer and Hahne, 1992) but the phenomenon is interesting as a manifestation of experimentally triggered cyto-differentiation and has been subject of several detailed studies (Barthou *et al.*, 1997; Caumont *et al.*, 1997; Petitprez *et al.*, 1995). Further attempts to raise the regeneration efficiency from sunflower protoplasts include selection for sub-populations with increased regeneration potential among the isolated protoplasts (Chibbar *et al.*, 1987; Laparra *et al.*, 1997a; Petitprez *et al.*, 1995; von Keller *et al.*, 1994), and electro-stimulation of populations of cultured protoplasts (Barth *et al.*, 1993). All of these studies have been instrumental for our understanding of the cellular processes during the early phases of sunflower protoplast development. Yet, regeneration frequencies have remained to date too low to be practically exploited. Transient expression in sunflower protoplasts of various origins (Kirches *et al.*, 1991) as well as the production of transgenic callus by protoplast-mediated transformation have been shown to be feasible (Moyne *et al.*, 1989) but plant regeneration from such callus has yet to be described.

It has become evident that the regeneration potential from sunflower protoplasts is extremely genotype dependent and that passage from one genotype to another requires the definition of experimental variables from scratch as the published protocols share few elements that they have in common. Recently, the heritability of the regeneration response of sunflower protoplasts has been investigated considering quantitative as well as qualitative aspects, *i.e.* the tendency to form the so-called pro-embryoids (Flores Berrios *et al.*, 2000b).

One may conclude from a detailed analysis of the published information on sunflower culture that no fundamental obstacle exists that would prevent the regeneration of plants from isolated protoplasts, since shoots and even plants have been obtained from several genotypes protoplasts types under a variety of conditions. However, in spite of intense efforts, the regeneration efficiency has remained too low to be practically exploited. Active research on sunflower protoplast regeneration systems seems to have been abandoned, at least in view of applications in genetic engineering.

2.3. Regeneration from cultured tissues

Most known approaches to plant regeneration from cultured tissue have been investigated for their applicability to sunflower. As has been found with protoplast systems, it was possible to regenerate plants in most instances but in many cases with low reproducibility or efficiency. Only few approaches could be developed into efficient, reliable protocols that work in several laboratories and are robust enough to be combined with gene transfer procedures. For practical purposes, it is useful to make a distinction between indirect and direct regeneration systems.

In the following, tissue culture systems that have been described in view of sunflower transformation will be reviewed in terms of their respective merits and problematic aspects. In order for a culture system to constitute a practically useful base for a transformation system, it must satisfy several requirements. (i) The cells from which adventitious meristems or proembryogenic masses will be formed must be accessible to the employed gene transfer technology; (ii) since gene transfer methods are generally characterized by low efficiency, an efficient selection agent for the cells expressing the introduced trait must be available; (iii) it must be possible to multiply putative transformants *in vitro* or *ex vitro*, either by vegetative multiplication or by seed production.

2.3.1. Indirect culture systems

Most sunflower tissues are suitable material for the production of callus- and cell suspension cultures. Such cultures grow vigorously and for prolonged periods on a variety of media, and constitute convenient sources for cell biological and gene transfer studies.

Normal, fertile sunflower plants have been regenerated from cultured callus *via* induction of adventitious meristems in several instances (Espinasse and Lay, 1989; Espinasse *et al.*, 1989; Greco *et al.*, 1984; Lupi *et al.*, 1987). The donor tissues that appear most useful for this purpose are hypocotyl segments and immature embryos although the regeneration of shoots has also been described from anther-derived callus (Gurel *et al.*, 1991). All these studies share a common result, *i.e.* that the efficiency and reproducibility of plant regeneration is extremely low. In addition, the morphogenic potential of such cultures is extremely volatile. It is lost after very short culture periods, often already after the first subculture; it is also highly genotype-specific (Espinasse and Lay, 1989).

The same explants that have been used for the induction of caulogenic callus have also given rise to embryogenic callus (Paterson and Everett, 1985; Paterson Robinson and Adams, 1987; Wilcox McCann *et al.*, 1988) and cell suspension cultures (Prado and Berville, 1990), depending on the culture conditions and media that were employed. The embryogenic potential of such callus cultures is subject to the same limitations as were described for the caulogenic potential: it is extremely genotype-specific, sometimes limited to specially selected proprietary genotypes, and the potential is very rapidly lost. In fact, it is not unusual to see somatic embryos emerging from callus that is still attached to the donor tissue during the callus induction phase (G. Hahne, unpublished). However, we have never been able to maintain this capacity once the callus was detached from the donor tissue. We are not aware of any callus or cell suspension culture of *H. annuus* that has been maintained for a prolonged period and has retained its embryogenic potential in a practically useful range.

All published information considered, all investigated indirect regeneration pathways from sunflower tissues are of too limited reproducibility and efficiency for an application in view of transgenic plant production. This is regrettable since callus and cell suspension cultures generally provide convenient and efficient tools for the enrichment of the cultures with cell populations expressing the introduced transgenic construct. The observation made about research on sunflower protoplast culture holds also true for research on indirect regeneration systems in a larger sense: The general difficulties have been recognized and this approach does not appear to be actively pursued any more.

2.3.2. Direct regeneration systems

In contrast to callus-based regeneration systems, direct regeneration systems are more easily mastered. Several culture systems have been described where the induction of either adventitious shoots or somatic embryos and subsequent production of normal plants proceeded reliably, rapidly and with high efficiency.

Frequently used explants for the induction of adventitious meristems include cotyledons of mature seedlings (Cariani *et al.*, 1992; Chraibi B *et al.*, 1992a; Chraibi B *et al.*, 1992b; Chraibi B *et al.*, 1992c; Chraibi B *et al.*, 1991; Knittel *et al.*, 1991; Nataraja and Ganapathi, 1989; Power, 1987; Sarrafi *et al.*, 1996a; Sarrafi *et al.*, 1996b) and immature embryos (Bronner *et al.*, 1994; Charrière and Hahne, 1998; Jeannin *et al.*, 1993; Jenanin *et al.*, 1995; Jeannin *et al.*, 1998; Wilcox McCann *et al.*, 1988; Witrzens *et al.*, 1988). Although the genotype of the explant may exert some influence on the regeneration efficiency, a genetically determined limitation does not appear to exist, or at least is much less pronounced for the direct regeneration systems than for the indirect ones. Judiciously chosen culture conditions may alleviate those variations that are introduced by various genotypes. In certain cases, regeneration efficiency may be modulated by ethylene concentration and reception (Chraibi B *et al.*, 1992b).

The induction of adventitious shoots in such direct regeneration systems is extremely rapid and may occur within a few hours after excision of the explant (Jeannin *et al.*, 1998). The overall time from the onset of the experiment to the harvest of seeds may be as short as 4 months. From a practical point of view, this short response time has not only advantages but also some drawbacks. The conditions for an optimal regeneration response and optimal gene transfer to the regenerating cells may not be the same (Laparra *et al.*, 1995) and the time window is too short for experimental adjustment of the conditions between the two key steps, represented by gene transfer and induction of the regeneration event. In addition, the short time available between a possible gene transfer and determination of the morphogenic response leaves little possibility for the selection of transformed cells. Consequently, regenerated structures will often be chimerical.

Similar to direct caulogenesis, direct somatic embryogenesis from sunflower tissues is distinguished by its rapidity. Embryo-specific features such as the presence of characteristic storage proteins may become detectable in somatic embryos only a few days after the explant was put into culture (Jeannin *et al.*, 1993). Plantlets can be separated from the explant as early as two weeks after induction and be transferred to the greenhouse 4-5 weeks after the begin of the experiment. Cotyledons (Fiore *et al.*, 1997), hypocotyl-derived epidermal thin layers (Pélissier *et al.*, 1990) and immature embryos (Finer, 1987; Freyssinet and Freyssinet, 1988; Jeannin and Hahne, 1991; Prado and Berville, 1990) have all been used with success for the direct induction of somatic embryos.

The embryogenic competence of the explant is in many cases limited to a confined, rather small cell population that may or may not be distinguishable from non-responding cells by particular cytological features. It has been shown that the exact cells responding in immature zygotic embryos are not determined by pre-existing competence but in response to clues perceived in function of the culture conditions (Jeannin *et al.*, 1995; Jeannin *et al.*, 1998). In this experimental system, the somatic embryos are of multicellular origin (Bronner *et al.*, 1994). Not much information is available concerning the early development of somatic embryos in other direct regeneration systems of *H. annuus,* in particular it is unknown whether somatic embryos of unicellular origin do occur in this species. However, a unicellular origin of directly as well as of indirectly forming somatic embryos has been demonstrated in the related species, *H. smithii* (Laparra *et al.*, 1997b).

Direct regeneration systems, whether via the caulogenic or the embryogenic pathway, do not suffer from the same problems of genotype dependence, reliability and efficiency as the indirect ones. Yield and reproducibility are quite satisfactory and have been demonstrated in a range of different genotypes. Fertile plants are also recovered in a very short time span, a feature, which may be considered as advantageous. However, the morphogenic zones are small and only a minute fraction of the explant will actually become involved in the formation of regenerating structures. This fact reduces the probability that a gene transfer event will indeed touch a cell that will give rise to a regenerated plant, and enrichment for transformed cell prior to the induction of the morphogenic response is impossible due to the short response times. Selection on a later stage is much less efficient and results in numerous false positives. It is thus not surprising that the vast majority of the tested direct regeneration system – as the indirect systems – has proven unsuitable for the production of transgenic sunflower plants under practical conditions.

However, the analysis of the existing possibilities allows defining a number of requirements a working transformation system for sunflower must fulfill. Considerations of reliability, efficiency and genotype dependence exclude all indirect culture systems that have been published to date. In contrast, several direct systems are satisfactory in terms of these criteria but the regeneration response originates from a small cell population only, which is difficult to select for. The challenge consists of identifying explants, and defining conditions, where available gene transfer techniques succeed in targeting the morphogenically competent cells. Furthermore, the resulting structures are likely to be chimerical. Suitable selection schemes will also have to be defined. At a later step, the produced chimeras can be 'purified" by passing through meiosis: the offspring is composed of genetically pure plants.

2.4. Properties of regenerated plants

Plantlets obtained from shoots or somatic embryos resulting from direct regeneration schemes are often vigorous and present little difficulty for root induction and subsequent transfer to the greenhouse (Chraibi B *et al.*, 1991; Fiore *et al.*, 1997; Freyssinet and Freyssinet, 1988; Jeannin and Hahne, 1991; Knittel *et al.*, 1991; Krasnyanski and Menczel, 1993). In contrast, shoots obtained from protoplast- or callus-based regeneration systems but also from certain direct systems, tend to be frail and small, and root induction and further development are difficult to obtain. In these cases, the success rate of transfer to the greenhouse and the vigor of the transferred plants can be enhanced in a quite spectacular way by grafting the shoots with the deficient root system onto greenhouse-grown or *in vitro-* grown sunflower root stocks (Fischer *et al.*, 1992). The formation of adventitious roots from sunflower cuttings is stimulated by the micro-nutrient boron (Josten and Kutschera, 1999) but the universality of this observation for all regenerated shoots is not clear. Particular culture regimes may also enhance the rooting capacity of regenerated shoots (Baker *et al.*, 1999).

The primary regenerants (R_0 generation) issued from *in vitro* culture, regardless of their origin, present in many cases a peculiar phenotype: they are severely stunted, often not taller than 20-50 cm but any size up to a normal plant may also occur. Regenerated plants tend to be branched, even in normally unbranched genotypes. This "regenerated" phenotype is typically observed in plants growing on their own roots but also frequently in grafted plants. The severity of the phenotype has a (rather loose) correlation with the time spent *in vitro* and whether a precocious floral induction (Henrickson, 1954; Ivanov *et al.*, 1997; Paterson, 1984; Trifi *et al.*, 1981) has taken place. The reasons for this altered phenotype are unknown but appear to be the consequence of physiological influences rather than genetic or epigenetic changes.

In spite of the unusual phenotype that regenerated plant tend to exhibit, seed set does not usually pose any problems. Even in small inflorescences consisting of but a few disk flowers a corresponding number of viable seeds can be obtained. It is not unusual to obtain more than 200 seeds from a single head of a regenerated plant. Plants obtained from such seeds (R_1 generation) are perfectly normal and indistinguishable from sunflower plant produced by conventional means.

Few investigations are concerned with the genetic stability of regenerated sunflower plants. Somaclonal variation and other (epi-)genetic alterations do not appear to be a common phenomenon. In various instances, no significant deviation from the parental phenotype has been detected over several generations (Freyssinet and Freyssinet 1988; Jeannin and Hahne, 1991), nor could karyotypic variations be detected (Jeannin and Hahne, 1991; Jeannin *et al.*, 1990). Likewise, no sign of somaclonal variation could be detected by analysis with microsatellite markers in protoplast-derived plants and their offspring (F. Charrière and G. Hahne, Unpublished). However, treatment with mercuric chloride of immature embryo-derived callus produced plants with altered coumarin contents (Roseland *et al.*, 1991). The possible occurrence of somaclonal variants among plants regenerated from a similar source has been mentioned although no details are available (Witrzens *et al.*, 1988).

For all practical purposes, genetic stability of sunflower plants regenerated from *in vitro* culture systems is not perceived as a major concern or a limiting factor.

3. GENE TRANSFER TO SUNFLOWER

The standard transformation technique for a great number of dicotyledoneous plants is the leaf disk approach (Horsch *et al.*, 1985). Unfortunately, this powerful technique is impracticable for sunflower since no leaf-based plant regeneration protocol has been established for this species. A wide range of alternative combinations of gene transfer methods and their compatibility with existing culture techniques has therefore been evaluated before transgenic sunflower plants could be produced in a reproducible manner.

3.1. Direct gene transfer

Three technical approaches are currently used in order to introduce foreign DNA into plant cells without the help of a biological vector: (i) direct gene transfer into protoplasts, mediated by either electroporation or treatment with polyethylene glycol (PEG), (ii) microinjection, and (iii) particle bombardment. The latter two techniques can, in principle, be applied to intact plant tissues while the former one requires isolated protoplasts where, consequently, transgenic plant production depends on the availability of an efficient regeneration protocol from isolated protoplasts.

3.1.1. Gene transfer to isolated protoplasts

The introduction of foreign DNA into isolated sunflower protoplasts has been shown to be feasible under conditions comparable to those frequently used with protoplasts isolated from other species, by either exposition to a short electric pulse (electroporation) or treatment with PEG.

Transient expression was observed as expected in extrapolation from results obtained with other species (Kirches *et al.*, 1991), and the production of transgenic calli from PEG-treated sunflower protoplasts did not present particular problems (Moyne *et al.*, 1989) except for the low efficiency of recovery of transgenic colonies. The combination of low gene transfer efficiency with the notoriously low and unreliable plant regeneration frequency from sunflower protoplast cultures creates a situation, which is extremely unfavorable for the production of transgenic sunflower plants. No other publications on the use of sunflower protoplasts for direct gene transfer have followed these two pioneering studies, and no published report relates to transgenic sunflower plants produced by direct gene transfer to protoplasts.

3.1.2. Microinjection

Microinjection, the preferred transformation technique for animal cells, does not enjoy much success for the production of transgenic plants. Plant cells are characterized by a large central vacuole barring access to the nucleus, and they are either surrounded by a rigid cell wall rendering the penetration of the injection needle difficult, or extremely fragile (isolated protoplasts). Success rates for the recovery of transformed cells per injected cell are therefore low in all species, a circumstance demanding an extremely efficient regeneration system from an accessible cell type if transgenic plants are to be produced. One of the few published examples concerns the production of transgenic rapeseed plants from by microinjection of microspore-derived embryoids

(Neuhaus and Spangenberg, 1990; Neuhaus *et al.*, 1987). A single conference abstract (Espinasse-Gellner) reports on the application of this time-consuming and complex procedure to sunflower embryos *in ovulo,* but detailed information on the recovered plants is not available. It is unlikely that a routine system for the transformation of sunflower based on microinjection will be developed.

3.1.3. Particle bombardment

Accelerated heavy metal particles of sub-cellular size can penetrate into plant cells even without prior removal of the cell wall. They can thus penetrate into cells still embedded in the natural context of their tissue, including tissues with the inherent capacity for shoot production such as meristems (McCabe and Christou, 1993; McCabe and Martinell, 1993; McCabe *et al.*, 1988). This technical approach has made possible the transformation of species for which transformation protocols based on other techniques had not been available, often due to limitations in the generation capacity from transformable tissues (McCabe and Martinell, 1993; McCabe *et al.*, 1988). The fact that the cells expressing the introduced transgene remain in their natural environment makes particle bombardment a preferred choice also for transient expression studies (Hamilton *et al.*, 1992; Van der Leede-Plegt *et al.*, 1992). However, such results must still be interpreted with caution since bombardment is often accompanied by death of the expressing cells (Hunold *et al.*, 1994).

Despite the apparent advantages, studies investigating particle bombardment for sunflower transformation are scarce. Bombardment of immature embryos has yielded good success for transient expression but did not result in transgenic plants (Hunold *et al.*, 1993; Laparra *et al.*, 1995). Bombardment of other tissues was less conclusive. The only published case where transgenic sunflower plants were obtained by a technique involving particle bombardment used the unusual combination of bombardment with naked particles (*i.e.,* without prior coating with DNA) *Agrobacterium* infection, bombardment being used to introduce the lesions favoring entry of the bacteria (Malone-Schoneberg *et al.*, 1994).

3.1.4. Conclusion

The survey of the existing literature on the use of direct gene transfer methods for the transformation of sunflower results in the view that no convincing combination of direct gene transfer technology with a suitable protocol for subsequent plant regeneration has been found to date. In the light of the low efficiency for stable transformation of direct techniques, and the difficulties in regenerating fertile plants from sunflower explants discussed above, this situation is hardly surprising and unlikely to change in the near future.

3.2. *Agrobacterium*-mediated gene transfer

Sunflower is one of the first species that have been used for studying the interaction of *Agrobacterium tumefaciens* with higher plants. Early as well as recent studies used virulent strains that resulted in tumor formation for the investigation of general physiological phenomena (*e.g.* De Ropp, 1946; Kutschera *et al.*, 2000; Matzke *et al.*, 1984; Ursic, 1985; Yao *et al.*, 1988). Regeneration of plants was not a objective in these studies. The combination of disarmed *Agrobacterium* strains with an indirect regeneration protocol would appear as an ideal approach to sunflower transformation. Unfortunately, the only publication investigating this strategy (Everett *et al.*, 1987) relates the production of a single transgenic plant. It has not been possible since to extend this strategy to a larger scale, and no further publication has followed on this subject.

A wide range of sunflower regeneration systems has been tested for their compatibility with *Agrobacterium*-mediated gene transfer, most based on direct regeneration systems or exploiting the natural function of the excised organs (*e.g.* shoot production of apical meristems). Hypocotyl explants (Escandon and Hahne, 1990; Everett *et al.*, 1987), seedling cotyledons and immature embryos (Laparra *et al.*, 1995; Laparra *et al.*, 1996; Lucas *et al.*, 2000), young seedlings (Grayburn and Vick, 1995; Sankara Rao and Rohini, 1999), and explants derived from mature embryos (Alibert *et al.*, 1999; Bidney *et al.*, 1991; Knittel *et al.*, 1994 Malone-Schoneberg *et al.*, 1994; Schrammeijer *et al.*, 1990) have all been investigated in this respect. However, all studies that did yield transgenic plants share a common feature in that they make use of explants with juvenile characteristics or involve tissue close to the apex of mature or immature embryos and young seedlings.

The use of embryonic axes, exploiting the inherent regeneration capacity of the region around the apical meristem in combination with *Agrobacterium*-mediated gene transfer was introduced by Schrammeijer *et al.* (Shcrammeijer *et al.*, 1990). It was later refined by using particle bombardment to introduce micro-wounds which increased the efficiency of the system to a practically useful range (Bidney *et al.*, 1992; Malone-Schoneberg *et al.*, 1994). Knittel *et al.*, 1994 obtained similar efficiencies and showed that the use of existing meristems did not abolish genotype dependence. This observation can be explained by the fact that transformed shoots do not necessarily arise from the existing meristem but the more uniformly transformed shoots originate from adventitious mersitems induced in the immediate vicinity of the apical meristem (Burrus *et al.*, 1996). It is possible to increase the regeneration potential from this type of explants by genetic selection (Weber *et al.*, 2000). Protocols utilizing other means of wounding the explants prior to co-culture with agrobacteria have also yielded transgenic plants, including vigorous shaking with glass beads (Grayburn and Vick, 1995) and partial digestion with pectolytic enzymes (Alibert *et al.*, 1999). A typical experimental plan, highlighting the key steps, is shown in Fig. 2. This overall plan is valid for most of the published approaches that have yielded transgenic offspring with the exception of Lucas *et al.* (Lucas *et al.*, 2000) who used immature embryos as explants. The individual protocols differ in modifications of one or more of the indicated steps, such as the age of the explants, the methods used for wounding of the tissue or for the induction of the *vir* genes of the employed agrobacteria. Timing of the individual steps and the indicated time scale must be adapted accordingly.

Whatever the treatment of the explant, the regenerated shoots tend to be weak and their survival and development after transfer to the greenhouse is one of the major limiting steps in the protocol. As was the case for protoplast-derived shoots, this problem can be satisfactorily solved by grafting of the weak shoots onto sunflower root-stocks, either *in vitro* or directly in the greenhouse (Hahne *et al.*, unpublished; for details see our web site, http://ibmp.u-strasbg.fr).

Although protocols are available that allow the production of transgenic sunflower plants in a number of laboratories, this species still belongs to those for which a universally applicable, efficient and reliable routine transformation system is not yet available. Apart from the fine-tuning that is necessary when introducing the technique to a new laboratory, due to the dependence of the overall efficiency on genotype dissection technique, bacterial strain, selection system, and *in vitro* culture and greenhouse conditions, the efficiency is still quite low even in those laboratories where the technique has been mastered for several years. The recovery rate of plants giving rise to transgenic offspring from embryonic axes is usually in or below the percent range, with a high degree of variability between individual experiments.

The use of direct regeneration protocols, including from pre-existing

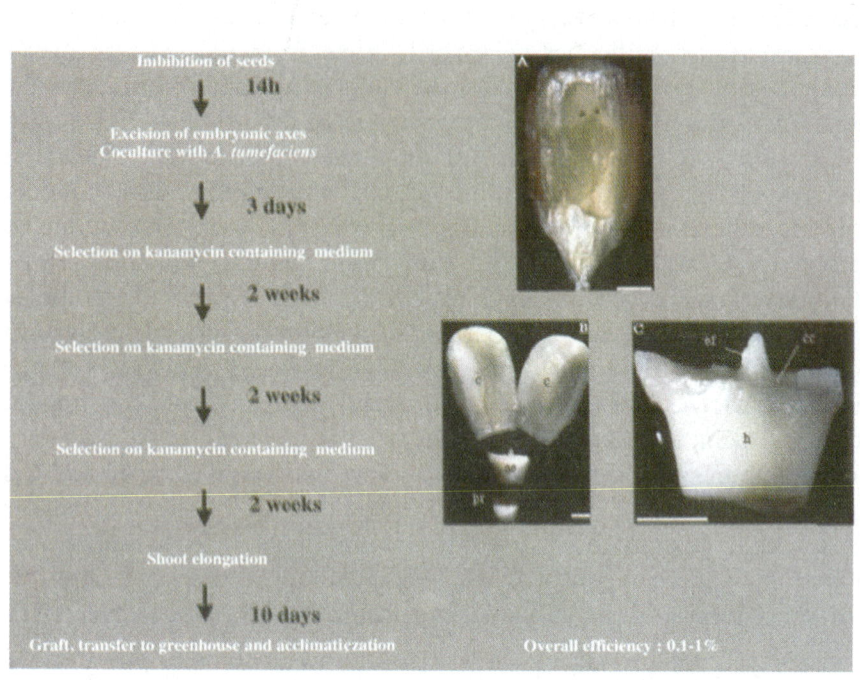

Figure 2 : Indicative representation of the key steps involved in the most frequently used protocol for sunflower transformation. Each individual step is subject to modification in order to adapt the protocol to the specific conditions and genotype, and to increase the overall efficiency. A,B,C: Preparation of explants. A. Mature embryo before dissection. B. Dissected embryo; cotyledons [c] and root pole [pr] have been removed from the embryonic axis [ae]. C. The explant consists of the hypocotyl [h] of the embryo and the apical region including leaf primordia [ef] and cotyledonal scar [cc] Bar: 1 mm.

meristems, leads to the production of primary transformants that are often chimerical in nature. This is not a serious limitation since the offspring is free of chimeras. However, even a preliminary analysis of primary transformants is rendered difficult by this fact and the segregation ratio of the transgene in the F_1 generation may deviate significantly from the values expected for Mendelian transmission.

3.3. Selection of transformed tissue

Whatever the method used for gene transfer, a crucial step in every transformation protocols is the enrichment of the cultured tissue for those cells that are capable of expressing the transferred genetic construct. This requirement is usually satisfied by integrating selectable genetic markers such as genes conferring resistance to antibiotics or herbicides in the genetic construct used for transformation. While such selection is usually straightforward with callus- and cell suspension cultures due to the high degree of accessibility of the target cells by the selective agent, selection among intact organs or even plantlets is much more prone to artifacts and escapes, due to the distortion of diffusion patterns and detoxification phenomena.

The *nptII* gene, coding for the enzyme, neomycin phosphotransferase II (NPT II), is among the oldest and most widely used selection markers in plant transformation. It is also the one that is most often used with sunflower tissues although this combination is not always unproblematic. The NPT II enzyme detoxifies a range of compounds belonging to the class of aminoglycoside antibiotics, including neomycin, kanamycin, geneticin, paromomycin etc. When used with sunflower tissues, the results obtained with these different compounds varied in function of the tissue type, the origin of the regenerated shoots, and the culture conditions (particularly media composition) (Escandon and Hahne, 1990; Escandón and Hahne, 1991). In some instances, cultured sunflower tissues were found to be resistant to these compounds, even when not expressing any genetic marker. In other cases, prolonged exposure to kanamycin allegedly caused morphogenic callus to loose its morphogenic capacity (Everett *et al.*, 1987). Despite these difficulties, selection of transformed tissue on kanamycin-containing media seems to represent the best compromise in conjunction with transformation systems using embryonic axes (Alibert el al., 1999; Bidney *et al.*, 1991; Knittel *et al.*, 1994; Malone-Schoneberg *et al.*, 1994; Schrammeijer *et al.*, 1990). Among the considerations to be taken into account, the rapidity of the regeneration systems, the direct origin of the regenerated shoots and their ensuing chimerical nature, and the necessity to conduct the selection on the level of whole shoots are the most important ones. In particular, the chimerical nature of the regenerating structures is of decisive importance for the choice of the selective agent, as the transgenic sectors likely to give rise to transgenic offspring are expected to be located in regions of the shoots distant from the culture medium, and thus to be separated from the source of the selective agent by non-transgenic tissue. Any selective agent resulting in cell death, as is the case for most herbicides and many antibiotics, are unsuitable for such samples for they cause the whole shoot to perish, including the transgenic sector. Suitable selective agents result in a visible alteration, but not destruction, of the untransformed tissue. Most published protocols for transformation of sunflower incorporate a selection step on non-toxic concentrations of kanamycin, which results in bleaching of the untransformed tissues. Although this approach is prone to a high level of escapes and false positives, in particular when using the low concentrations that are compatible with the objective of not loosing chimerical samples, it still represents a reasonable compromise.

Further available visual markers for the selection of transformed tissue include luciferase, an enzyme responsible for a chemiluminescent phynotype (Barnes, 1990). The use of this marker is limited by the requirement for sophisticated optical equipment capable to detect and visualize the low level of the emitted light. Green fluorescent protein (GFP) causes fluorescence of the cells expressing this protein. Under suitable exciting light, this fluorescence is easily detected with standard equipment (dissecting or compound microscopes, McCormac *et al.*, 1998; Molinier *et al.*, 2000). However, many plant tissues are characterized by pronounced autofluorescence over a wide range of wavelengths, and this feature is the major reason why GFP so far is not widely used for the selection of transgenic sunflower tissues.

4. INTRODUCED CHARACTERS AND PROPERTIES OF TRANSGENIC PLANTS

Long-term studies of transgenic sunflower plants have not been reported in published form. Little information is available about structural properties of the introduced transgenes, their genetic stability and expression properties of the introduced constructs over several generations under laboratory and field conditions. A due appreciation of these aspects must await publication of further studies.

In spite of this lack of published information, transgenic sunflower has been subject to numerous field trials in several countries. The earliest trials date back to 1991 (Table 1). It is noteworthy that the number of field trials decreased towards the end of the decade and no field trials were registered after 1999 worldwide (at least in those countries for which such information is included in the international databases; see Table 1). The traits involved in the registered field trials range from marker genes obviously intended to gather information about transgenic stability and expression data, to genes matching the challenges presented by the difficulties and constraints of large-scale cultivation (*e.g.,* disease resistance) as well as market demand for improved product quality and novel products (Table 1). It may be concluded from this table that to

Table 1. Field trials with transgenic sunflowers

Character	Year	Country	Declared Main Trait	Company	Reference
Seed quality	1991	USA	Altered seed storage protein	Pioneer	91-067-01R
	1992	USA	Altered seed storage protein	Pioneer	92-042-02R
	1993	USA	Altered seed storage protein	Pioneer	93-033-02R
	1995	USA	Altered seed storage protein	Van der Have	95-033-01R
	1995	NL	Albumin synthesis	Van der Have	B/NL/95/16
	1995	NL	Asparagine synthesis	Van der Have	B/NL/95/16
	1995	NL	Threonine synthesis	Van der Have	B/NL/95/16
	1997	F	Alteration of oil composition (low stearate content)	Rustica Prograin Génétique	B/FR/97/11/29
	1999	F	Reduction of stearic acid content	Rustica Prograin Génétique-Biogemma	B/FR/99/05/03
Fungal disease resistance	1995	USA	Sclerotinia resistance	Van der Have	95-033-01R
	1995	NL	Chalcone synthesis	Van der Have	B/NL/95/16
	1995	NL	Chitinase synthesis	Van der Have	B/NL/95/16
	1995	NL	Fungal resistance	Van der Have	B/NL/95/16

Table 1. Continued

Character	Year	Country	Declared Main Trait	Company	Reference
	1995	NL	Glucanase synthesis	Van der Have	B/NL/95/16
	1995	NL	Oxalate synthesis	Van der Have	B/NL/95/16
	1996	F	Oxalate oxidase synthesis	SES France, Van der Have France	B/FR/96/02/12
	1997	USA	Sclerotinia resistant	Pioneer	97-224-01n
	1997	USA	Sclerotinia resistant	Pioneer	97-307-01n
	1997	USA	Sclerotinia resistant	Pioneer	97-307-02n
	1997	F	Oxalate oxidase synthesis	Pioneer Génétique France	B/FR/97/05/29
	1998	E	Oxalate oxidase synthesis	Semillas Pioneer	B/ES/98/14
	1998	USA	Sclerotinia resistance	Pioneer	98-44-11N
	1998	USA	Sclerotinia resistance	Pioneer	98-078-10n
	1998	USA	Sclerotinia resistance	Pioneer	98-201-05N
	1998	USA	Sclerotinia resistance	Pioneer	98-329-06n

Table 1. Continued

Character	Year	Country	Declared Main Trait	Company	Reference
	1998	USA	Sclerotinia resistance	Pioneer	98-34-04n
	1999	F	Oxalate oxidase synthesis	Pioneer Genetique	B/FR/99/03/07-CON
	1999	USA	Sclerotinia resistance	Pioneer	99-070-04N
	2000	F	Oxalate decarboxylase synthesis	Pioneer Génétique SARL	B/FR/00/03/03-CON
	2000	F	Oxalate oxidase sysnthesis	Pioneer Génétique SARL	B/FR/00/03/07-CON
Viral disease resistance	1997	USA	Virus (CMV/TMV) resistance	Pioneer	97-029-03R
Insect resistance	1995	NL	Bt-derived insect resistance (Cowpea trypsin inhibitor synthesis)	Van der Have	B/NL/95/16
	1996	USA	Insect resistant	Van der Have	96-071-02R
	1997	USA	Lepidopteran resistant	Pioneer	97-211-01n
	1997	USA	Lepidopteran resistant	Mycogen	97-160-02n
	1998	USA	Lepidopteran resistant	Indiana U	98-195-02n

Table 1. Continued

Character	Year	Country	Declared Main Trait	Company	Reference
	1998	USA	Lepidopteran resistant	Indiana U	98-110-06n
	1998	USA	Lepidopteran resistant	Pioneer	98-320-05n
	1998	USA	Lepidopteran resistant	Pioneer	98-343-06n
	1998	USA	Lepidopteran resistant	Pioneer	98-349-05n
	1999	USA	Insect resistance	Pioneer	99-07-05n
Herbicide resistance	1994	NL	Glufosinate tolerance	Van der Have	B/NL/94/18
	1995	NL	Chlorsulphuron tolerance	Van der Have	B/NL/95/16
	1995	NL	Glufosinate tolerance	Van der Have	B/NL/95/16
Marker genes	1994	F	Marker	Limagrain Genetics	B/FR/94/02/13
	1994	F	Marker	Van der Have France	B/FR/94/02/19
	1995	USA	Visual marker (GUS)	Pioneer	95-031-02R
	1997	USA	Visual marker (GUS)	Van der Have	97/126-01R
Other traits	1994	E	Drought tolerance; Levan sucrase synthesis	Van der Have Cubain	B/ES/94/10

Table 1. Continued

Character	Year	Country	Declared Main Trait	Company	Reference
	1994	USA	Confidential business information	Van der Have	94-025-01R
	1995	NL	Fructosyltransferase synthesis	Van der Have	B/NL/95/16
	1995	NL	Levan sucrase synthesis	Van der Have	B/NL/95/16
	1995	NL	Male sterility/fertility restoration	Van der Have	B/NL/95/16
	1995	Nl	Nitrate reductase synthesis	Van der Have	B/NL/95/16
	1995	NL	Nitrite reductase synthesis	Van der Have	B/NL/95/16
	1996	USA	Confidential business information	Van der Have	96-071-04R
	1998	E	Broomrape control	Semillas Pioneer	B/ES/98/13

Sources :
Robert-Koch-Institut (RKI) (http://www.rki.de/GENTEC/GENTEC.HTM)
European Commission Joint Research Centre (jrc) (htt://biotech.jrc.it/gmo.htm),
USDA APHIS (http://www.aphis.usda.gov/biotech/),
OECD (http://www.olis.oecd.org/biotrack.nsf)

date, the developed transgenic sunflower varieties have not yet reached a point of maturity that makes their introduction into the market desirable.

5. CONCLUSIONS AND FUTURE PROSPECTS

Considered individually, each step involved in the production of transgenic sunflower plants does not present fundamental difficulties that would explain why this species is so recalcitrant to the widespread and efficient application of genetic engineering. Plant regeneration is possible through several mechanisms from a range of tissues, gene transfer has been demonstrated to be feasible by several approaches that are routinely employed with other species, and the existence of a number of transgenic sunflowers that have been submitted to field trials validates the feasibility in principle of the protocol most currently used today. Low efficiencies for stable integration of transgenes are not characteristic of sunflower, as they are the norm for most gene transfer systems. The explanation for the exceptionally low overall efficiency of sunflower transformation systems is most probably found in the combination of a number of unfavorable characteristics: very high genotype dependence of the regeneration efficiency, combined with the impossibility for vegetative propagation (itself resulting in the difficulty to maintain a particular genotype with *e.g.* high regeneration potential faithfully); absence of an indirect regeneration system allowing for efficient enrichment for transgenic cell populations prior to induction of the morphogenic process; the difficulty to target gene transfer to cells that become involved in morphogenic activities in the direct regeneration systems; chimerical nature of putative primary tansfromants and resulting difficulties for selection schemes. It is probable that a number of other, as yet unidentified, parameters contribute to the observed recalcitrance.

The practical result of this combination of unfavorable circumstances is that today, all steps and parameters must be fine-tuned with high precision in order for a protocol to yield any transgenic plants at all. Consequently, the time and experience needed for the adaptation of the protocol to a new laboratory is longer than usual for other species and the result subject to stochastic variations in function of a complex set of parameters. Personal experience of the experimenter is still an important parameter.

In spite of all these limitations, genetic engineering of sunflower has become a reality and has already been practiced on industrial scale for several years (see Table 1). The technology is available to a growing number of private and public laboratories, as demonstrated by the field trials that have been conducted by a number of companies in several countries. However, the performance level of genetic engineering of sunflower does not yet match the economic importance of this species, and it is obvious that substantial further efforts in fundamental research are needed in order to understand the phenomena better that limit the efficiency of genetic engineering techniques, and to devise more robust transformation systems.

With increased understanding of the mechanisms governing the *in vitro* development of sunflower, and in-depth knowledge of the physiological processes involved in the desirable traits to be introduced in this important oil crop, it can be expected that genetic engineering of the complex biochemical pathways involved in resistance to biotic and abiotic stresses as well as product quality trait will become a reality in the near future also for *Helianthus annuus*.

REFERENCES

Alibert B, Lucas O, Le Gall V, Kallerhoff J and Alibert G (1999) Pectolytic enzyme treatment of sunflower explants prior to wounding and cocultivation with *Agrobacterium tumefaciens* enhances efficiency of transient GUS expression. *Physiol. Plant.*, **106** : 232-237.

Baker CM, Moñoz-Fernandez N and Carter CD (1999) Improved shoot development and rooting from mature cotyledons of sunflower. *Plant Cell Tiss. Org. Cult.*, **58** : 39-49.

Barnes WM (1990) Variable patterns of expression of luciferase in transgenic tobacco leaves. *Proc Natl. Acad. Sci. USA.*, **87** : 9183-9187.

Barth S, Voeste D, Wingender R and Schnabl H (1993) Plantlet regeneration from electrostimulated protoplasts of sunflower (*Helianthus annuus* L.). *Bot. Acta*, **106** : 220-222.

Barthou H, Briere C, Caumont C, Petitprez M, Kallerhoff J, Borin C, Souvré A and Alibert G (1997) Effect of atmospheric pressure on sunflower (*Helianthus annuus* L.) protoplasts division. *Plant Cell Rep.*, **16** : 310-314.

Bechtold N, Ellis J and Pelletier G (1993) *In planta Agrobacterium*-mediated gene transfer by infiltration of adult *Arabidopsis thaliana* plants. *Comptes Rendus Acad. Sci. Paris,* **316** : 1194-1199.

Bidney DL, Scelonge CJ and Malone-Schonenberg JB (1991) Transformed progeny can be recovered from chimeric plants regenerated from *Agrobacterium tumefaciens* treated embryonic axis of sunflower. *Proceedings of the 13th International Sunflower Conference, Pisa, Italy,* pp. 1408-1412.

Bidney DL, Scelonge CJ and Malone-Schoneneberg JB (1992) Transformed progeny can be recovered from chimeric plants regenerated from *Agrobacterium tumefaciens* treated embryonic axes of sunflower. *Proceedings of the 13th International Sunflower Conference, Pisa,* pp. 1408-1412.

Bohorova NE, Cocking EC and Power JB (1986) Isolation, culture and callus regeneration of protoplasts of wild and cultivated *Helianthus species. Plant Cell Rep.,* **5** : 256-258.

Bolandi AR, Branchard M, Alibert G, Genzbittel L, Berville A and Sarrafi A (2000) Combining-ability analysis of somatic embryogenesis from epidermal layers in the sunflower (*Helianthus annuus* L.). *Theor. Appl. Genet.*, **100** : 621-624.

Borisjuk NV, Borisjuk LG, Logendra S, Peterson F, Gleba Y and Raskin I (1999) Production of recombinant proteins in plant root exudates. *Nature Biotech.*, **17** : 466-469.

Bronner R, Jeannin G and Hahne G (1994) Early cellular events during organogenesis and somatic embryogenesis induced on immature zygotic embryos of sunflower (*Helianthus annuus*). *Can. J. Bot.*, **72** : 239-248.

Burrus M, Chanabe C, Alibert G and Bidney D (1991) Regeneration of fertile plants from protoplasts of sunflower (*Helianthus annuus*). *Plant Cell Rep.*, **10** : 161-166.

Burrus M, Molinier J, Himber C, Hunold R, Bronner R, Rousselin P and Hahne G (1996) *Agrobacterium*-mediated transformation of sunflower (*Helianthus annuus* L.) shoot apices : transformation patterns. *Mol. Breed.*, **2** : 329-338.

Carter JF (1987) Sunflower science and technology. In: *Agronomy.* American Society of Agronomy, Crop Science Society of America, Soil Science of America, Inc., Madison.

Caumont C, Petitprez M, Woynaroski S, Barthou H, Briére C, Kellerhoff J, Borin C, Souvré A and Alibert G (1997) Agarose embedding affects cell wall regeneration and microtubule organization in sunflower protoplasts. *Physiol. Plant.*, **99** : 129-134.

Ceriani MF, Hopp HE, Hahne G and Escandon AS (1992) Cotyledons: an explant for routine regeneration of sunflower plants. *Plant Cell Physiol.*, **33** : 157-164.

Chanabé C, Burrus M and Alibert G (1989) Factors affecting the improvement of colony formation from sunflower protoplasts. *Plant Sci.*, **64** : 125-132.

Charrière F and Hahne G (1998) Induction of embryogenesis versus caulogenesis on *in vitro* cultures sunflower (*Helianthus annuus* L.) immature zygotic embryos: role of plant growth regulators. *Plant Sci.*, **127** : 63-71.

Chibbar RN, Shyluk J, Georges F and Constabel F (1987) Biochemical aspects of sunflower protoplast culture. *Plant Physiol.*, 83-74.

Chraibi BKM, Castelle JC, Latche A, Roustan JP and Fallot J (1992a) Enhancement of shoot regeneration potential by liquid medium culture from mature cotyledons of sunflower (*Helianthus annuus* L.). *Plant Cell Rep.*, **10** : 617-620.

Chraibi BMM, Castelle JC, Latche A, Roustan JP and Fallot J (1992b) A genotype-independent system of regeneration from cotyledons of sunflower (*Helianthus annuus* L.) The role of ethylene. *Plant Sci.*, **86** : 215-221.

Chraibi BKM, Castelle JC, Latche A, Roustan JP and Fallot J (1992c) Influence de l'acid 2-chloroethylphosphonique et des olyamines sur la caulogenese du touresol (*Helianthus annuus* L.) a partir de cotyledons. *Comptes Rendus Acad. Sci. Paris,* **315** : 459-462.

Chraibi BKM, Latche A, Roustan JP and Fallot J (1991) Stimulation of shoot regeneration from cotyledons of *Helianthus annuus* by ethylene inhibitors, silver and cobalt. *Plant Cell Rep.*, **10** : 204-207.

Datta K, Muthukrishnan S and Datta SK (1999) Expression and function of PR-protein genes in transgenic plants. In: *Pathogenesis-related proteins in plants.* (Eds Datta SK and Muthukrishnan S) CRC Press, Boca Raton, pp. 261-277.

De Ropp RS (1946) The isolation and behaviour of bacteria-free crown gall tissue from primary galls of *Helianthus annuus. Phytopathol.*, **37** : 201-206.

Deglene L, Lesignes P, Alibert G and Sarrafi A (1997) Genetic control organogenesis in cotyledons of sunflower (*Helianthus annuus*). *Plant Cell Tiss. Org. Cult.*, **48** : 127-130.

Diepenbrock W (1987) Die Ertragsbildung der Sonnenblume. Kali-Briefe (Büntehof), **18** : 639-659.

Dunwell JM (1999) Transgenic crops: The next generation, or an example of 2020 vision. *Ann. Bot.,* **84** : 269-277.

Dupuis JM, Pean M and Chagvardieff P (1990) Plant donor tissue and isolation procedure effect on early formation of embryoids from protoplasts of *Helianthus annuus* L. *Plant Cell Tiss. Org. Cult.,* **22** : 183-190.

Escandon AE and Hahne G (1990) Sunflower transformation: a study of selectable markers. In: *The impact of Biotechnology in Agriculture.* (Eds Sangwan RS and Sangwan-Norreel BS) Kluwer Acad. Publ., Dordrecht (NL), pp. 345-353.

Escandón AS and Hahne G (1991) Genotype and composition of culture medium are factors important in the selection for transformed sunflower (*Helianthus annuus*) callus. *Physiol. Plant.,* **81** : 367-376.

Espinasse A and Lay C (1989) Shoot regeneration of callus derived from globular to torpedo embryos from 59 sunflower gentotypes. *Crop Sci.,* **29** : 201-205.

Espinasse A, Lay C and Volin J (1989) Effects of growth regulator concentrations and explant size on shoot organogenesis from callus derived from zygotic embryos of sunflower (*Helianthus annuus* L.) *Plant Cell Tiss. Org. Cult.,* **17** : 171-181.

Espinasse-Gellner A (1990) A simple and direct technique of transformation in sunflower. Sunflower Research Workshop Fargo ND. National Sunflower Associated, USA,

Everett NP, Paterson-Robinson KE and Mascarenhas D (1987) Genetic engineering of sunflower (*Helianthus annuus* L.). *Bio/Technology,* **5** : 1201-1204.

Finer JJ (1987) Direct somatic embryogenesis and plant regeneration from immature embryos of hybrid sunflower (*Helianthus annuus* L.) on a high sucrose-containing medium. *Plant Cell Rep.,* **6** : 372-374.

Fiore MC, Trabace T and Sunseri F (1997) High frequency of plant regeneration in sunflower from cotyledons *via* somatic embryogenesis. *Plant Cell Rep.,* **16** : 295-298.

Fischer C and Hahen G (1992) Structural analysis of colonies derived from sunflower (*Helianthus annuus* L.) protoplasts cultured in liquid and semi-solid media. *Protoplasma,* **169** : 130-138.

Fischer C, Klethi P and Hahne G (1992) Protoplasts from cotyledon and hypocotyl of sunflower (*Helianthus annuus*) : shoot regeneration and seed production. *Plant Cell Rep.,* **11** : 632-636.

Fischer C, Laparra H, Charriere F, Jung JL and Hahne (1996) Regeneration of plants from protoplasts of *Helianthus annuu* L. (Sunflower). In: *Biotechnology in Agriculture and Forestry.* (Ed Bajaj YPS) Springer Verlag. Berlin Heidelberg, pp. 48-63.

Fischer R, Drossard J, Commandeur U, Schillber S and Emans N (1999) Towards molecular farming in the future: moving from diagnostic protein and antibody production in microbes to plants. *Biotechnol. Appl. Biochem.,* **30** : 101-108.

Flores Berrios E, Gentzbittel L, Kayyal H, Alibert G and Sarrafi A (2000a) AFLP maping of QTLs for *in vitro* organogenesis traits using recombinant inbred lines in sunflower (*Helianthus annuus* L.). *Theor. Appl. Genet.,* **101** : 1299-1306.

Flores Berrios E, Gentzbittel L, Mokrani L, Alibert G and Sarrafi A (2000b) Genetic control of early events in protoplast division and regeneration pathways in sunflower. *Theor. Appl. Gent.,* **101** : 606-612.

Flores Berrios E, Sarrafi A, Fabre F, Alibert G and Gentzbittel L (2000c) Genotypic variation and chromosomal location of QTLs for somatic embryogenesis revealed by epidermal layers culture of recombinant inbred lines in the sunflower (*Helianthus annuus* L.). *Theor. Appl. Genet.,* **101** : 1307-1312.

Franken E, Teuschel U and Hain R (1997) Recombinant proteins from transgenic plants. *Curr. Opin. Biotechnol.,* **8** : 411-416.

Freyssinet M and Freyssinet G (1988) Fertile plant regeneration from sunflower (*Helianthus annuus* L.) immature embryos. *Plant Sci.,* **56** : 177-181.

Grayburn WS and Vick BA (1995) Transformation of sunflower (*Helianthus annuus* L.) following wounding with glass beads. *Plant Cell Rep.,* **14** : 285-289.

Greco B, Tanzarella OA, Carrozzo G and Blanco A (1984) Callus induction and shoot regeneration in sunflower (*Helianthus annuus* L.). *Plant Sci. Lett.,* **36** : 73-77.

Gruber V and Theisen M (2000) Genetically modified crops as a source for pharmaceuticals *Annual Reports in Medicinal Chemistry.* Academic Press, Orlando, pp. 357-364.

Guilley E and Hahne G (1989) Callus formation from isolated sunflower (*Helianthus annuus*) mesophyll protoplasts. *Plant Cell Rep.,* **8** : 226-229.

Gürel A, Nichterlein K and Friedt W (1991) Shoot regeneration from anther culture of sunflower (*Helianthus annuus*) and some interspecific hybrids as affected by genotype and culture procedure. *Plant Breed.,* **106** : 68-76.

Hamilton DA, Roy M, Rueda J, Sindhu RK, Sanford J and Mascarenhas JP (1992) Dissection of a pollen-specific promoter from maize by transient transformation assays. *Plant Mol. Biol.,* **18** : 211-218.

Heiser C, JR (1978) Taxonomy of *Helianthus* and origin of domesticated sunflower. In: *Sunflower science and technology.* (Ed Carter JF) American society of Agronomy, Crop Science Society of America, Soil Science Society of America, Inc., Madison, pp. 31-53.

Heiser CB, JR, Smith DM, Clvenger SB and Martin WC, JR (1969) The North American Sunflowers (*Helianthus*). In: *Memoirs of the torrey botanical club.* (Ed Delevoryas T). The Seeman Printery, Durham, pp. Vol. 22.

Heiser J, Charles B (1976) The Sunflower. Press UOO (ed), Norman, pp. 11-184.

Henderson JHM, Durrell ME and Bonner J (1952) The culture of normal sunflower stem callus. *Am. J. Bot.,* **39** : 467-473.

Henn HJ, Wingender R and Schnabl H (1995) Plant regeneration from sunflower (*Helianthus annuus* L.) protoplasts and intespecific hybridization. *3rd European Symposium on Sunflower Biotechnology,* pp. 4.

Henrickson CE (1954) The flowering of sunflower explants in aseptic culture. *Plant Physiol.,* **29** : 536-538.

Hiei Y, Ohta S, Komari T and Kumashiro T (1994) Efficient transformation of rice (*Oryza sativa* L.) mediated by *Agrobacterium* and sequence analysis of the boundaries of the T-DNA. *Plant J.,* **6** : 271-282.

Hilderbrandt AC, Riker AJ and Duggar BM (1946) The influence of the composition of the medium on growth *in vitro* of excised tobacco and sunflower tissue cultures. *Am. J. Bot.,* **33** : 591-597.

Horsch RB, Fry JE, Hoffmann NL, Eichholtz D, Rogers SG and Fraley TT (1985) A simple and general method for transferring genes into plants. *Science,* **227** : 1229-1231.

Hunold R, Bronner R and Hahne G (1993) GUS expression in sunflower following microprojectile bombardment. *Biotechnol. Biotechnol. Eq.,* **4** : 91-95.

Hunold R, Bronner R and Hahne G (1994) Early events in microprojectile bombardment: cell viability and particle location. *Plant J.,* **5** : 593-604.

Ivanov P, Encheva J and Ivanova I (1997) A protocol to avoid precocious flowering of sunflower plantlets *in vitro. Plant Breed.,* **117** : 582-584.

Jeannin G, Bronner R and Hahne G (1993) Early cytological discrimination between organogenesis and somatic embryogenesis induced on immature zygotic embryos of sunflower (*Helianthus annuus* L.). *Biotechnol. Biotechnol. Eq.,* **4** : 96-99.

Jeannin G, Bronner R and Hahne G (1995) Somatic emrbyogenesis and organogenesis induced on the immature zygotic embryo of sunflower (*Helianthus annuus* L.) cultivated *in vitro*: role of the sugar. *Plant Cell Rep.,* **15** : 200-204.

Jeannin G, Charrière F, Bronner R and Hahne G 1998) Is predetermined cellular competence required for alternative embryo or shoot induction on sunflower zygotic embryos? *Bot. Acta,* **111** : 280-286.

Jeannin G and Hahne G (1991) Donor plant growth conditions and regeneration of fertile plants from somatic embryos induced on immature zygotic embryos of sunflower (*Helianthus annuus* L.). *Plant Breed.,* **107** : 280-287.

Jeannin G, Poirot M and Hahne G (1990) Régénération de plantes fertiles á partir d'embryons zygotiques immatures de tournesol. I: Doré C. (ed) Cinquantenaire de la culture *in vitro*. Ed. INRA, Versailles, France, pp. 275-276.

Josten P and Kutschera U (1999) The micronutrient boron causes the development of adventitious roots in sunflower cuttings. *Ann. Bot.,* **84** : 337-342.

Kandler O (1952) Uber eine physiologische Umstimmung von Sonnenblumenstengelgewebe durch Dauereinwirkug von β-Indolylessingsäure. *Planta,* **40** : 346-349.

Kirches E, Frey N and Schnabl H (1991) Transient gene expression in sunflower. *Bot. Acta,* **104** : 212-216.

Knittel N, Escandon AS and Hahne G (1991) Plant regeneration at high frequency from mature sunflower cotyledons. *Plant Sci.,* **73** : 219-226.

Knittel N, Gruber V, Hahne G and Lénée P (1994) Transformation of sunflower (*Helianthus annuus* L.): a reliable protocol. *Plant Cell Rep.,* **14** : 81-86.

Konov A, Bronner R, Skryabin K and Hahne G (1998) Formation of epiphyllus buds in sunflower (*Helianthus annuus* L.): induction and cellular origin. *Plant Sci.,* **135** : 77-86.

Koornneef M, Bade J, Hanbart C, Horsman K, Schel J, Soppe W, Verkerk R and Zabel P (1993) Characterization and mapping of a gene controlling shoot regeneration in tomato. *Plant J.,* **3** : 131-141.

Krasnyanski S and Menczel L (1993) Somatic embryogenesis and plant regeneration from hypocotyl protoplasts of sunflower (*Helianthus annuus* L.) *Plant Cell Rep.,* **12** : 260-263.

Kutschera U, Bahrami M and Grotha R (2000) Sucrose metabolism during *Agrobacterium tumefaciens*-induced tumor growth in sunflower. *J. Plant Physiol.,* **157** : 1-6.

Laparra H, Bronner R and Hanhe G (1997a) Amyloplasts as a possible indicator of morphogenic potential in sunflower protoplasts. *Plant Sci,* **122** : 183-192.

Laparra H, Bronner R and Hahne G (1997b) Histological analysis of somatic embryogenesis induced in leaf explants of *Helianthus annuus* Heiser. *Protoplasma*, **196** : 1-11.

Laparra H, Burrus M, Hunold R, Damm B, Bravo-Angel AM, Bronner R and Hahne G (1995) Expression of foreign genes in sunflower (*Helianthus annuus* L.) – evaluation of three gene transfer methods. *Euphytica,* **85** : 63-74.

Laparra H, Burrus M, Hunold R, Himber C, Damm B, Bravo-Angel AM, Knittel N, Bronner R and Hahne G (1996) Approaches to the genetically engineered sunflower (*Helianthus annuus* L.). In: *Compositae: Biology & Utilization.* (Eds, Caligari PDS and Hind DJN) Kew Royal Botanical Gardens, Kew (GB), pp. 593-601.

Leclercq P (1969) Une stérilité mâle cytolasmique chez le tournesol. *Ann. Amélior. Plantes,* **19** : 99-106.

Lenee P and Chupeau Y (1986) Isolation and culture of sunflower protoplasts (*Helianthus annuus* L.): factors influencing the viability of cell colonies derived from protoplasts. *Plant Sci,* **43** : 69-75.

Lenée P and Chupeau Y (1989) Development of nitrogen assimilating enzymes during growth of cells derived from protoplasts of sunflower and tobacco. *Plant Sci.,* **59** : 109-117.

Lucas O, Kallerhoff J and Alibert G (2000) Production of stable transgenic sunflowers (*Helianthus annuus* L.) from wounded immature embryos by particle bombardment and co-cultivation with *Agrobacterium tumefaciens. Mol. Breed.,* **6** : 479-487.

Lupi MC, Bennici A, Locci F and Gennai D (1987) Plantlet formation from callus and shoot-tip culture of *Helianthus annuus* (L.). *Plant Cell Tiss. Org. Cult.,* **11** : 47-55.

Malone-Schoneberg J, Scelonge CJ, Burrus M and Bidney DL (1994) Stable transformation of sunflower using *Agrobacterium* and split embryonic axis explants. *Plant Sci.,* **103** : 199-207.

Marchant R, Davey MR, Lucas LA, Lamb CJ, Dixon RA and Power JB (1998) Expression of a chitinase transgene in rose (*Rosa hybrida* L.) reduces development of blackspot disease (*Diplocarpon rosae* Wolf). *Mol. Breed.,* **4** : 187.

Matzke MA Susani M, Binns AN, Lewis ED, Rubenstein I and Matzke AJM (1984) Transcription of a zein gene introduced into sunflower using a Ti plasmid vector. *EMBO J.,* **3** : 1525-1531.

McCabe D and Christou P (1993) Direct DNA transfer using electric discharge particle acceleration (ACCELL^{TM}technology). *Plant Cell Tiss. Org. Cult.,* **33** : 227-236.

McCabe DE and Martinell BJ (1993) Transformation of elite cotton cutivars via particle bombardment of meristems. *Bio/Technology,* **11** : 596-598.

McCabe DE, Swain WF, Martinell BJ and Christou P (1988) Stable transformation of soybean (*Glycine max*) by particle acceleration. *Bio/Technology,* **6** : 923-925.

McCormac AC, Wu H, Bao M, Wang Y, Xu R, Elliott MC and Chen DF (1998) The use of visual marker genes as cell-specific reporters of *Agrobacterium* mediated T-DNA delivery to wheat (*Triticum aestivum* L.) and barley (*Hordeum vulgare* L.) *Euphytica,* **99** : 17-25.

Molinier J, Himber C and Hahne G (2000) Use of green fluorescent protein for detection of transformed shoots and homozygous offspring. *Plant Cell Rep.,* **19** : 219-223.

Moyne AL, Tagu D, Thor V, Bergounioux C, Freyssinet G and Gadal P (1989) Transformed calli obtained by direct gene transfer into sunflower protoplasts. *Plant Cell Rep.,* **8** : 97-100.

Moyne AL, Thor V, Pelissier B, Bergounioux C, Freyssinet G and Gadal P (1988) Callus and embryoid formation from protoplasts of *Helianthus annuus. Plant Cell Rep.,* **7** : 437-440.

Nataraja K and Ganapathi TR (1989) *In vitro* plantlet regeneration from cotyledons of *Helianthus annuus* cv. Morden (sunflower). *Indian J. Exp. Biol.* **27** : 777-779.

Neuhaus G and Spangenberg G (1990) Plant transformation by microinjection techniques. *Physiol. Plant.,* **79** : 213-217.

Nehauhs G, Spangenberg G, Mittelsten Scheid O and Schweiger HG (1987) Transgenic rapeseed plants obtained by the microinjection of DNA into microspore-derived embryoids. *Theor. Appl. Genet.,* **75** : 30-36.

Paterson KE (1984) Shoot tip culture of *Helianthus annuus*-flowering and development of adventitious and multiple shoots. *Amer. J. Bot.,* **71** : 925-931.

Paterson KE and Everett NP (1985) Regeneration of *Helianthus annuus* inbred plants from callus. *Plant Sci.,* **42** : 125-132.

Paterson Robinson KA and Adams DO (1987) The role of ethylene in the regeneration of *Helianthus annuus* (sunflower) plants from callus. *Physiol. Plant.,* **71** : 151-156.

Pelissier B, Bouchefra O, Pepin R and Freyssinet G (1990) Production of isolated somatic embryos from sunflower thin cell layers. *Plant Cell Rep.,* **9** : 47-50.

Petitprez M, Briere C, Borin C, Kallerhoff J, Souvre A and Alibert G (1995) Characterization of protoplasts from hypocotyls of *Helianthus annuus* in relation to their tissue origin. *Plant Cell Tiss. Org. Cult.,* **41** : 33-40.

Power CJ (1987) Organogenesis from *Helianthus annuus* inbreds and hybrids from the cotyledons of zygotic embryos. *Amer. J. Bot.,* **74** : 497-503.

Prado E and Berville A (1990) Induction of somatic embryo development by liquid culture in sunflower (*Helianthus annuus* L.). *Plant Sci.,* **67** : 73-82.

Rogers C (1992) Inscet pests and strategies for their management in cultivated sunflower. *Field Crops Res.,* **30** : 301-332.

Roseland CR, Espinasse A and Grosz TJ (1991) Somaclonal variants of sunflower with modified coumarin expression under stress, *Euphytica,* **54** : 183-190.

Sackston WE (1992) On a treadmill : breeding sunflowers for resistance to disease. *Annu. Rev. Phytopathol.,* **30** : 529-551.

Sankara Rao K and Rohini VK (1999) *Agrobacterium*-mediated transformation of sunflower (*Helianthus annuus* L.): A simple protocol. *Ann. Bot.,* **83** : 247-254.

Sarraffi A, Bolandi AR, Serieys H, Bervillé A and Alibert G (1996a) Analysis of cotyledon culture to measure genetic variability for organogenesis parameters in sunflower (*Helianthus annuus* L.). *Plant Sci.,* **121** : 213-219.

Sarrafi A, Roustan JP, Fallot J and Alibert G (1996b) Genetic analysis of organogenesis in the cotyledons of zygotic embryos of sunflower (*Helianthus annuus* L.). *Theor. Appl. Genet.,* **92** : 225-229.

Schmitz P and Schanbl H (1989) Regeneration and evacuolation of protoplasts from mesophyll, hypocotyl and petioles from *Helianthus annuus* L. *J. Plant Physiol.,* **135** : 223-227.

Schrammeijer B, Sijmons PC, Van den Elzon PJM and Hoekema A (1990) Meristem transformation of sunflower via *Agrobacterium. Plant Cell Rep.,* **9** : 55-60.

Schuster W (1993) Die Züchtung der Sonnenblume. Paul Parey Scientific Publishers, Berlin, Hamburg.

Service NE (1995) Sunflower Production. North Dakota State University, http://www.ext.nodak.edu/extpubs/plantsci/rowcrops/eb25w-3.htm

Steeves TA, Hicks MA, Naylor JM and Rennie P (1969) Analytical studies on the shoot apex of *Helianthus annuus. Can. J. Bot.,* **47** : 1367-1375.

Steeves TA and Sussex IM (1989) Alternative patterns of development. In: *Patterns in plant development.* (Eds Steeves TA and Sussuex IM) Cambridge University Press, Cambridge, pp. 348-369.

Tabei Y, Kitade S, Nishizawa Y, Kikuchi N, Kayano T, Hibi T and Akutsu K (1998) Transgenic cucumber plants harboring a rice chitinase gene exhibit enhanced resistance to gray mold (*Botrytis cinerea*). *Plant Cell Rep.,* **17** : 159.

Trabace T, Vischi M, Fiore MC, Sunseri F, Vanadia S, Marchetti S and Olivieri AM (1995) Plant regeneration from hypocotyl protoplast in sunflower (*Helianthus annuus* L.). *J. Genet. Breed.,* **49** : 51-54.

Trifi M, Mezghani S and Marrakchi M (1981) Multiplication végétative du tournesol. *Physiol. Vég.,* **19** : 99-102.

Ursic D (1985) Eight DNA insertion events of *Agrobacterium tumefaciens* Ti-plasmids in isogenic sunflower genomes are all distinct. *Biochem. Biophys. Res. Comm.,* **131** : 152-159.

Van der Leede-Plegt LM, Van de Ven BCE, Bino RJ, Van der Salm TPM and Van Tunen AJ (1992) Introduction and differential use of various promoters in pollen grains of *Nicotiana glutinosa* and *Lilium logniflorum. Plant Cell Rep.,* **11** : 20-24.

von Keller A, Coster HGL, Schnabl H and Mahaworasilpa TL (1997) Influence of electrical treatment and cell fusion on cell proliferation capacity of sunflower protoplasts in very low density culture. *Plant Sci.,* **126** : 79-86.

von Keller A, Frey-Koonen N, Wingender R and Schnabl H (1994) Ultrastructure of sunflower protoplasts derived calluses differing in their regenerative potential. *Plant Cell Tiss. Org. Cult.,* **37** : 277-285.

Weber S, Horn R and Friedt W (2000) High regeneration potential *in vitro* of sunflower (*Helianthus annuus* L.) lines derived from interspecific hybridization. *Euphytica,* **116** : 271-280.

Wilcox McCann A, Cooley G and Van Dreser J (1988) A system for routine plantlet regeneration of sunflower (*Helianthus annuus* L.) from immature embryo-derived callus. *Plant Cell Tiss. Org. Cult.,* **14** : 103-110.

Wingender R, Henn HJ, Barth S, Voeste D, Machlab H and Schnabl H (1996) A regeneration protocol for sunflower (*Helianthus annuus* L.) protoplasts. *Plant Cell Rep.,* **15** : 742-745.

Witrzens B, Scowcroft WR, Downes RW and Larkin PJ (1988) Tissue culture and plant regeneration from sunflower (*Helianthus annuus*) and interspecific hybrids (*H. tuberosus* × *H. annuus*). *Plant Cell Tiss. Org. Cult.,* **13** : 61-76.

Yao X, Jingfen J and Kuochang C (1988) Transfer and expression of the T-DNA harboured by *Agrobacterium tumefaciens* in cultured explants of *Helianthus annuus. Act. Bot. Yunnan.,* **10** : 159-166.

Chapter 6

BIOTECHNOLOGY OF SAFFLOWER (*CARTHAMUS TINCTORIUS* L.) : STRATEGIES FOR GENE TRANSFER

K Sankara Rao★ and VK Rohini

Department of Biochemistry, Indian Institute of Science, Bangalore - 560 012, INDIA

Summary

Safflower (Carthamus tinctorius L.) is known for its importance as a non-conventional oil seed crop. The oil is valued for its high degree of polyunsaturation and elevated levels of α-tocopherol. India has a vast varietal wealth of safflower and is a major producer of safflower. Very little work has been done to create broad-based genetic improvement of safflowers. Recourse to in vitro techniques and genetic engineering could go a long way towards genetic upgrading of this crop including herbicide tolerance, disease resistance and oil enrichment. In the last eighteen years, in vitro regeneration of several Indian and American cultivars of safflower was attempted. To date, direct and adventitious shoot bud regeneration, axillary bud proliferation or somatic embryogenesis were obtained using mostly primary seedling explants and immature embryos and leaf explants. It was observed that seedling explants of safflower are comparatively more amenable to callus induction and/or shoot bud regeneration. However, an efficient in vitro plant regeneration system applicable to a wide group of genotypes/cultivars is still lacking. Sensitivity of regenerated shoots to media water content and high relative humidity in the culture vessels, differential rooting response among cultivars to the auxin source have been the major constraints.

★Corresponding author : E-mail : baradwaj@biochem.iisc.ernet.in

Agrobacterium-mediated gene transfer has been attempted in some of the Indian and American cultivars of safflower. Direct and callus-mediated regeneration from primary explants was examined following Agrobacterium cocultivation with little success. A transformation system has been recently established to overcome the constraints of in vitro plant regeneration from Agrobacterium-treated explants that exploits the use of the entire embryo axes from the germinating seeds as the target tissue for transformation and for its subsequent growth directly into a transformed plant. The strategy has shown that transgenic safflower plants can be obtained by Agrobacterium-mediated transformation of cells of the plumule and cotyledonary node of germinating seeds. The possibility of some of the progeny resulting in fully transformed plants for the introduced genes has been demonstrated in this study. No tissue culture or regeneration necessary with this method and amazingly little labour is involved. Thus far, this is the only procedure available for safflower that could successfully be used to generate whole plant transformants, and in principle, can be applied to all those genotypes and cultivars of safflower which are susceptible to Agrobacterium tumefaciens infection.

Keywords : *Agrobacterium tumefaciens*, non-tissue culture-based transformation, Safflower (*Carthamus tinctorius* L.), tissue culture, transgene integration.

1. INTRODUCTION

Safflower (*Carthamus tinctorius* L.) is an important oil seed crop in India, North America, and Mexico. The plant is known only under cultivation and is believed to have originated either from *C. lanatus* or *C. oxyacantha* (Weiss, 1971; Knowles and Schank, 1964). It is cultivated in almost all states in India and has attained considerable importance. India has a vast varietal wealth and is a major producer of safflower.

Safflower oil is valued for its high degree of polyunsaturation and elevated levels of α-tocopherol (Furuya *et al.*, 1987). Genetic variability has allowed successful breeding of cultivars with widely varying oil content and quality ranging from high oleic varieties used for human consumption to high linoleic varieties used for industrial coatings and lubricants.

Several cultivars of safflower under cultivation are adapted to low humidity and low soil water content and can withstand prolonged drought conditions. The primary agronomical problem associated with safflower is the lack of an effective herbicide. Among the diseases affecting Indian safflowers, the most serious are wilt caused by *Fusarium carthami* and leaf spot caused by *Alternaria carthami* (Sankara Rao and Rohini, 1999). The most serious pest of safflower in India is *Acanthiophilus helianthii,* the maggot of a fruit fly, which causes serious damage to the florets. A green leaf-eating caterpillar, *Perigaea capensis* G., is recorded from south India, which appears sometimes in large numbers and defoliates the plants (Wealth of India).

Very little work has been done to create broad-based genetic improvement of safflower. Recourse to *in vitro* techniques and genetic engineering could go a long way towards genetic upgrading including herbicide and disease resistance and oil enrichment.

2. GENE TRANSFER AND GENETIC ENGINEERING OF SAFFLOWER

2.1. *In vitro* regeneration of safflower

For successful genetic transformation, an efficient *in vitro* plant regeneration system for the crop of interest is essential. In the last 18 years, *in vitro* regeneration of several Indian and American cultivars of safflower was attempted. To date, direct and adventitious shoot bud regeneration, axillary bud proliferation or somatic embryogenesis were obtained using mostly primary seedling explants and immature embryos and leaf explants. It was observed that seedling explants of safflower are comparatively more amenable to callus induction and/or shoot bud regeneration. Overall, the most effective basal medium used for regeneration was Murashige and Skoog (1962) medium.

2.1.1. Bud regeneration

The potential of shoot bud regeneration from the explants was studied under conditions either to allow direct regeneration or through callus stages in the Indian and American safflowers. Media containing BAP alone or in combination with NAA encouraged shoot formation from all safflowers investigated, but other cytokinins used *viz.,* kinetin, 2-isopentyl adenine, and zeatin in combination with NAA did not promote shoot formation. Thidiazuron (TDZ) was also used in the case of American cultivar 'Centennial'. The *in vitro* grown seedlings provided the explants.

The age of the seedling when explants are excised varied with the cultivar.

Regeneration from cotyledons **:** Shoot buds were induced directly on the cotyledons of cultivars A-300, and Centennial whereas shoot buds as well as inflorescence were obtained from the cotyledons of cvs. Manjira and A-1.

In the cv. A-300, cotyledons from 10-12 days old seedlings were used for culture. They were excised close to their attachment to the axis and were incubated with their abaxial surface in contact with medium. Shoot buds emerged directly from the proximal cut end after 7-10 days. Bud regeneration was observed in all combination treatments of BAP and NAA tried although BAP at 0.1 mg/l and NAA 0.01mg/l induced high level of regeneration. The optimal response, however, was obtained when the explants were initially cultured for 7 days in medium with higher BAP level (0.5 mg/l) and later shifted to a lower level of 0.1 mg/l of BAP, the auxin, NAA in the combinations remaining constant at 0.01 mg/l throughout. Increased thiamine HCl (4 mg/l) in the medium further helped recovery of healthy-looking shoots. Small amounts of callus growth accompanied shoot bud regeneration (Anil Kumar and Sankara Rao, 1996).

In another Indian safflower cultivar, cotyledons produced callus and shoots on Murashige and Skoog (MS) medium containing 0.5 to 2.0 mg/l BAP alone and in combination with 0.1 or 0.5 mg/l NAA. Direct regeneration of shoot buds occurred from cotyledons of 7-day-old seedlings. About 50% of the explants formed 8-10 shoot buds each (George and Rao, 1982).

In the cvs. 'Manjira' and A-1, direct induction of shoots as well as inflorescence was obtained on MS medium containing 0.5 mg/l BAP or kinetin plus 0.1 mg/l NAA. The capitula arise from the inner surface of the cotyledons. Complete blooming of florets in a capitulum was observed and a few seeds were also recovered (Tejovathi and Anwar, 1984).

In the American cv., 'Centennial', cotyledons from 3-7-day-old seedlings produced leafy structures on medium containing 0.1 mg/l NAA and 0.5 mg/l BAP or 0.1 mg/l thidiazuron (TDZ). Number of regenerated shoots were comparable on media containing BAP or TDZ, although TDZ medium was superior in reducing shoot hyperhydricity and permitting multiple harvests of regenerated shoots from primary explants (Orlikowska and Dyer, 1993).

Cotyledons of A-300 produced callus on media containing BAP and NAA in different concentrations and ratios. However, callus that developed shoot buds later was obtained with 1 mg/l BAP and 0.1 mg/l NAA. Shoot buds developed on the same BAP-NAA combination as the one used for callus induction. Four weeks after subculture, buds appeared as small green protuberances on the non-green friable callus and elongated eventually into leafy shoots (Fig. 1) (Sankara Rao and Rohini, 1999).

Regeneration from hypocotyl : In 'Centennial' and 'Montola', older seedlings (10-20-day old) produced leafy structures from hypocotyls (Orlikowska and Dyer, 1993) on medium containing 0.1 mg/l NAA and 0.5 mg/l BAP or 0.1 mg/l TDZ. Leaf primordia appeared after 7-10 days on hypocotyl sections closest to the apical meristem (Orlikowska and Dyer, 1993). For conditions, refer to second para of results in the paper.

Regeneration from leaves : Leafy structures are produced directly on leaves of 10-20-day-old seedlings of cv. 'Centennial'. Regeneration of leafy structures was observed on most NAA/cytokinin combinations. Number of regenerated leafy structures increased with increase in BAP concentration in combination with NAA and regeneration was more efficient overall on media containing BAP rather than TDZ. Addition of 0.1 or 0.5 mg/l NAA to BAP medium induced the highest levels of regeneration, while NAA at 1.0 mg/l, decreased direct regeneration and stimulated callus growth. However, leafy structures regenerated on BAP media suffered from hyperhydricity. On TDZ media, regeneration frequency increased slightly with increase in TDZ concentrations and leafy structures grew vigorously. Overall, the most effective medium for direct shoot regeneration with a minimum of hyperhydricity contained MS basal salts with 0.1 mg/l NAA and 0.1 mg/l TDZ (Orlikowska and Dyer, 1993).

Callus with bud forming capability was obtained from leaf segments of the American cv. 'Centennial' on MS medium containing 1 mg/l BAP and 1 mg/l NAA. Shoot buds were regenerated from this leaf-derived calli with a frequency of about 26%. Most buds were obtained on the same medium used for callus growth. Small buds (1-2 mm) were regenerated from friable, light green callus as well as from dark green sectors after about six weeks. Some buds elongated into small shoots (1-1.5 cm long) at a frequency of about 5%. However, many regenerated buds failed to elongate and instead became swollen and vitrified-(Ying *et al.*, 1992). Leaf explants from one-month-old greenhouse plants produced only callus (Orlikowska and Dyer, 1993).

Induction of morphogenetic calli from the leaf explants of cv. Co-1 of safflower was obtained from three-week-old *in vitro* grown seed lines. Callusing occurred in MS medium with NAA (1.0-1.5 mg/l) and BAP (1.0-0.5 mg/l). Shoot bud primordia surfaced when the level of BAP was raised to 5.0 mg/l in combination with 0.25 mg/l NAA after fourth week of subculturing. For leafy shoot development, shoot buds were transferred to half strength basal MS with adenine sulfate (20 mg/l) (Chatterji and Singh, 1993). Similarly, leaf segments of cv. A-1 callused on medium containing BAP and NAA at 1 mg/l, one week after inoculation. The proximal (basal) halves of the leaves callused effectively and gave out shoot buds on the callus. At this stage, *i.e.,* when shoot buds appear as green spots on the callus, the calluses were separated from the explants and on subculture to MS medium with BAP and NAA at 3 mg and 1 mg/l respectively, the shoot buds opened up (Fig. 2). Elongation of the shoots took place on media containing IPA (Sankara Rao and Rohini-unpublished data).

Shoot regeneration from immature embryo : Shoot regeneration from immature embryos was observed in cv. 'Centennial' on MS medium containing NAA and TDZ. Immature embryos regenerated leafy structures on almost all media tested, accompanied by callus growth especially at higher TDZ concentrations. Minimal regeneration was observed on media containing TDZ and 10 mg/l NAA (Orlikowska and Dyer, 1993). Hyperhydricity was observed in some leafy structures regenerating from primary explants and many structures regenerating from callus. Transfer of explants with regenerating leafy structures onto elongation medium after 2-3 weeks limited callusing and hyperhydricity and permitted normal shoot development. When regenerated leafy structures were subcultured repeatedly on media with low concentrations of cytokinins (0.25 mg/l BAP), 0.01 mg/l TDZ or 1 mg/l kinetin, shoots grew as rosettes and flower buds were formed which eventually flowered. Eventhough some regenerated shoots were obtained from immature embryos, direct regeneration from seedling explants was judged superior because of easier explant preparation and better vigor of regenerated shoots.

2.1.2. Axillary bud proliferation and Multiple shoots

Adventitious shoots were produced on 1cm long internodal shoot sections from 20-day–old seedlings on MS media supplemented with varying amounts of BAP, TDZ, 2-iP or kinetin. None of the media tested stimulated shoot proliferation from 1 cm-long shoot apical meristems. Most shoot proliferation was observed on media containing BAP or 2-iP, but shoots on BAP medium became yellow and eventually necrotic. Low numbers of shoots were formed on media containing TDZ and did not elongate. Kinetin was not effective in inducing useful numbers of adventitious shoots. Numerous shoots were obtained on 2-iP medium especially at 1 and 2 mg/l and the shoots remained healthy and elongated readily (Orlikowska and Dyer, 1993).

Adventitious shoots were produced in both A-1 and A-300 from cotyledonary axils as well as from the node above (Fig. 3). Shoot primordia appeared in about two weeks when the seedling shoot segments were cultured on MS media supplemented with 1 mg/l each of NAA and BAP.

2.1.3. Somatic embryogenesis

Somatic embryos were induced directly on adaxial surface of cotyledonary leaves of safflower cultivar 'Girna' within 8-10 days of culture on Murashige and Skoog medium containing 5.37 to 10.74 μM NAA and 2.22 μM BAP. Germinated embryos with shoot axes developed into complete plants after transfer onto half-strength Murashige and Skoog medium containing 1.07 μM NAA. Histological studies suggested direct origin of somatic embryos with broad-base attachment (Mandal *et al.*, 1995).

2.1.4. Regeneration from anther (haploid callus induction)

Prasad *et al.* 1991 reported that MS medium was the most effective of five basal salt combinations tested in inducing haploid callus production from safflower anthers. Shoot regeneration from haploid callus was observed on MS medium containing 2.0 mg/l BAP and 0.5 mg/l NAA.

2.1.5. Constraints of regeneration

Safflower is well adapted to low humidity and soil water conditions, and can withstand extended drought conditions. Because of these characteristics, safflower explants and regenerated shoots seem to be particularly sensitive to media water content and high relative humidity in culture vessels. An early symptom of excess water content is hyperhydricity (Deberg *et al.*, 1992). Hyperhydricity of regenerated safflower shoots was successfully prevented by early transfer to medium containing 0.5X MS salts, weak cytokinin (1 mg/l kinetin) and 2.5 mg/l $AgNO_3$. Ag^{2+} ions inhibit ethylene action (Biddington, 1992) and $AgNO_3$ has been successfully used to enhance regeneration efficiency from sunflower (Chraibi *et al.*, 1991) and brassica cotyledon (Chi *et al.*, 1990) cultures. The presence of $AgNO_3$ in safflower elongation medium generally improved shoot health and growth. Hyperhydricity also decreased during direct shoot regeneration by the substitution of TDZ for BAP in regeneration media. TDZ was superior to BAP in allowing extended direct shoot regeneration from primary explants, even after transfer of primary explants with shoots to elongation medium.

Extended periods of callus-mediated regeneration led to hyperhydricity and vitrification of the shoots in the cultivars A-300 and A-1. Shoots regenerated from explants directly had, however, a normal appearance (Sankara Rao and Rohini, 1999).

2.1.6. Rooting of regenerated shoots and constraints

Rooting of regenerated shoots was achieved on one-half strength MS medium containing 0.1 mg/l NAA and 1% sucrose in the cv. "Manjira" (Prasad *et al.*, 1991). Shoot buds obtained from the leaf callus of safflower cv. 'Co-1' were rooted on medium added with 2, 4, 5 Cl_3 POP (Chatterji and Singh, 1993). However, attempts to root regenerated shoots from leaf-derived callus in 'Centennial' were not successful (Ying *et al.*, 1992). Rooting was not observed on any combination of MS salt strength, sucrose levels, or IAA levels tested, including those reported by Prasad *et al* (1991) as successful for the Indian cv. 'Manjira'. Likewise, small shoots transferred to a peat-pearlite (1:1) mixture in a mist tent did not root, although this treatment is successful for rooting shoots directly regenerated from leaf explants. A similar lack of success in rooting Turkish and other Indian safflower varieties has been reported (George and Rao, 1982; Tejovathi and Anwar, 1987).

Nevertheless, shoots regenerated from primary seedling explants of the American 'Montola' and 'Centennial' were rooted on 0.5X MS medium containing 1 mg/l NAA and successfully transferred to greenhouse.

Successful rooting of safflower shoots has been consistently problematic, thus reducing the overall efficiency of whole plant regeneration and the likelihood of obtaining transgenic plants after transformation. Less than 10% of shoots regenerated from cotyledons of Indian cultivars rooted on hormone-free MS basal salts medium containing 6-8% sucrose (George and Rao, 1982) or 9% sucrose (Tejovathi and Anwar, 1987). In both cases, shoots formed multiple slender roots instead of the normal tap root morphology characteristic of safflower. Of the several rooting media tested on cotyledon callus-derived shoots of the cv. A-300, only medium containing NAA at 0.1 mg/l formed roots occasionally. Axillary shoots of cv. A-1 did not respond to the NAA-containing and other root-inducing media used (Sankara Rao and Rohini, 1999). The regenerated shoots of both the cvs. A-1 and A-300 showed a high degree of vitrification and/or formed callus at the base when treated with IBA in the medium. These differences in the rooting behavior observed with auxin treatments could be related to varying endogenous levels of auxins among cultivars (Sankara Rao and Rohini, 1999). TIBA (2,3,5-Triiodobenzoic acid) used to overcome the apparent auxin sensitivity of the shoots of cvs. A-300 and A-1 did not result in root formation. Nevertheless, these shoots showed elongation when treated with 0.5 mgl^{-1} TIBA in half-strength MS medium prior to root induction treatment. Further, the symptoms of hyperhydricity and vitrification of shoots were overcome with this treatment. These differences in response may at least partially explain the inhibitory effect of high endogenous auxin levels that might exist in some cultivars (Sankara Rao and Rohini, 1999).

For the American cultivars, 'Centennial' and 'Montola', shoots directly regenerated from primary explants were first elongated on MS medium containing 1 mg/l kinetin and 0.05 mg/l NAA for 21 d and then rooted on MS medium containing 1 mg/l NAA and 1 mg/l riboflavin (Orlikowska and Dyer, 1993). Rooted shoots were transferred to soil, with better survival of larger (4 cm) long shoots. In subsequent attempts to improve rooting efficiency using this protocol, it was observed that many shoots did not elongate on elongation medium and the percentage of shoots successfully rooted was unacceptably low. It was further shown (Baker and Dyer, 1996) that an encouraging number of regenerated shoots rooted soon after removal from primary explants. Therefore, the use of elongation medium was discontinued which shortened the rooting procedure and protocols to optimize rooting of shoots taken directly from explants were tried. It was claimed (Baker and Dyer, 1996) that the optimal protocol for inducing root formation consisted of a 7 d exposure. Shoots directly regenerated from seedling explants to 10 mg/l IBA in root induction media followed by incubation in media containing 15 g/l sucrose and 1 g/l activated charcoal for 21 d. The culture period has been reduced by

21 d from previous protocols (Orlikowska and Dyer, 1993), and the incidence of hyperhydricity symptoms has been substantially reduced.

2.2. *Agrobacterium*-mediated transformation systems

Since safflower is known for its importance as an oil seed crop, genetic engineering of safflower for traits such as herbicide tolerance and disease resistance would help in the improvement of the crop. For genetic engineering of safflower, *Agrobacterium*-mediated transformation has been tried in some of the American and Indian cultivars.

2.2.1. Tissue culture-based transformation

Ying *et al.*, in 1992 showed *Agrobacterium*-mediated transformation of safflower cv. 'Centennial' via callus-mediated regeneration from cotyledon, stem and leaf explants. The different explants inoculated with *Agrobacterium tumefaciens* containing *npt*II and *uid*A marker genes produced kanamycin resistant calli on MS medium supplemented with 1.0 mg/l of BAP and NAA. Shoot buds were regenerated from 26% of leaf derived calli on the callus induction medium. Transgenic plants could not be recovered as rooting of regenerated shoots was unsuccessful. Nevertheless, transformation and stable integration of transgenes was confirmed by GUS assay in regenerated shoots and Southern analysis of kanamycin resistant calli.

Subsequently direct as well as callus-mediated regeneration of transgenic shoot buds expressing the *uid*A and *npt*II reporter genes following *Agrobacterium tumefaciens* co-cultivation of primary seedling explants was accomplished in two of the Indian cultivars of safflower (*Carthamus tinctorius* L.), A-1 and A-300 (Sankara Rao and Rohini, 1999). The procedure yielded 23 and 34 transformation events per 100 co-cultivated explants with direct and callus-mediated shoot recovery, respectively. The use of *uid*A gene in pKIWI105 that lacks a bacterial ribosome-binding site precluded *uid*A expression in residual *Agrobacterium* cells. Two different *Agrobacterium tumefaciens* strains *viz.,* At459 and LBA4404, both harbouring the same binary vector were used. There was no significant difference between *Agrobacterium* strains in their infectivity and transformation efficiency as determined by the number and extent of GUS-expressing sectors observed per explant 4 days later. This is contrary to the observation that transformation efficiency varied with the use *of Agrobacterium tumefaciens* strains on the American cv. 'Centennial' (Orlikowska *et al.*, 1995). The susceptibility of a cultivar to a particular bacterial strain might have been the reason for the difference in the transformation efficiency. Consistently high levels of GUS activities were detected in selected putative transgenic calli (Fig. 4a) and in shoot regenerants (Fig. 4b) by histochemical assay. The *vir* gene induction treatments to enhance the transformation efficiency involved the use of tobacco leaf extract and acetosyringone. The use of *Agrobacterium* previously treated with acetosyringone (100 μM) did not have any effect on transformatiom efficiency. A similar observation was also made by Orlikowska *et al.*, 1995, with safflower 'Centennial' using the bacterial strains EHA105 and LBA4404. Acetosyringone treatment further increased hyperhydricity of the explant that would hamper the regeneration process. The use of wounded tobacco leaf extract added to the AB induction medium improved transformation efficiency by reducing the symptoms of hypersensitivity and increasing number of embryos expressing GUS. Western blot analysis using GUS antiserum and the NPTII expression assays confirmed the expression of marker genes in the putative transformants. Transgene integration was examined by PCR and dot blot hybridization of the transformants. Compared to controls, the efficiency of regeneration was markedly decreased subsequent to co-cultivation. Extended periods of callus–mediated regeneration led to hyperhydricity and vitrification of the shoots. Shoots regenerated from explants directly, however had a normal appearance. The rooting response of regenerated shoots was poor and remains a continuing obstacle for safflower plant regeneration and transformation.

2.2.2. Embryo transformation and *ex vitro* plant regeneration

Earlier transformation experiments have merely served the purpose of demonstrating the transformability of safflower as recovery of whole plant transformants was not successful. A method has been developed to obtain transformants which is independent of the problems inherent to tissue culture of safflower (Rohini and Sankara Rao, 2000). The strategy is based on *Agrobacterium*-mediated gene transfer wherein the *in vitro* plant regeneration step has been totally eliminated. The procedure adapted is essentially an *in planta* infection similar to those used earlier for other crops *viz.,* DNA uptake by whole embryos of rice by imbibition (Junhi and Guhung, 1995), transformation by injection of *Agrobacterium* into seeds of cauliflower (Eimert *et al.*, 1992), co-cultivation of the whole or the split embryo apices with *Agrobacterium* in sunflower (Schoneberg *et al.*, 1994), co-cultivation of germinating seeds of *Arabidopsis thaliana* with *Agrobacterium* (Feldman and Marks, 1987) and *Agrobacterium* infection of the cotyledonary node of the germinating

Figure 1 : Callus-mediated regeneration in safflower cultivar A-300. The callus was obtained from the cotyledonary explant (bar = 3 cm).
Figure 2 : Callus-mediated regeneration in safflower cultivar A-1. The callus was obtained from the leaf explant (bar = 2.5 cm).
Figure 3 : Multiple shoots from the apex of the decapitated epicotyl of safflower cultivar A-1 (bar = 0.9 cm).
Figure 4 : GUS histochemical assay in safflower cultivars. (a) Callus showing GUS expression in safflower cv. A-300 (bar = 0.1 cm); (b) Shoot segments of cv.A-1: one showing GUS expression and the other (shoot of uninfected plant-negative control) lacking GUS expression (bar = 0.13 cm).
Figure 5 : Localization of GUS expression in primary transformants (T_0) of safflower A-1.

seeds by excision of one of the cotyledons in sunflower (Sankara Rao and Rohini, 1999) and peanut (Rohini and Sankara Rao, 2000).

Surface decontaminated and pre-incubated embryo axes of germinating seeds of safflower cvs. 'A-1' and A-300 with one of the cotyledons removed were pricked with a sterile sewing needle at the cotyledonary node and infected by gentle agitation for 10 min in a suspension of *Agrobacterium tumefaciens,* LBA4404/pKIWI105. Winans AB medium (Winans *et al.*, 1988) added with wounded tobacco leaf extract (2 g in 2 ml sterile water) was used for *vir* gene induction before infection. Following a 24 h co-cultivation and decontamination with cefotaxime for 1 h, they were placed on soilrite moistened with water to allow germination to progress under growth room conditions for atleast 10 days before they were transferred to the greenhouse. In the greenhouse, they grew into normal healthy plants. The histochemical assay of an *uid*A gene that expresses only in plant tissues (Fig. 5) and PCR amplification of *uid*A and *npt*II marker genes (Fig. 6a & 6b) were used for early determination of putative transformants. In the experiments performed initially to determine the tolerance to kanamycin, it was observed that uninfected embryo axes (control) did not root beyond 150 $\mu g\ ml^{-1}$ of kanamycin. There was a reduction in the survival rate when wounded uninfected as well as infected embryos were germinated in the presence of kanamycin. Also, it is possible that the transformed sectors would be eliminated when embryo organs begin to die when subjected to kanamycin selection. For these reasons, selection on kanamycin was eliminated. Southern analysis of T_0 (Fig. 7a) and T_1 (Fig. 7b) plant DNA was used to confirm integration of the transgenes.

The combined results indicated that the frequency of transformation was 5.3 % in safflower cv. A-1 and 1.3 % in A-300. Four T_0 plants of cv. A-1 yielded transformed T_1 progeny. The strategy, in principle, should be applicable to all cultivars and genotypes of safflower, which are susceptible to *Agrobacterium tumefaciens* infection. Thus far, this is the only procedure available for safflower that could successfully be used to generate whole plant transformants.

3. CONCLUSIONS AND FUTURE PROSPECTS

Cultivated safflower (*Carthamus tinctorius* L.) is a promising non-conventional oil seed and occupies a unique position among the oil seed crops for the high degree of polyunsaturation and elevated α-tocopherol of its oil. Very little attempt has been made to have a broad-based

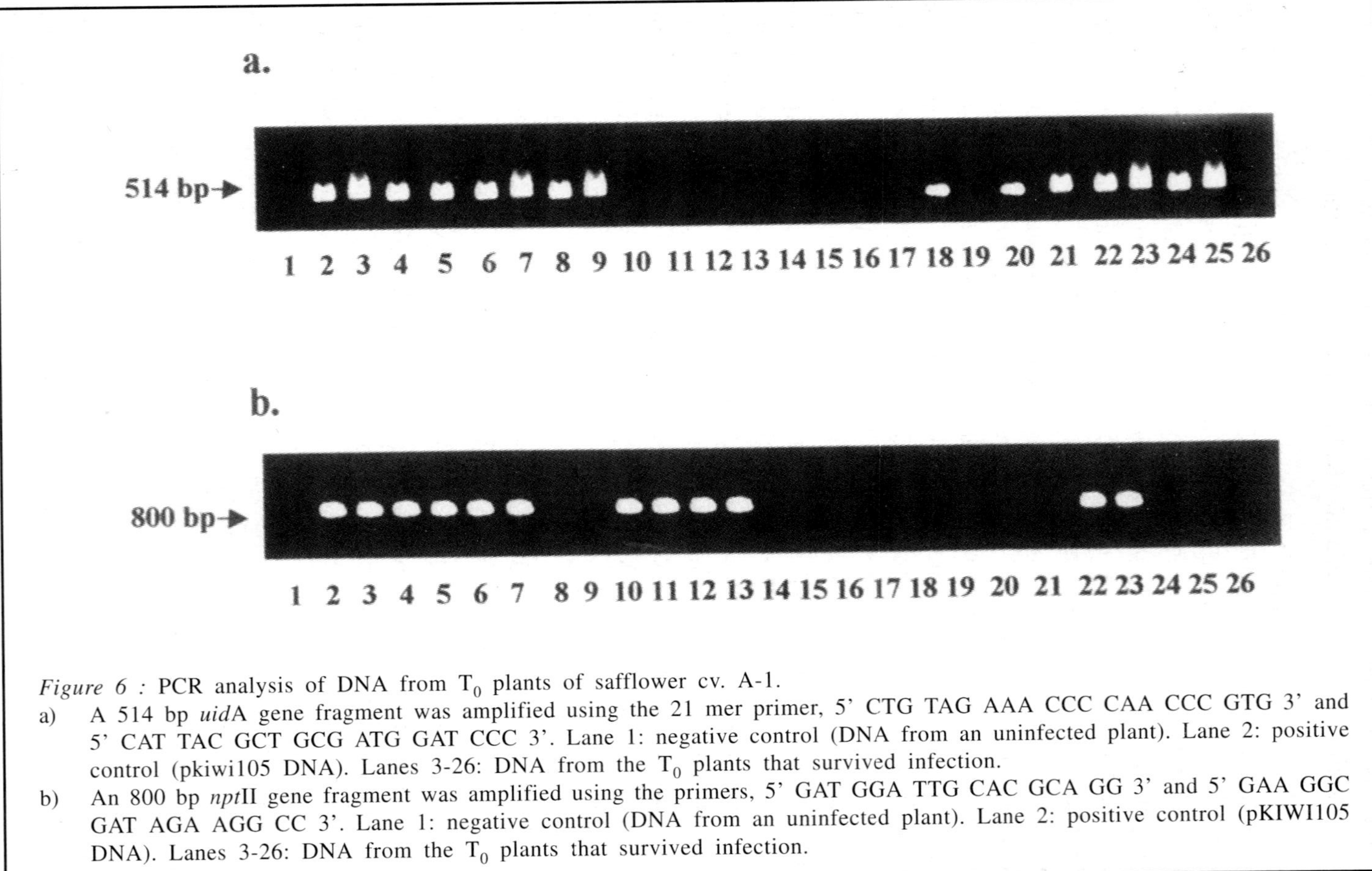

Figure 6 : PCR analysis of DNA from T_0 plants of safflower cv. A-1.

a) A 514 bp *uid*A gene fragment was amplified using the 21 mer primer, 5' CTG TAG AAA CCC CAA CCC GTG 3' and 5' CAT TAC GCT GCG ATG GAT CCC 3'. Lane 1: negative control (DNA from an uninfected plant). Lane 2: positive control (pkiwi105 DNA). Lanes 3-26: DNA from the T_0 plants that survived infection.

b) An 800 bp *npt*II gene fragment was amplified using the primers, 5' GAT GGA TTG CAC GCA GG 3' and 5' GAA GGC GAT AGA AGG CC 3'. Lane 1: negative control (DNA from an uninfected plant). Lane 2: positive control (pKIWI105 DNA). Lanes 3-26: DNA from the T_0 plants that survived infection.

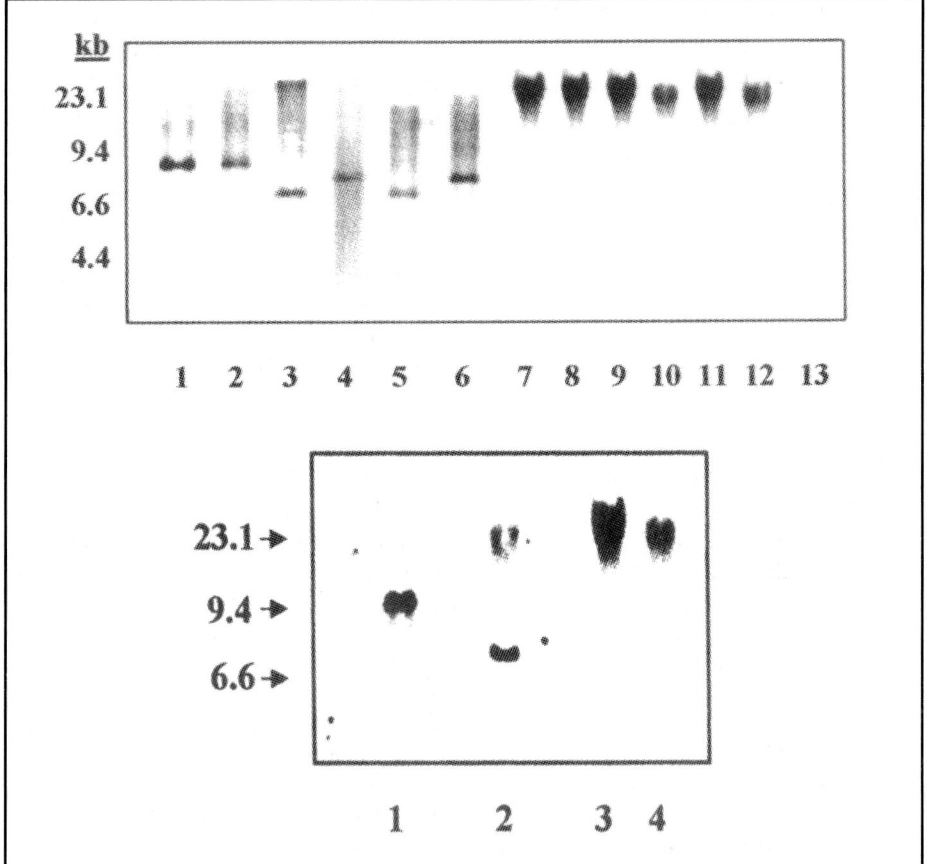

Figure 7 : (a) Southern analysis of safflower cv. A-1 T_0 plants for the integration of the *uid*A gene. 10 µg DNA was digested with *Hind*III and probed with a 2.1 kb *uid*A gene fragment. Lanes 1-6: digested DNA from PCR positive T_0 plants. Lanes 7-12: Uncut DNA. Lane 13: DNA from uninfected plant (negative control); (b) Southern analysis of T_1 plants from safflower cv. A-1 for the integration of the *npt*II gene. Lanes 1 & 2: *Hind*III digested DNA. Lanes 3 & 4: uncut DNA.

genetic improvement of safflower. Genetic engineering of safflower for such traits as herbicide tolerance and disease resistance would help in the improvement of the crop. Genetic engineering strategies have so far employed *Agrobacterium*-mediated transformation in some of the American and Indian cultivars. The most crucial step however has been to optimize the tissue culture regeneration system. Although several reports of *in vitro* culture of safflower have been published, an efficient plant regeneration system applicable to a wide group of genotypes/ cultivars is still lacking. An *Agrobacterium*-based transformation system that has recently been established (Rohini and Sankara Rao, 2000) renders genetic improvement via gene transfer a reality. As more information about specific traits becomes available, it will become possible to integrate important genes into the breeding programme of this crop and to design cultivars with specific traits.

REFERENCES

Anil Kumar PS and Sankara Rao K (1996) Genetic transformation of Safflower (*Carthamus tinctorius* L.) and Niger (*Guizotia abyssinica* Cass). *In Agricultural Biotechnology, 2nd Asia Pacific Conference.* 173-180.

Baker CM and Dyer WE (1996) Improvements in rooting regenerated safflower (*Carthamus tinctorius L.)* shoots. *Plant Cell Rep.,* **16** : 106-110.

Biddington NL (1992) The influence of ethylene in plant tissue culture. *Plant Growth Regul.,* **11** : 173-187.

Chatterji AK and Singh HP (1993) Plant regeneration from leaf calli of safflower. *Proceedings, Third Intl. Safflower Conference,* June 14-18. 139-143.

Chi GL, Barfield DG, Sim GE and Ch. Pua E (1990) Effect of $AgNO_3$ and aminoethoxyvinylglycine on *in vitro* shoot and root organogenesis from seedling explants of recalcitrant Brassica genotypes. *Plant Cell Rep.,* **9** : 195-198.

Chraibi KM, Latche A, Roustan JP and Fallot J (1991) Stimulation of shoot regeneration from cotyledons of *Helianthus annuus* by the ethylene inhibitors, silver and cobalt. *Plant Cell Rep.,* **10** : 204-207.

Debergh P, Aitken-Christie J, Cohen D, Grout B, von Arnold S, Zimmerman R and Ziv M (1992) Reconsideration of the term 'vitrification' as used in micropropogation. *Plant Cell Org. Cult.,* **30** : 135-140.

Eimert K, Schroder C and Siegemund F (1992) Expression of the NPT II-sequence in cauliflower after injection of *Agrobacterium* into seeds. *J. Plant Physiol.,* **140** : 37-40.

Feldmann KA and Marks MD (1987) *Agrobacterium*-mediated transformation of germinating seeds of *Arabidopsis thaliana*: a non-tissue culture approach. *Mol. Gen. Genet.,* **208** : 1-9.

Furuya T, Yoshikawa T, Kimura T and Kancko H (1987) Production of tocopherols by cell culture of safflower. *Phytochemistry,* **26** : 2741-2747.

George L and Rao PS (1982) *In vitro* multiplication of safflower (*Carthamus tinctorius* L) through tissue culture. *Proc. Ind. Natl. Sci. Acad.,* **B48** : 791-794.

Junhi Y and Guhung J (1995) DNA uptake by imhibition and expression of a foreign gene in rice. *Physiol. Plant.,* **94** : 453-459.

Knowles TF and Schank SC (1964) Artificial hybrids of *Carthamus nitidus* Boiss and *C. tinctorius* L. (Compositae). *Crop Sci.,* **4** : 596-599.

Mandal AKA, Chatterji AK and Dutta Gupta S (1995) Direct somatic embryogenesis and plantlet regeneration from cotyledonary leaves of safflower. *Plant Cell Tiss. and Org. Cult.,* **43** : 287-289.

Murashige T and Skoog F (1962) A revised medium for rapid growth and bioassays with tobacco tissue cultures. *Physiol. Plant.,* **15** : 473-497.

Orlikowska TK and Dyer WE (1993) *In vitro* regeneration and multiplication of safflower (*Carthamus tinctorius* L.). *Plant Sci.,* **93** : 151-157.

Orlikowska TK, Cranston HJ and Dyer WE (1995) Factors influencing *Agrobacterium tumefaciens*-mediated transformation of the safflower cultivar 'Centennial'. *Plant Cell Tiss. Org. Cult.,* **40** : 85-91.

Prasad BR, Khadeer MA, Seeta P and Anwar SY (1991) *In vitro* induction of androgenic haploids in safflower (*Carthamus tinctorious* L.). *Plant Cell Rep.,* **10** : 48-51.

Rohini VK and Sankara Rao K (2000) Transformation of peanut (*Arachis hypogaea* L.): a non-tissue culture based approach for generating transgenic plants. *Plant Sci.,* **150** : 41-49.

Sankara Rao K and Rohini VK (1999) Gene transfer into Indian cultivars of safflower (*Carthamus tinctorius* L.) using *Agrobacterium tumefaciens. Plant Biotechnol.,* **16** : 201-206.

Sankara Rao K and Rohini VK (1999) *Agrobacterium*-mediated transformation of sunflower (*Helianthus annuus* L.): A simple protocol. *Ann. Bot.,* **83** : 347-354.

Schoneberg JM, Scelonge CJ, Burris M and Bidney DL (1994) Transformation of sunflower using embryonic axis explants. *Plant Sci.,* **103** : 199-207.

Tejovathi G and Anwar SY (1984) *In vitro* induction of capitula from cotyledons of *Carthamus tinctorious* (Safflower). *Plant Sci. Lett.,* **36** : 165-168.

Tejovathi G and Anwar SY (1987) Plantlet regeneration from cotyledonary cultures of safflower: (*Carthamus tinctorius* L.). *In:* (Ed. Reddy GM) *Proc. Symp. Plant Cell and Tissue culture of Economically Important Plants*, 347-353.

The Wealth of India. Vol II Raw materials (1950) *Carthamus. Publications and Information Directorate*, CSIR, New Delhi. pp 83-88.

Winans SC, Kerstetter RA and Nester EW (1988) Transcriptional regulation of the *vir* A and *vir* G genes of *Agrobacterium tumefaciens. J. Bacteriol.,* **170** : 4047-4054.

Weiss EA (1971) Castor, sesame and safflower. Leonard Hill Books, University Press, Aberdeen, London.

Ying M, Dyer WE and Bergman J (1992) *Agrobacterium tumefaciens*-mediated transformation of safflower (*Carthamus tinctorius* L.) cv. 'Centennial'. *Plant Cell Rep.,* **11** : 581-585.

Chapter 7

IN VITRO REGENERATION AND GENETIC TRANSFORMATION IN NIGER (*GUIZOTIA ABYSSINICA*)

PB Kavi Kishor[1★], A Sarvesh[2], Amitha S Katta[1], TP Reddy[1], DVSSR Prakash[3] and K Vaidyanath[1]

[1]*Department of Genetics, Osmania University, Hyderabad - 500 007*
[2]*147 Kiely Boulevard, Santa Clara, California, USA*
[3]*Department of Biotechnology, Administrative Management College, Post Graduate Centre, Bannerghatta Main Road, Bangalore - 560 083, India*

Summary

In vitro organogenesis, somatic embryogenesis and subsequent plant regeneration in niger is dependent on genotype, choice of explant and concentrations of plant growth regulators used in the medium. Protocols were developed for high frequency induction of embryos from cultured anthers and flowering from in vitro developed plantlets. Isolated shoots could be induced to flower with 40% frequency in the variety Ootacamund. Anther induced callus not only produced embryos but also haploid shoots with 85% frequency. Plants regenerated from diploid tissues were transferred to the soil, selfed and seeds were collected separately. The progeny of these plants were evaluated for somaclonal variations and all the plant characters with a negligible reduction or increase bred true in the second generation. Seed raised and also non-variant tissue culture derived plants served as controls. Non-variant tissue culture derived plants displayed traits like that of seed raised plants. Embryogenic callus obtained from anthers developed into embryos and shoots initiated from non-embryogenic callus produced while plants upon transfer into the Chaleff's R-2 and MS basal media, respectively. One hundred

★Corresponding author : E-mail : pbkavi@yahoo.com

and fifty plants derived from anther callus of the niger variety Ootacamund were transferred to pots and grown to maturity. Out of these, 8 plants were fertile diploids (2n=30) and showed significant variations in agronomical characters like plant height, leaf length, number of branches per plant, width of recepticle and number of seeds per capitulum in both first and second generations. Dwarfs, large flower head types and self-compatible plants obtained in this study are useful for the improvement of this neglected crop plant. Plantlets regenerated from various explants after co-cultivation with A. tumefaciens exhibited resistance to 100 mg/l kanamycin and gus activity. However, these plants need to be confirmed further by Southern blots for their transgenic nature. There is an urgent need to develop transgenics for altering the qualitative and quantitative profiles of seed oil through metabolic engineering and niger into a model oil-yielding crop plant.

Keywords : *Guizotia abyssinica,* haploids, Niger, somatic embryogenesis, somaclonal variation, transgenic plants.

1. INTRODUCTION

1.1. Geographic distribution of niger

Guizotia abyssinica, commonly known as niger belongs to the family Asteraceae. The leaves are generally opposite, but the upper leaves are alternate. Heads are peduncled, axiallary or terminal and heterogamous. Achenes (seeds) are glabrous, dorsally compressed with no pappus. Niger, originally is a native of tropical Africa, and is also grown in Ethiopia, East African countries, India and the West Indies. This crop is sown on light lateritic or sandy soils with very simple preparation of the land. It is a rainfed crop, which receives very little attention by way of manuring or irrigation. The crop is recognized for its edible and industrial oil. In India, the crop is grown over an area of 0.6 million hectares with 0.17 million tons of production, and is widely grown in Andhra Pradesh, Bihar, Gujarat, Karnataka, Madhya Pradesh, Maharashtra, Orissa and West Bengal as a marginal crop. The crop accounts for almost 2% of total oil seed crops with an annual production of 1,73,200 tons and an average yield of 285 kg/ha. But, as a pure crop niger yields 337-450 kg of seed per hectare. Niger occupies the fifth position in total production of oilseeds, and India accounts for 79% of total niger cultivation and 75% of niger production in the world (Joshi, 1990). India exports niger seed to United Kingdom, Germany, Netherlands, Belgium, USA, Canada, France etc.

1.2. Economic importance

Niger is an oil yielding herbaceous annual crop. Niger seed resembles sunflower seed in shape, but is much smaller in size (3.9 - 4.1 mm long). Niger seed contains 1.8% moisture, 19.4% protein, 31.3% fatty oil, 39.7% carbohydrate, 1.8% ash, 0.05% calcium and 0.18% phosphorous. The seeds of niger are used for extraction of oil which is about 30 - 50% of the total seed weight (Seegler, 1983). In some places, the seeds are fried with oil or ghee and eaten. The oil content reaches the maximum value at the end of about 45 days after opening of flowers. Niger oil contains 1.1% myristic acid (along with some capric, caprylic and lauric acids), 5% palmitic acid, 2% stearic acid, 38.9% oleic acid and 51.6% linoleic acid along with traces of arachidic, behenic and lignoceric acids. The oil is clear, pale-yellow or orange and is used as a substitute for gingelly oil, which often contains an admixture of niger seed oil. The oil is usually limpid, pale and used in pickles and for culinary purposes (Nikam and Shitole, 1993). The extracted oil is bluish-white in colour with a faint odour and slightly sweet taste. The oil when refined, is fairly good for lighting purpose and used for painting and cleaning machinery. It is also used for anointing the body (applying to the body), in paints and soaps. Niger oil absorbs flower fragrances making it useful as a base-oil in perfumes (Sharma, 1990). In India, niger oil is used mostly in cooking, either in pure form or mixed with other vegetable oils. The oil cake is used as a cattle feed, especially for dairy cows. The cake contains about 4.5% nitrogen. In certain regions, it is used both for manure and as cattle feed. At the flowering stage, the niger crop is cut and fed as green fodder, particularly to sheep. Silage prepared from niger plants is used as cattle-feed. Niger oil is also used for pharmaceutical purposes as sesamum oil.

1.3. Need for *in vitro* studies

Keeping in view its high economic value, attempts have been made to improve the yields of this plant through conventional breeding methods. The primary aim of breeding programmes of niger both in India and Ethiopia is to improve the yield and oil content (Riley and Belayneh, 1989). Low yield potential and high instability is the main barrier of niger yields. A sharp increase in oil content could probably be achieved through selection of thin hull types. The fatty acid composition of the oil is similar to that of sunflower except, that the sunflower contains 2% lignoceric acid (Seegler, 1983). Hence, selection for improved fatty acid balance has a good potential in this crop. Some of the problems hampering the realization of the full potentialities of niger in India are low seed yields due to the self-incompatible nature of the crop and inherent low yielding capacity of the local cultivars, susceptibility to diseases, pests and abiotic stresses. When conditions for vegetative growth are favourable, niger is prone to lodging, hence shorter plants with stronger stems and better harvest index are needed for these situations (Riley and Belayneh, 1989).

2. GENOTYPIC DIFFERENCES FOR CALLUS INDUCTION AND ORGANOGENESIS

Induction of callus and shoot organogenesis, and subsequent plant regeneration in niger seem to be dependent on the choice of genotype (Table 1) and composition of media growth regulators (Sarvesh *et al.*, 1994). A distinct regeneration protocol was developed for organogenesis by culturing immature zygotic embryos of different genotypes on Linsmaier and Skoog's (LS) medium (1965) containing various combinations of hormones and the results obtained are shown in Table 1. Of different hormones tested, 2 mg/l 2,4-D and 0.5 mg/l kinetin (KN) were good for callus initiation. Per cent frequency of callus as well as organogenetic response was highest in Ootacamund (93 and 77) and least in the variety RCR-18 (82 and 62).

3. GENOTYPIC DIFFERENCES FOR SOMATIC EMBRYOGENESIS

The per cent frequency of embryogenic callus initiation from cotyledons of the genotype Ootacamund was higher in Murashige and Skoog's (MS 1962), LS and B_5 (Gamborg *et al.*, 1968) media (incorporated with 2 mg/l 2,4-D and 0.3 mg/l kinetin) when compared to other genotypes (Table 2). Embryogenic callus with 26 and 25% frequencies was noticed only in the cultivar Ootacamund (not in other genotypes) in MS and LS media respectively but not in B_5. This could be because of differences in the nitrogen content of these media, especially the ammonium ion. Histological sections of the embryogenic callus revealed a large number of globular-shaped embryos and a few heart-shaped embryos.

Cotyledons were reported to be efficient for embryogenesis in other oil-yielding plants like *Carthamus* (Tejovathi, 1986) and *Arachis* (Ozias-Atkins, 1989). Auxins, especially 2,4-D, are known to induce somatic

Table 1. Genotypic differences for callus induction and organogenesis from immature zygotic embryos

Genotype	Hormonal concentration (mg/l)	No. of explants inoculated	No. of explants showing callus	Callusing frequency (%)	Nature of response*
Ootacamund	2,4-D (2) + KN (0.5)	60	56	93	C
	NAA (2) + KN (0.5)	45	33	73	CSR
	NAA (3) + IAA (1) + KN (0.2)	48	37	77	CSR
	NAA (1.5) + IAA (0.5) + KN (0.5)	34	23	67	CSR
No.5	2,4-D (2) + KN (0.5)	54	47	87	C
	NAA (2) + KN (0.5)	41	28	68	CSR
	NAA (3) + IAA (1) + KN (2)	48	35	73	CSR
	NAA (1.5) + IAA (0.5) + KN (1)	46	30	65	CSR
RCR-18	2,4-D (2) + KN (0.5)	50	41	82	C
	NAA (2) + KN (0.5)	49	32	65	CSR
	NAA (3) + IAA (1) + KN (2)	37	26	70	CSR
	NAA (1.5) + IAA (0.5) + KN (1)	43	27	62	CSR
UN-4	2,4-D (2) + KN (0.5)	43	31	85	C
	NAA (2) + KN (0.5)	40	27	67	CSR
	NAA (3) + IAA (1) + KN (2)	42	31	73	CSR
	NAA (1.5) + IAA (0.5) + KN (1)	42	27	64	CSR

*Data were scored after 4 weeks of culture; C = Callus; S = Shoot buds; R = Roots; Callus induction medium = LS + hormones + 3% sucrose

Table 2. Embryogenic and non-embryogenic callus initiation from cotyledons of different genotypes of niger

Genotype	MS		LS		B_5	
	Embryogenic callus	Non-embryogenic callus	Embryogenic callus	Non-embryogenic callus	Embryogenic callus	Non-embryogenic callus
Ootacamund	26 (± 1.0)	73 (± 1.0)	25 (± 1.0)	73 (± 1.0)	Nil	60 (± 3.0)
RCR-18	Nil	75 (± 2.3)	Nil	88 (± 1.4)	Nil	56 (± 2.6)
IET-12	Nil	79 (± 2.2)	Nil	87 (± 1.8)	Nil	54 (± 2.4)
GA-1	Nil	70 (± 1.1)	Nil	75 (± 0.6)	Nil	42 (± 1.2)
IGP-76	Nil	69 (± 1.1)	Nil	85 (± 1.5)	Nil	40 (± 1.2)
No-71	Nil	75 (± 1.2)	Nil	90 (± 0.6)	Nil	65 (± 1.1)

*Two milligram per litre 2,4-D and 0.3 mg per litre KN were used in all the three media. Numbers in parenthesis indicate standard error.

embryogenesis in many plants (Stuart and McCall, 1992). Kamada and Harada (1979) reported that somatic embryogenesis can be induced using auxins with the following order of effectiveness: IAA<NAA<4-chlorophenoxyacetic acid<2,4-D<2,4,5-T. These results indicate that in addition to 2,4-D, the ring modified forms of phenoxy acids are also effective in inducing embryogenesis, although 2,4-D is the preferred synthetic auxin.

4. *IN VITRO* REGENERATION FROM DIFFERENT SEEDLING EXPLANTS

Several explants were used for direct plant regeneration in niger. Plantlet regeneration was observed from explants such as cotyledons (Sarvesh *et al.*, 1993), mature zygotic embryos (our unpublished data), hypocotyl segments (Nikam and Shitole, 1993), zygotic embryos (Bir Bahadur *et al.*, 1992) and shoot apices (Malathi *et al.*, 1989). Mature zygotic embryos of niger cultivar Ootacamund cultured on Linsmaier and Skoog's medium supplemented with 1 mg/l 2,4-D and 0.2 mg/l kinetin exhibited over 70% somatic embryogenesis (Fig. 1a). Prolonged culture on the same medium resulted whole plant regeneration within 30-40 days (Fig. 1b). Immature embryos isolated after 15-18 days of pollination were also used as explants for direct regeneration of plants.

In general, callus initiation was observed from cut ends of the explants within a week when cultured on LS media, containing 3 mg/l NAA + 1 mg/l IAA + 2 mg/l KN, 1.5 mg/l NAA + 0.5 mg/l IAA + 1 mg/l KN and 2 mg/l NAA + 0.5 mg/l KN. Calli were green, friable and non-embryogenic on all the above three media. Some of the callus clumps showed shoot bud initiation after 2 to 3 weeks of culture. When the callus clumps containing shoot buds were subcultured onto MS basal medium, multiple shoot buds were observed within 15 days, and shoots with well developed roots were observed after 30 days of culture. Three explants *viz.*, immature zygotic embryos, cotyledons and immature leaves were used in the present study for direct regeneration.

4.1. Immature zygotic embryos

Immature zygotic embryos, 10 to 15 days after anthesis were isolated and used in the present study for the induction of somatic embryogenesis. Immature zygotic embryos of different varieties of niger were tried for *in vitro* morphogenic response on MS medium supplemented with different combinations and concentrations of growth regulators. Callus initiation was observed from cut ends of the explants within a week when cultured on LS medium supplemented with 2 mg/l 2,4-D and 0.5 mg/l KN. However, it took 8-10 days for callusing on LS medium containing 3 mg/l NAA, 1 mg/l IAA and 2 mg/l KN and also about the same in other combinations. The effect of various growth regulators on morphogenetic response and genotypic differences are shown in Table 3. MS medium supplemented with 0.5 mg/l BAP exhibited the highest percentage of regeneration as well as maximum number of shoot buds per culture. Addition of GA_3 (0.01 mg/l) to the MS medium, containing 0.5 mg/l BAP and 0.1 mg/l NAA, resulted in the total suppression of shoot-bud formation but caused the development of green callus. Total inhibition of shoot buds by GA_3 was reported earlier in tobacco callus (Thorpe and Meier, 1975). Among different genotypes tested for morphogenetic potential, the variety Ootacamund showed the highest percentage (88%) of regeneration followed by variety No.5 (85%), while it was the least (69%) in the variety UN-4 (Venkatesham and Reddy, 1996).

In another cultivar No.5, soft, fragile, transluscent, creamy, compact, globular, white calli were observed from immature embryos on LS medium supplemented with 2 mg/l 2,4-D and 0.5 mg/l KN. Embryogenic, compact, globular, white callus was separated and cultured on MS medium, fortified with KN, IAA, NAA and 2,4-D either alone or in combination, to induce somatic embryos (Table 4). Maximum number of embryos was observed on MS medium containing 2 mg/l IAA and 0.3 mg/l KN while it was least on the medium supplemented with 0.5 mg/l 2,4-D and 0.5 mg/l NAA. However, somatic embryogenesis was not observed on MS medium containing 2,4-D (2 mg/l) or kinetin (0.1 mg/l) alone. Conversion of embryos into plantlets occurred on MS medium lacking growth regulators. Regeneration of plants through embryogenic pathway is of special significance since the embryos arise from single cells and, therefore, the regenerated plants are generally regarded as genetically stable (Heinz and Mee, 1971). Somatic embryogenesis was reported in the tissue cultures of many oil crops like *Brassica* (Mehta *et al.*, 1993), groundnut (Venkatachalam *et al.*, 1999), sunflower (Finer, 1987), sesame (Rajender Rao and Vaidyanath, 2000) etc. Plants produced through callus cultures and somatic embryogenesis are successfully established under field conditions after hardening.

4.2. Cotyledons

Cotyledons from the cultivar Ootacamund exhibited non-embryogenic callus with 65, 88 and 79% frequencies respectively when LS medium

Table 3. Genotypic differences and effect of hormonal concentrations on morphogenetic response of immature zygotic embryo derived callus

Variety	Hormonal concentration (mg/l)	No. of calli inoculated	No. of calli with shoots	Frequency of response	No. of shoots/ culture*
Ootacamund	BAP (0.5)	60	53	88	6-8
	BAP (0.5) + NAA (0.1)	54	45	83	4-6
	BAP (0.5) + NAA (0.1) + GA_3 (0.01)	53	-	-	-
	BAP (1) + NAA (0.1)	48	32	67	3-6
No. 5	BAP (0.5)	60	51	85	5-7
	BAP (0.5) + NAA (0.1)	48	35	73	3-6
	BAP (0.5) + NAA (0.1) + GA_3 (0.01)	36	-	-	-
	BAP (1) + NAA (0.1)	45	29	64	3-5
RCR-18	BAP (0.5)	55	42	76	4-6
	BAP (0.5) + NAA (0.1)	47	33	70	3-5
	BAP (0.5) + NAA (0.1) + GA_3 (0.01)	45	-	-	-
	BAP (1) + NAA (0.1)	48	29	60	3-4
UN-4	BAP (0.5)	36	25	69	5-7
	BAP (0.5) + NAA (0.1)	45	28	62	3-6
	BAP (0.5) + NAA (0.1) + GA_3 (0.01)	-	-	-	-
	BAP (1) + NAA (0.1)	31	18	58	2-4

*Data were scored after 30 days of culture; Regeneration medium = MS + hormones + 3% sucrose

Table 4. Effect of growth regulators on somatic embryogenesis from immature zygotic embryo derived callus of niger variety No. 5.

Hormonal concentration (mg/l)	No. of calli cultured	No. of calli showing somatic embryos	No. of somatic embryos/ culture
MS + KN (0.1)	7	-	-
MS + IAA (2) + KN (0.3)	15	9	13-21
MS + NAA (2) + KN (0.3)	15	7	10-15
MS + 2,4-D (2)	11	-	-
MS + 2,4-D (0.5) + NAA (0.5)	13	3	4-7
MS + 2,4-D (0.2) + NAA (0.5)	11	3	7-11

Embryo induction medium = MS + hormones + 3% sucrose

was supplemented with 2 mg/l 2,4-D or 2,4,5-trichlorophenoxyacetic acid or 2,4,5-trichlorophenoxypropionic acid (Table 5). Addition of 0.3 mg/l kinetin to 2,4-D or its analogues promoted embryogenic callus induction. Embryogenic callus induction was better (35%) on 2,4,5-T (0.5 mg/l) but 2,4,5-TP totally suppressed the induction of embryogenic callus at lower concentrations. Cotyledons were superior to hypocotyls for callus induction but root explants were still better than cotyledons (data not shown). The number of embryos produced directly from cotyledonary explants varied with the concentration of 2,4,5-T and 2,4,5-TP. Direct embryogenesis obtained from cotyledons could probably be used for artificial seed production.

The development of embryos from the initial globular stage to fully mature stage from cotyledons occurred in the presence of 2,4,5-T and 2,4,5-TP within 25 days. However, induction of somatic embryogenesis from cotyledonary callus required two separate media with different hormonal concentrations. Callus could be initiated and proliferated on LS medium containing 2 mg/l 2,4-D and 0.3 mg/l kinetin, but embryogenic clumps containing globular shaped embryos could be induced from the callus in MS medium containing 2,4-D. When these embryogenic clumps were transferred to MS medium fortified with 10.7 μM NAA and 2.3 μM kinetin, they developed into mature embryos. The embryogenic callus development appeared to be the result of the omission of 2,4-D from the medium (Reinert, 1973). These results are in agreement with the

Table 5. Effect of auxins and KN on somatic embryogenesis in cotyledons of niger variety Ootacamund

Hormonal concentration (mg/l)	Per cent of cotyledons showing non-embryogenic callus	Per cent of cotyledons showing callus with somatic embryos
2,4-D (0.25)	42	0
2,4-D (0.5)	50	0
2,4-D (1.0)	59	10
2,4-D (2.0)	65	25
2,4,5-T (0.25)	88	0
2,4,5-T (0.5)	15	35
2,4,5-T (1.0)	67	15
2,4,5-T (2.0)	88	0
2,4,5-TP (0.25)	30	0
2,4,5-TP (0.5)	67	0
2,4,5-TP (1.0)	71	0
2,4,5-TP (2.0)	79	12
2,4-D (2.0) + KN (0.3)	73	25
2,4,5-T (2.0) + KN (0.3)	93	0
2,4,5-TP (2.0) + KN (0.3)	84	9

LS medium was used in all the above treatments

observations of Wochok and Wetherell (1971) who suggested that the 2,4-D induced suppression of embryo maturation may be mediated through endogenous ethylene production.

Plant growth regulators such as benzyladenine (BA), indole 3-acetic acid (IAA), naphthaleneacetic acid (NAA), kinetin (KN) and GA_3 were used at different concentrations for morphogenetic response of cotyledonary explants of the cultivar Ootacamund and the results are shown in Table 6. In general, callusing response was good excepting at 2.2 µM concentration of BA. Maximum frequency of callus (80%) was recorded at 17.8 µM BA. The per cent frequency of shoot regeneration was 81 when 11.4 µM IAA was incorporated along with

BA. On this medium, multiple shoots with an average of 28.2 shoots per explant were observed. Lower or higher levels of BA plus NAA, BA plus GA_3, BA plus kinetin and BA plus NAA and GA_3 reduced the number of shoots from cotyledons. On the other hand, a combination of KN (0.93 μM) plus NAA (10.7 μM) exhibited excellent root differentiation (Table 6). Whenever GA_3 was used, shoot regeneration was totally suppressed. Suppression of shoots by GA_3 was observed earlier in callus cultures of tobacco (Thorpe and Meier, 1975) and rice (Kavi Kishor, Unpublished).

The effect of auxins and cytokinins on organogenesis from cotyledonary derived callus of niger cultivar Ootacamund is shown in the Table 7. Mean number of shoot buds produced from callus cultures was highest (8.7 per culture) in presence of 2.2 μM BA, but a combination of 4.4 μM BA plus 0.54 μM NAA reduced the number to 4.8 per culture (Table 7). Replacing BA with kinetin only reduced the number of shoots formed per culture. Shoots regenerated from cotyledonary callus also produced roots in NAA/IAA and kinetin combinations and also on MS basal medium (data not shown). This investigation clearly demonstrated the optimal cultural conditions for regeneration of plantlets from cotyledonary callus of niger (Sarvesh *et al*., 1993).

4.3. Immature leaves

Multiple shoot bud and somatic embryo induction was achieved from the immature leaf segments of niger cultivar Ootacamund with very high frequency. Shoot-bud induction was maximum (65%) on the LS medium supplemented with 2 mg/l BAP and 0.5 mg/l NAA. But, mean number (12.88) of shoot buds per explant was higher when BAP concentration was reduced to 1 mg/l while maintaining the same concentration of NAA (Table 8). Somatic embryos were induced on LS medium fortified with 1-2 mg/l 2,4-D alone or in combination with 0.1-0.3 mg/l kinetin. The frequency of explants producing somatic embryos and the mean number of embryos per explant were maximum on the medium containing 1 mg/l 2,4-D and 0.2 mg/l kinetin (Table 9). LS medium devoid of growth regulators proved ideal for further growth of shoot buds and somatic embryos. Jadimath *et al.* (1998) reported earlier direct adventitous shoot bud differentiation but not somatic embryogenesis in *G. abyssinica* and *G. scabra.* Direct regeneration observed in this study helps in the clonal propagation of elite species of niger like male sterile lines, high yielding varieties (Venkatesham and Reddy, 1996) and also in genetic transformation.

Table 6. Morphogenic response of cotyledonary explants of niger variety Ootacamund

Growth regulators (μM)	Callus (%)	Shoot (%)	Root (%)
BA (2.2)	-	-	-
BA (4.4)	45	-	-
BA (8.9)	60	-	-
BA (17.8)	80	-	-
BA (2.2) + NAA (0.54)	15	-	-
BA (4.4) + NAA (2.7)	25	-	-
BA (8.9) + NAA (5.4)	40	-	18
BA (8.9) + NAA (10.7)	38	-	23
BA (17.8) + NAA (2.7)	55	20	3
BA (17.8) + NAA (5.4)	70	16	30
BA (17.8) + NAA (10.7)	63	10	38
BA (4.4) + IAA (5.7)	60	40	15
BA (4.4) + IAA (11.4)	43	81	15
BA (8.9) + IAA (5.7)	58	73	-
BA (8.9) + IAA (11.4)	40	55	-
IAA (5.7) + GA_3 (2.9)	58	-	45
IAA (5.7) + GA_3 (5.8)	63	3	70
IAA (11.4) + GA_3 (2.9)	65	10	50
IAA (11.4) + GA_3 (5.8)	70	-	65
NAA (5.4) + KN (0.93)	75	10	83
NAA (5.4) + KN (2.3)	78	-	75
NAA (10.7) + KN (0.93)	63	-	95
NAA (10.7) + KN (2.3)	73	38	88
BA (2.2) + NAA (0.54) + GA_3 (0.29)	20	-	-
BA (0.44) + NAA (0.54) + GA_3 (0.29)	25	-	-

Data were scored after 4 weeks of culture from 40-50 replicates

Table 7. Effect of auxins and cytokinins on organogenesis from cotyledon derived callus of niger variety Ootacamund

MS + growth regulators (µM)	Mean no. of shoots/culture	Nature of response
BA (2.2)	8.7	Green callus with shoots
BA (4.4)	2.1	-
BA (8.9)	Nil	-
BA (4.4) + NAA (0.54)	4.8	Green callus with shoots
BA (4.4) + NAA (1.1)	4.3	-
BA (4.4) + NAA (2.7)	2.3	Shoots and callus
BA (8.9) + NAA (0.54)	3.3	Shoots and callus
BA (8.9) + NAA (1.1)	2	Shoots and callus
BA (8.9) + NAA (2.7)	1.6	Shoot buds and callus
BA (17.8) + NAA (0.54)	Nil	Callus
BA (17.8) + NAA (2.7)	Nil	-
BA (4.4) + NAA (2.7) + KN (2.3)	1.8	Green thread-like structures
BA (8.9) + NAA (2.7) + KN (2.3)	1	Shoot buds with callus
BA (17.8) + NAA (2.7) + KN (2.3)	Nil	-

Data represent an average of 40 replicates/treatment

5. *IN VITRO* FLOWERING

Protocols were developed for high frequency induction of embryos from cultured anthers and also a highly organized stage of differentiation *i.e.,* flowering from *in vitro* developed plantlets of niger in the present study (Sarvesh *et al*., 1996). Capitula differentiation could be obtained consistently from isolated shoots of niger with a very high percent frequency. But, sessile capitula that were formed directly from the callus did not contain equal number of organs. Induction of flowers was observed in other oil yielding crop plants like groundnut (Narasimhulu and Reddy, 1984), *Brassica* (Mehta *et al*., 1993) and sunflower (our unpublished results).

Table 8. Effect of growth regulators on multiple shoot-bud induction from leaf segments of niger variety Ootacamund

Growth regulators (mg/l)	% Frequency of shoot-bud regeneration	Mean number of shoot-buds/explant
BA (0.5)	62	4.25
BA (1.0)	57	5.14
BA (2.0)	53	5.40
BA (3.0)	42	5.71
BA (4.0)	30	4.00
BA (1.0) + NAA (0.2)	61	6.60
BA (1.0) + NAA (0.5)	47	12.88
BA (1.0) + NAA (1.0)	42	8.00
BA (2.0) + NAA (0.5)	65	8.90
BA (2.0) + NAA (1.0)	57	6.80
BA (2.0) + NAA (2.0)	45	5.90
BA (3.0) + NAA (0.5)	47	7.51
BA (3.0) + NAA (1.0)	51	8.74
BA (3.0) + NAA (2.0)	44	6.20
BA (4.0) + NAA (0.5)	32	4.71
BA (4.0) + NAA (1.0)	38	5.10
BA (4.0) + NAA (2.0)	27	4.20

Data were scored after 4 weeks of culture from 40-50 replicates

6. SOMACLONAL VARIATIONS

Regenerated plants of Ootacamund were transferred to the soil along with the seed raised plants, selfed and seeds were collected separately. The progeny of these plants (30 individual clones) were evaluated for six important plant traits and all the plant characters with a negligible reduction or increase bred true in the second generation. Seed raised and also non-variant tissue culture derived plants served as controls. Non-variant tissue culture derived plants displayed traits like that of

Table 9. Effect of growth regulators on callus induction and somatic embryogenesis from leaf segments of niger cultivar Ootacamund

Growth regulators (mg/l)	Callusing frequency (%)	% Frequency of explants with embryos	Mean number of embryos/ explant
2,4-D (0.5)	62	-	-
2,4-D (1.0)	68	12	2.8
2,4-D (1.5)	75	9	2.5
2,4-D (2.0)	82	6	1.7
2,4-D (1.0) + 0.1 KN	70	19	3.5
2,4-D (1.0) + 0.2 KN	77	24	4.3
2,4-D (1.5) + 0.2 KN	83	18	4.0
2,4-D (1.5) + 0.3 KN	88	15	2.4
2,4-D (2.0) + 0.3 KN	92	-	-
2,4-D (2.0) + 0.5 KN	92	-	-

Data were scored after 4 weeks of culture from 40-50 replicates

seed raised plants. The range of variation in seed raised plants of the variety Ootacamund for some important characters like plant height was 155-170 cm and for number of branches was 11 to 16. Except clones SC2 and SC4, the remaining clones showed a decrease in plant height, particularly SC5 and SC6. Number of branches, flowers and seed per plant were significantly higher in clone SC2 compared to the seed raised plants (Table 10). Clone SC5 is a dwarf one with a mean number of 20.1 seeds per plant. Though morphological variations in other varieties were observed, they were not evaluated.

7. ANTHER CULTURE AND ANDROCLONAL VARIATIONS

Haploid plants derived from anther and microspore cultures have considerable potential in plant breeding programmes because of the time saved by the reduction of the classical selection cycle period and the genetic value of isogenic lines or homozygous diploids. Also, the presence of a single set of chromosomes in haploids allow the detection of mutations controlled by recessive genes. Androgenic haploids have

Table 10. Somaclonal variations for some plant traits in the variety Ootacamund in the second generation[a]

Genotype	Plant height (cm)	Leaf length (cm)	Leaf width (cm)	No. of branches/ plant	No. of flowers/ plant	No. of seeds/ capitulum
Seed raised plants (control)★★	165.4 (± 8.0)	10.6 (± 0.6)	2.2 (± 0.2)	13.9 (± 2.0)	11.4 (± 1.3)	43.1 (± 6.4)
SC1	120.6★ (± 7.2)	9.9 (± 0.8)	1.7 (± 0.2)	10.5 (± 1.6)	9.2 (± 1.0)	38.5 (± 5.8)
SC2	176.9★ (± 9.0)	16.9★ (± 1.1)	2.6 (± 0.3)	19.8★ (± 3.4)	17.5★ (± 1.9)	59.0★ (± 8.4)
SC3	110.5★ (± 6.6)	9.0 (± 0.4)	1.5★ (± 0.2)	11.7 (± 1.8)	9.7 (± 1.0)	35.2★ (± 5.7)
SC4	169.1 (± 8.7)	14.6★ (± 1.0)	2.4 (± 0.3)	13.6 (± 2.2)	11.1 (± 1.8)	48.9★ (± 7.9)
SC5	49.4★ (± 4.2)	6.9★ (± 0.4)	1.0★ (± 0.2)	8.0★ (± 1.0)	6.3★ (± 0.8)	20.1★ (± 4.2)
SC6	81.7★ (± 5.1)	10.2 (± 0.7)	1.6★ (± 0.2)	9.8★ (± 1.5)	8.0★ (± 1.0)	29.3★ (± 4.8)
SC7	101.3★ (± 6.3)	8.6★ (± 0.7)	1.8 (± 0.3)	11.2 (± 1.9)	9.5 (± 1.2)	32.0★ (± 5.1)

[a]Data represent an average of 30 individual clones for all the above traits.
★significant at p = 0.05. Other values are not significantly different.
★★Ten seed raised plants were scored for control. Non-variant tissue culture derived plants displayed traits like that of seed raised Ootacamund plants (data not shown).

been induced mainly in cereals and vegetable crops (Foroughi-Wehr and Wenzel, 1989) and a few oil seed crops like *Brassica campestris* (Keller *et al.*, 1975), *Arachis hypogaea* (Bajaj *et al.*, 1981) and *Helianthus* sps (Bohorova *et al.*, 1986). No reports are available on *in vitro* culture of anthers of niger, when the work was initiated on this plant. Therefore, an attempt was made to produce haploids and to recover androclonal variations, if any, for the breeding programmes in niger.

Anthers of niger were inoculated onto five different media differing mainly in their inorganic and organic constituents and plant growth regulators to study their influence on callus induction (embryogenic/ non-embryogenic) and plant regeneration. Around 200 to 292 anthers from the variety Ootacamund were inoculated into each medium and the percent frequency of response varied in between 16 to 90 (Table 11). Of the five media tested, LS was superior followed by N_6, MS, B5 and Chaleff's R-2 (Chaleff and Stolarz, 1981) for callus induction from anthers of Ootacamund. Time taken for callus induction also varied significantly. While callus appeared in 10 days in LS, it took 20 days in Chaleff's R-2. Embryogenic callus was observed only when 2,4-D was used in the medium but in presence of 2 mg/l NAA and 0.3 mg/l KN, non-embryogenic callus was noticed.

Considerable variation was observed in the degree of response to callus induction when different growth regulators were used (Table 12). The combination of 2,4-D plus KN was more effective for callus induction compared to BAP plus KN. But, others failed to show any response when 2,4-D was used alone. This points out that cytokinin is essential for triggering cell division. BAP alone (1 and 2 mg/l) yielded very little callus, and the frequency ranged from 4 to 8% (Sarvesh *et al.*, 1993). Wang *et al.* (1978) and Huang *et al.* (1986) demonstrated that a combination of hormones rather than in single had a profound influence on callusing of rice anthers. This may be due to their role in increasing the survival rate and establishment of the state of synchrony among embryogenic pollen grains (Rout, 1989). In the variety Ootacamund, callus formed with 89% frequency on LS medium fortified with 2 mg/ l 2,4-D plus 0.3 mg/l KN and this was shiny, loosely packed and appeared in 7-9 days. Prolonged culture in the same medium resulted in the production of a few globular embryos directly from callus. Transfer of this callus onto Chaleff's R-2 medium (with the same hormonal combination), also resulted in the production of embryos with 80% frequency. Callus formed on the LS medium supplemented with BAP (2 mg/l) and KN (0.2 mg/l) was green, friable, non-embryogenic and showed a high frequency of callusing (69%) compared to other BAP containing media. Higher concentrations of BAP (3 mg/l) considerably suppressed the callus induction (4 to 8%).

Response to callusing from anthers of many genotypes such as Ootacamund, No. 71, GA-1, IET-11, IET-19, RCR-18, RCR-140, RCR-317 and IGP-76 is shown in the Table 13. Anthers were inoculated onto LS medium supplemented with 2 mg/l 2,4-D and 0.3 mg/l KN and the per cent frequency of embryogenic callus was 90, 80, 49, 67, 45, 67, 55, 33 and 30 respectively. For embryo germination, Chaleff's R-

Table 11. Influence of different media and plant growth regulators on anther callus induction in the variety Ootacamund

Type of the media + hormones (mg/l)	No. of anthers inoculated	No. of anthers responded	Callus induction (%)	No. of days taken for callus induction	Nature of response
MS					
2,4-D(2) + KN (0.3)	292	152	52	18	Embryogenic callus, roots and shoots
NAA (2) + KN (0.3)	200	36	18	18	Non-embryogenic callus
B_5					
2,4-D(2) + KN (0.3)	284	121	43	17	Green friable callus
NAA(2) + KN(0.3)	294	95	32	17	Callus with shoots and roots
Chaleff's R-2					
2,4-D(2) + KN (0.3)	240	39	29	20	Embryogenic callus
NAA(2) + KN(0.3)	228	66	16	20	
N_6					
2,4-D(2) + KN (0.3)	286	170	59	12	Friable callus
NAA(2) + KN(0.3)	270	45	17	12	
LS					
2,4-D(2) + KN (0.3)	275	248	90	10	Friable callus
NAA(2) + KN(0.3)	240	110	46	10	

Immature capitula were pretreated for one day at 4° ± 2 °C.

Table 12. Effect of various plant growth regulators on callus induction from anthers of the niger variety Ootacamund*

Hormonal concentrations (mg/l)	No. of anthers inoculated	No. of anthers forming callus	Frequency of callus formation (%)	Nature of response
2,4-D (1)	130	-	-	-
2,4-D(1) + KN(0.1)	128	13	10	Embryogenic callus
2,4-D(1) + KN(0.3)	124	16	13	Embryogenic callus
2,4-D(1) + KN(0.5)	124	8	7	Embryogenic callus
2,4-D(2)	140	-	-	-
2,4-D(2) + KN(0.1)	125	44	35	Embryogenic callus
2,4-D(2) + KN(0.3)	134	119	89	Embryogenic callus
2,4-D(2) + KN(0.5)	142	94	66	-
2,4-D(3) + KN(0.1)	140	30	21	-
2,4-D(3) + KN(0.3)	142	13	9	-
2,4-D(3) + KN(0.5)	138	8	6	-
BAP(1)	140	6	4	Green friable callus
BAP(1) + KN(0.2)	136	21	15	Green friable callus
BAP(2)	126	10	8	Green friable callus
BAP(2) + KN(0.2)	138	92	69	Green friable callus
BAP(2) + KN(0.5)	134	74	54	Green friable callus
BAP(3) + KN(0.2)	128	38	30	Green friable callus
BAP(3) + KN(0.5)	124	10	8	-

*Data were scored at the end of 30 days. Immature capitula were pretreated for one day at 4° ± 2 °C, Callus induction medium = LS + hormones + 2% sucrose.

2 medium devoid of growth regulators was used and the highest frequency of 80% was observed in the variety Ootacamund and the least in IGP-76 (Table 13).

Callus cultures of the above genotypes initiated on LS medium containing 2 mg/l 2,4-D and 0.3 mg/l KN were transferred to MS media supplemented either with 1 mg/l BAP plus 0.1 mg/l NAA or 2 mg/l BAP plus 0.2 mg/l NAA for shoot regeneration. Lower concentration of BAP (1 mg/l) was found to be better than a higher concentration (2 mg/l) for shoot regeneration as well as for the number of shoots formed per callus mass in all the genotypes. Of many genotypes tested, Ootacamund exhibited the highest frequency of response (36.6%) followed by NO. 71 (34.2%) and IGP-76 showed the lowest.

A large number of adventitious shoot bud primordia and shoots were differentiated from anther callus on MS medium containing 1 mg/l BAP and 0.1 mg/l NAA in the genotypes tested (Table 14). Differentiation of capitula was also observed on this medium, mainly from the callus of the variety Ootacamund. When the callus of Ootacamund with shoot

Table 13. Genotypic variation for embryogenic callus induction and embryo germination

Genotype	No. of anthers cultured	% Frequency of embryogenic callus*	% Frequency of embryo germination**
Ootacamund	206	90	80
No.71	162	80	55
GA-1	172	49	40
IET-11	248	67	60
IET-19	168	45	40
RCR-18	160	67	55
RCR-140	180	55	50
RCR-317	150	33	30
IGP-76	140	30	25

*LS medium containing 2 mg/l 2,4-D and 0.3 mg/l KN was used.
**Chaleff's R-2 medium without any growth regulators was used for embryo germination.

buds was subcultured after 30 days onto the MS basal medium without growth regulators, it produced more tiny shoot buds, shoots and roots. These shoots grew further up to 10 to 15 cm long in 15 days and 23% of these shoots developed small capitula after 30 to 35 days. Six small capitula arising from a single stalk could also be seen on this medium. Capitula differentiation was seldom observed in other genotypes. These cultures would be valuable for analyzing the interaction between environmental and genetic factors controlling flower bud differentiation as pointed out by Greco *et al.* (1984). A basal medium without growth regulators was found to be the best for further growth and rooting of the shoots. The shoot forming potential of the callus on MS medium containing 1 mg/l BAP and 0.1 mg/l NAA could be maintained for 5-6 subcultures, but thereafter the morphogenetic capacity was lost. Shoot bud differentiation was also noticed on MS medium containing 2 mg/l BAP and 0.2 mg/l NAA, but was reduced invariably in all the varieties (Table 14).

In niger, regeneration could be obtained by both embryogenic and organogenic pathways. The anthers of variety Ootacamund gave high frequency of embryos via callus formation. It was observed that lower or higher than 2 mg/l NAA and 5 mg/l ABA reduced the number of embryos formed per callus and also the percent frequency of callus showing embryos. Chaleff's R-2 medium fortified with 2 mg/l NAA and 0.3 mg/l KN appreciably influenced the formation of embryos of various stages, which could be germinated upon transfer to the same medium but devoid of growth regulators (Table 15). KN (0.3 mg/l) in combination with 2 mg/l NAA was highly effective for embryogenesis than alone. The ability of callus to form embryogenic cultures on Chaleff's R-2 medium was found to be dependent on the concentration of auxin, cytokinin and ABA. This underlines the critical concentration and the role of growth regulators for efficient induction of embryogenesis from anthers of the genotype Ootacamund. Similar results were observed with other genotypes but the results are not shown here.

A total of 19.5 embryos per callus mass could be obtained when the callus of Ootacamund containing embryogenic cell clusters was cultured onto Chaleff's R-2 medium incorporated with 5 mg/l ABA. In other species, ABA was used to aid maturation of embryos, for example in the case of *Pennisetum americanum* (Vasil and Vasil, 1981) and *Picea abies* (Arnold and Hakman, 1988). Higher or lower than 5 mg/l ABA in the medium reduced the number of embryos formed per callus and also the frequency of callus showing embryogenesis. The concentration

Table 14. Shoot forming ability of anther callus tissues of different niger genotypes★

Genotype	Hormonal concentration (mg/l)	No. of calli inoculated	No. of calli with shoots	Frequency of response (%)	No. of shoots/ callus mass
NO. 71	BAP(1) + NAA(0.1)	38	13	34	7-9
	BAP(1) + NAA(0.2)	42	10	24	5-7
IGP-76	BAP(1) + NAA(0.1)	38	6	16	3-5
	BAP(2) + NAA(0.2)	42	5	12	2-4
Ga-1	BAP(1) + NAA(0.2)	40	7	18	3-6
	BAP(2) + NAA(0.2)	40	5	13	3-5
Ootacamund	BAP(1) + NAA(0.1)	41	15	37	8-10
	BAP(2) + NAA(0.2)	38	11	29	5-8
RCR-18	BAP(1) + NAA(0.1)	43	10	23	4-7
	BAP(2) + NAA(0.2)	40	7	18	3-5
IET-12	BAP(1) + NAA(0.1)	42	9	21	5-7
	BAP(2) + NAA(0.2)	41	6	15	3-6

★Data were scored at the end of 30 days on regeneration medium = MS + 2% sucrose.

$$\text{Frequency of response (\%)} = \frac{\text{No. of calli with plantlets}}{\text{No. of calli inoculated}} \times 100$$

Table 15. Effect of hormones on pollen embryogenesis in anther callus of Ootacamund

Hormones (mg/l)	Frequency of embryogenesis (%)	Mean number of embryos/callus
KN (0.1)	-	-
KN (0.1)	8	4.42
KN (0.1)	5	3.71
NAA (3) + KN (0.3)	24	5.60
NAA (1) + KN (0.3)	72	12.00
NAA (2) + KN (0.3)	80	15.33
ABA (2)	32	7.42
ABA (4)	56	12.16
ABA (5)	84	19.50
ABA (7)	16	4.00

Embryo induction medium = Chaleff R-2 + hormones + 2% sucrose

of ABA appeared to be critical for inducing globular to heart shaped embryos. Analysis of chromosomes from the root tips of the regenerated plants obtained via embryogenesis revealed the haploid nature in 85% of the plants. The remaining 15% included diploids (5%) and polyploids (10%). About 100 plants derived from anther callus of the niger variety Ootacamund were successfully transferred to pots and grown to maturity in the green house. Self-incompatibility is the main constraint for increasing the yields in niger, using conventional mutagenesis or somaclonal variation, and no new variants for the S-locus governing self-incompatibility could be obtained till now (Sree Ramulu, 1982). But, in the present study, a single self-compatible plant from the variety Ootacamund was generated through anther culture. Our studies therefore, indicate that anther culture could be exploited for generating useful androclones such as self-compatible lines which are expected to increase the yields in niger.

Anther culture of 9 genotypes of niger was studied for haploid plant production, so as to use derived haploids in breeding programmes. Both genotype and treatment affected the induction of embryogenic and non-

Table 16. Adroclonal variation for some plant traits of Ootacamund in the first generation*

Genotype	Plant height (cm)	Leaf length (cm)	Leaf width (cm)	No. of branches/ plant	Width of recepticle (mm)	Depth of recepticle (mm)	No. of seeds/ capitulum
Seed raised plants (control)**	162.2 (± 8.0)	10.4 (± 0.7)	2.2 (± 0.2)	14.2 (± 2.0)	9.4 (± 1.3)	9 (± 1.2)	45.2 (± 2.4)
AC1	78.2	11.5	1.4	11.7	8.2	8.7	42.2
AC2	103.5	12.4	2.0	11.3	7.6	7.1	28.2
AC3	165.1	11.1	1.4	8.5	8.2	7.8	29.4
AC4	158.1	19.2	2.5	16.0	13.8	12.9	40.3
AC5	146.3	10.3	2.2	19.3	8.2	7.9	49.4
AC6	36.2	10.2	2.3	12.6	8.4	7.8	31.8
AC7	145.6	15.1	1.6	18.0	18.2	17.6	50.8
AC8	108.7	12.2	1.8	13.1	8.3	7.2	36.5

*Eight individual clones were scored for the above traits.
**Ten seed raised plants were scored for control. Non-variant tissue culture derived plants displayed traits like that of seed raised Ootacamund plants (data not shown).

embryogenic callus from anthers of niger. Embryos developed from embryogenic callus and shoots initiated from non-embryogenic callus produced whole plants upon transfer into the Chaleff's R-2 and MS basal media, respectively. One hundred and fifty plants derived from anther callus of the niger genotype Ootacamund were successfully transferred to pots and grown to maturity. Out of these 150 plants, 8 plants were fertile, diploid (2n = 30) and showed significant variations in agronomical characters like plant height, leaf length, number of branches per plant, width of recepticle and number of seeds per capitulum in both first and second generations (Table 17). Dwarfs, large flower head types and self-compatible plants recovered in this study are extremely useful for the improvement of this crop plant (Sarvesh *et al.*, 1994).

8. RESPONSE OF DIFFERENT EXPLANTS TO GENETIC TRANSFORMATION

Genetic transformation is one of the emerging areas whereby specific traits can be added with minimal alteration of the target plant genome. Though many methods are currently in use, *Agrobacterium*-mediated genetic transformation is highly efficient. In niger, protocols developed for somatic embryogenesis and organogenesis have been utilized for transformation experiments. Zygotic embryos, cotyledons and immature leaf segments were co-cultivated with *Agrobacterium tumefaciens* for 8 to 10 days. The explants were later transferred onto the selective medium containing auxins and/or cytokinins along with 500 mg/l cefotaxime and 100 mg/l kanamycin. Somatic embryo formation (5 embryos per explant) was observed from mature embryos after 4 weeks of culture on LS medium fortified with 1 mg/l 2,4-D (Fig.1c). But no response was noticed on other media tested (Table 18). Plantlets regenerated from various explants after co-cultivation with *A. tumefaciens* exhibited resistance to 100 mg/l kanamycin (Fig.1e). Infected leaf segments showed higher shoot bud formation when compared to cotyledons. Further growth of somatic embryos and shoot buds into whole plants from explants was noticed on LS medium devoid of growth regulators. The roots produced in the basal medium were slender and weak. However, transferring the shoots on to LS medium supplemented with 2 mg/l IBA, 0.05 mg/l KN and 6% sucrose resulted in the regeneration of stout and healthy roots. Activity of the β-glucuronidase (GUS) in niger transformants was determined by histochemical assay after co-cultivation with *A. tumefaciens* (strain LBA4404) containing the plasmid pBI121. Two to four-leafed transgenic plants of the same

Table 17. Androclonal variation for some plant traits of Ootacamund in the second generation[#]

Genotype	Plant height (cm)	Leaf length	Leaf width	No. of branches/ plant	Width of recepticle (mm)	Depth of recepticle (mm)	No. of seeds/ capitulum
Seed raised plants★★	165.4 (± 8.0)	10.6 (± 0.6)	2.2 (± 0.2)	13.9 (± 2.0)	8.9 (± 1.2)	7.8 (± 1.0)	43.1 (± 2.3)
AC1	50.3★ (± 3.1)	12.1★ (± 0.3)	1.2★ (± 0.2)	10.8 (± 0.9)	8.2 (± 0.3)	7.4 (± 0.5)	40.4 (± 1.5)
AC2	112.3★ (± 4.9)	12.0★ (± 0.5)	1.6 (± 0.2)	9.8★ (± 0.7)	7.3 (± 0.5)	6.8 (± 0.7)	29.5★ (± 2.5)
AC3	172.2 (± 2.3)	10.7 (± 0.4)	1.2★ (± 0.2)	8.6★ (± 1.0)	8.4 (± 0.5)	8.0 (± 0.5)	34.0★ (± 3.1)
AC4	124.4 (± 2.6)	18.6★ (± 0.5)	2.2 (± 0.4)	16.0 (± 1.2)	14.5★ (± 1.2)	13.5★ (± 1.0)	44.5 (± 3.8)
AC5	139.5★ (± 4.0)	11.3 (± 0.5)	2.3 (± 0.3)	20.8★ (± 0.3)	9.8 (± 0.9)	8.5 (± 0.6)	42.0 (± 1.6)
AC6	39.0 (± 1.9)	9.8 (± 0.2)	2.3 (± 0.3)	10.3★ (± 1.0)	9.3 (± 0.9)	7.8 (± 0.9)	28.5★ (± 2.2)
AC7	122.3★ (± 2.6)	13.6★ (± 0.4)	1.3★ (± 0.3)	14.0 (± 0.4)	17.6★ (± 0.6)	16.0★ (± 1.0)	52.0★ (± 1.4)
AC8	111.8★ (± 3.0)	10.0 (± 0.3)	1.5★ (± 0.3)	10.3★ (± 1.3)	8.8 (± 0.5)	7.3 (± 0.5)	39.3 (± 1.9)

[#]Data represent an average of 30 replicates for all the traits shown above.
★significant at $p = 0.05$. Other values are not significantly different.
★★Ten seed raised plants were scored for control. Non-nariant tissue culture derived plants displayed traits like that of seed raised Ootacamund plants (data not shown).

Figure 1 : (a) Embryogenesis from mature zygotic embryo of niger cultured on Linsmear and Skoog's medium supplemented with 1 mg/l 2,4-D and 0.2 mg/l KN (b) Regeneration from somatic embryos of zygotic embryo on LS medium enriched with 1 mg/l 2,4-D and 0.2 mg/l KN (c) Shoot buds being regenerated from *Agrobacterium* infected cotyledons. (d) Infected leaf segments showing shoot-bud induction on LS medium enriched with 2 mg/l BAP and 0.5 mg/l NAA and 100 mg/l kanamycin (blue colour with X-gluc) (e) Regeneration of transgenic plants on LS medium supplemented with 1 mg/l 2,4-D and 100 mg/l kanamycin.

Table 18. Callusing and morphogenetic response of different explants of niger after *Agrobacterium* infection*

LS medium + hormones (mg/l)	% Frequency of explants callusing	% Frequency of explants with shoot buds	Mean no. of shoot buds/ explant	% Frequency of explants with somatic embryos	Mean no. of somatic embryos/ explant
Embryo					
2,4-D (1)	32	-	-	11	5
2,4-D (1) + KN(0.1)	21	-	-	-	-
2,4-D (1) + KN(0.2)	23	-	-	-	-
NAA(0.5) + BAP(2)	17	-	-	-	-
Cotyledon					
2,4-D (1)	-	-	-	-	-
2,4-D (1) + KN(0.1)	-	-	-	-	-
2,4-D (1) + KN(0.2)	-	-	-	-	-
NAA(0.5) + BAP(2)	20	14	1.7	-	-
Leaf					
2,4-D (1)	13	-	-	-	-
2,4-D (1) + KN(0.1)	42	-	-	-	-
2,4-D (1) + KN(0.2)	42	-	-	-	-
NAA(0.5) + BAP(2)	47	20	2.5	-	-

*A total of 52 explants were inoculated per treatment.

age group (35-40 days) were selected for GUS assay. High frequency of GUS expression was detected in the plants regenerated from mature embryos followed by leaf segment and cotyledon derived plants. After co-cultivation, mature zygotic embryos produced maximum number of somatic embryos per explant, followed by leaf and cotyledon. Within 4 weeks, 80% of embryos germinated into plantlets (Table 18). Blue colouration of transformed shoot buds indicated successful integration of the *gus* gene into the host genome and its expression (Fig.1d).

In our transformation experiments, co-cultivation for longer time generally produced higher frequency of transformed tissues. *Agrobacterium*-mediated transformation is advantageous because of prolonged co-culture (beyond two days) without any damage to the tissues. Similar results were obtained by Settu (1999) with *Glycine max* (L.) Merr. The transformed explants showed delayed regeneration response presumably because of the presence of kanamycin and cefotaxime in the regeneration medium (Kar *et al.*, 1996).

Activity of GUS as indicated by the blue colouration of cells/tissues with x-gluc indicates the primary evidence for transformation. Induction of somatic embryos is advantageous over organogenesis in genetic transformation as the frequency of somaclonal variations will be less. The other advantage of this system is the induction of fully developed or mature somatic embryos on the primary medium itself.

Table 19. Frequency of genetic transformation in the variety Ootacamund

LS + hormonal concentration (mg/l)	No. of kanamycin resistant somatic embryos/ shoot buds	Total number of plants regenerated	No. of plants exhibiting GUS activity	% Frequency of trans-formants
Embryo				
2,4-D(1)	40	32	28	87.5
Cotyledon				
BAP(2) + NAA(0.5)	25	20	14	70.0
Leaf				
BAP(2) + NAA(0.5)	25	17	15	88.2

The per cent frequency of genetic transformation from different explants of the variety Ootacamund is shown in the Table 19. GUS activity was observed in 88.2 per cent of plants derived from leaf explants, while it was 70% in cotyledon derived plants. The genetic transformation protocols optimized in the present study facilitate the transfer of beneficial alien genes into niger across sexual barriers. Further studies are in progress to confirm the transgenic nature of the regenerated plants by Southern blotting technique.

9. CONCLUSIONS AND FUTURE PROSPECTS

Recombinant DNA technology of niger is at the nascent stage. However, in view of its amenability for *in vitro* manipulations and the ease with which the somatic embryogenesis, plant regeneration, *in vitro* flowering and genetic transformation can be achieved, it can be developed as a model system for basic molecular biological research on par with *Brassica*. Besides, the transgenics capable of high yield potential with reduced incompatibility possessing high levels of resistance to biotic and abiotic factors can be evolved. The qualitative and quantitative profiles of seed oil can be altered through metabolic engineering to augment the supply of much needed quality oil for human as well as for industrial consumption. There is an immense potential to develop this plant into a model oil- yielding crop.

ACKNOWLEDGEMENTS

Dr. A. Sarvesh is grateful to the UGC, New Delhi, and Dr. Amita S. Katta is thankful to the CSIR, New Delhi, for financial support during the investigation.

REFERENCES

Arnold SV and Hakman I (1988) Regulation of somatic embryo development in *Picea abies* by abscisic acid. *J. Plant Physiol.,* **132** : 164-169.

Bir Bahadur, Malathi N, Reddy KRK and Rao GP (1992) High frequency plant regeneration from embryo cultures of niger (*Guizotia abyssinica* Cass). *Adv. Plant Sci.,* **5** : 216-220.

Bajaj YPS, Ram AK, Labana KS and Singh H (1981) Regeneration of genetically variable plants from anthers derived callus of *Arachis hypogaea* and *Arachis villosa*. *Plant Sci. Lett.,* **23** : 35-39.

Bohorova NE, Cocking EC and Power JB (1986) Isolation, culture and callus regeneration of protoplasts of wild and cultivated *Helianthus* species. *Plant Cell Rep.,* **5** : 256-258.

Chaleff RS and Stolarz A (1981) Factors influencing the frequency of callus formation among cultured rice (*Oryza sativa*) anthers. *Physiol. Plant.,* **51** : 201-206.

Finer JJ (1987) Direct somatic embryogenesis and plant regeneration from immature embryos of hybrid sunflower (*Helianthus annuus* L.) on a high sucrose containing medium. *Plant Cell Rep.,* **8** : 372-374.

Foroughi-Wehr B and Wenzel G (1989) Androgenic haploid production. *IAPTC Newsletter No.,* **58** : 11-18.

Gamborg OL, Miller RA and Ojima K (1968) Nutrient requirements of suspension cultures of soybean root cells. *Exp. Cell Res.,* **50** : 151-158.

Greco B, Tanzarella OA, Carrozzo G and Blanco A (1984) Callus induction and shoot regeneration in sunflower (*Helianthus annuus* L.) *Plant Sci.,* **36** : 73-77.

Heinz DJ and Mee GWP (1971) Morphogenic, cytogenic and enzymatic variation in *Saccharum* species hybrid clones derived from callus tissues. *Amer. J. Bot.,* **58** : 257-262.

Huang CR, Wu YH and Chen CC (1986) Effects of plant growth substances on callus formation and plant regeneration in anther culture of rice. *In : Rice Genetics,* International Rice Research Institute, Manila, Phillippines. pp 763-771.

Jadimath VG, Murthy HN, Pyati AN, Ashok Kumar HG and Ravishankar BV (1998) Plant regeneration from leaf cultures of *Guizotia abyssinica* (niger) and *Guizotia scabra. Phytomorphology,* **48** : 131-135.

Joshi AV (1990) Present production, research and future strategy for niger in Maharashtra. In: *Oil Crops : Proceedings of the Three Meetings Held at Pantnagar and Hyderabad,* (Ed. Abbas Omran) International Development Research Centre, Ottawa, pp 171-175.

Kamada H and Harada H (1979) Studies on organogenesis in carrot tissue cultures. I. Effect of growth regulators on somatic embryogenesis and root formation. *Z. Pfanzenphysiol.,* **91** : 255-266.

Kar S, Johnson TM, Nayak P and Sen SK (1996) Efficient transgenic plant regeneration through *Agrobacterium* mediated transformation of chickpea (*Cicer arietinum* L.). *Plant Cell Rep.,* **16** : 32-37.

Keller WA, Rajhathy and Lapra J (1975) *In vitro* production of plants from pollen in *Brassica campestris. Can J. Genet. Cytol.,* **17** : 655-666.

Linsmaier EM and Skoog F (1965) Organic growth factor requirements of tobacco tissue cultures. *Physiol. Plant.,* **18** : 100-127.

Malathi N, Reddy KRK, Rao GP and Bir Bahadur (1989) Growth and development of niger (*Guizotia abyssinica* L.F.Cass) shoot apices in *in vitro. Adv. Plant Sci.,* **2** : 294-296.

Mehta UJ, Hazra S and Mascarenhas AF (1993) Somatic embryogenesis and *in vitro* flowering in *Brassica nigra. In Vitro Cell Dev. Biol.,* **29** : 1-4.

Murashige T and Skoog F (1962) A revised medium for rapid growth and bioassays with tobacco tissue cultures. *Physiol. Plant.,* **15** : 473-497.

Narasimhulu SB and Reddy GM (1984) *In vitro* flowering in groundnut (*Arachis hypogaea L.*). *Theor. Appl. Genet.,* **69** : 87-90.

Nikam TD and Shitole MG (1993) Regeneration of niger (*Guizotia abyssinica* Cass.) cv. Sahyadri from seedling explants. *Plant Cell Tiss. Org. Cult.,* **32** : 345-349.

Ozias-Atkins P (1989) Plant regeneration from immature embryos of peanut. *Plant Cell Rep.,* **8** : 217-218.

Rajender Rao K and Vaidyanath K (2000) Biotechnology of sesame-problems and prospects, In: *Plant Biotechnology-Recent Advances,* (Ed. Trivedi PC) Panima Publishing Corporation, New Delhi, Bangalore, pp. 106-120.

Reinert J (1973) Aspects of organization: organogenesis and embryogenesis. In: *Plant Tissue and Cell Culture,* (Ed. Street HE) Blackwell Scientific Publications, London, pp 338-355.

Riley KW and Belayneh P (1989) Niger seed (*Guizotia abyssinica* Cass) In: *Oil Crops of the World: Their Breeding and Utilization,* (Eds. Grobbelen G, Downy RK and Ashri A) MacMillan Publishing Co., New York, pp. 394-403.

Rout JR (1989) Anther culture studies in *Oryza sativa* Linn. and interspecific hybrid *O. sativa* L. X *O. rufipogan* Griff. *Ph.D. Thesis,* Utkal University, Bhubeneswar, India.

Sarvesh A (1990) Genetic variability for callus induction, plantlet regeneration and stress among varieties of niger (*Guizotia abyssinica* (L.f) Cass). *Ph.D. Thesis* Osmania University, Hyderabad.

Sarvesh A, Reddy TP and Kavi Kishor PB (1993) Embryogenesis and organogenesis in cultured anthers of an oil yielding crop niger (*Guizotia abyssinica.* Cass) *Plant Cell Tiss. Org. Cult.,* **35** : 75-80.

Sarvesh A, Reddy TP and Kavi Kishor (1994a) Somatic embryogenesis and organogenesis in *Guizotia abyssinica. In Vitro Cell. Dev. Biol.,* **30** : 104-107.

Sarvesh A, Reddy TP and Kavi Kishor PB (1994b) Androclonal variation in niger (*Guizotia abyssinica* Cass). *Euphytica,* **79** : 59-64.

Sarvesh A, Reddy TP and Kavi Kishor PB (1994c) Plant regeneration from cotyledons of niger. *Plant Cell Tiss. Org. Cult.,* **32** : 131-135.

Sarvesh A, Reddy TP and Kavi Kishor PB (1996) *In vitro* anther culture and flowering in *Guizotia abyssinica* Cass. *Indian J. Exp. Biol.,* **34** : 565-568.

Seegler CJP (1983) *Oil Plants in Ethiopia, their Taxonomy and Agricultural Significance.* Centre for Agricultural Publishing and Documentation, PUDOC, Wageningen, Netherlands.

Sharma SM (1990) Niger seed in India: Present status of cultivation, research achievements and strategies. In: *Oil crops: Proceedings of the Three Meetings held at Pantnagar and Hyderabad,* (Ed. Abbas Omran) International Development Research Centre, Ottawa, Canada, pp. 159-165.

Settu A (1999) Organogenesis, embryogenesis, salt tolerance and *Agrobacterium* mediated transformation studies in soybean (*Glycine max* (L.) Merr. *Ph.D. Thesis*, Bharathidasan University, Tiruchirapalli, India.

Sreeramulu K (1982) Genetic instability at the S-locus of *Lycopersicon peruvianum* plants regenerated from *in vitro* culture of anthers. Generation of new S-specificities and S-allele resources. *Heredity,* **49** : 319-330.

Stuart DA and McCall CM (1992) Induction of somatic embryogenesis using chain and ring modified forms of phenoxy acid growth regulators. *Plant Physiol.,* **99** : 111-118.

Tejovathi G (1986) Studies on morphogenesis and *in vitro* flowering in *Carthamus tinctorious L. Ph.D. Thesis,* submitted to the Osmania University, Hyderabad, India.

Thorpe TA and Meier DD (1975) Effect of gibberellic acid on starch metabolism in tobacco callus cultured under shoot forming conditions. *Phytomorphology,* **25** : 238-245.

Vasil V and Vasil IK (1981) Somatic embryogenesis and plant regeneration from suspension cultures of pearl millet (*Pennisetum purpureum* schum). *Z. Pflanzenphysiol,* **111** : 233-239.

Venkatachalam P, Kavi Kishor PB, Geetha N, Thangavelu M and Jayabalan N (1999) A rapid protocol for somatic embryogenesis from immature leaflets of groundnut (*Arachis hypogaea* L.) *In Vitro Cell. Dev. Biol.,* **35** : 409-412 (1999).

Venkatesham G and Reddy TP (1996) Plant regeneration from immature zygotic embryo cultures of niger (*Guizotia abyssinica* Cass) *Adv. Plant Sci.,* **9(1)** : 15-20.

Wang CC, Sun CS, Chu CC and Wu SC (1978) Studies on the albino pollen plantlets of rice. In : *Proc. Plant Tissue Culture, Science Press, Peking, China,* pp 149-160.

Wochok ZS and Wetherell DF (1971) Suppression of organized growth in cultured wild carrot tissue by 2-chloroethylphosphonic acid. *Plant Cell Physiol.,* **12** : 771-774.

Chapter 8

BIOTECHNOLOGY OF SESAME

K Rajender Rao★[1] **and K Vaidyanath**

Department of Genetics, Osmania University, Hyderabad - 500 007, India

Summary

Sesame (Sesamum indicum) is an ancient oil seed crop and grown in India from antiquity. The primary aim of sesame breeding is to improve yield and oil contents. Both biotic and abiotic stress factors are hampering the realization of the full yield potential of sesame. Plant breeding programes include genetic variation, selection, mating and population propagation, whereas, molecular techniques ranging from cell and tissue culture, vegetative propagation, genetic mapping and gene transfer. The integration of molecular and classical genetics have paved the way for modern crop improvement programes. Molecular techniques are unlikely to replace classical breeding but instead will enhance the capabilities of breeders to solve practical problems.

Keywords : Embryo rescue, genetic transformation, haploids, protoplast culture, *Sesamum indicum,* somatic embyogenesis.

Abbrevations : BAP – benzylamino purine, 2,4-D-2,4-dichloro-phinoxy acetic acid, GA_3 – gibberellic acid, KN – kinetin, NAA – naphalene acetic acid.

---

[1]Present address : Institute National de la Research Agronomique (INRA), Research Unit on Breeding, Genetics and Physiology of Forest Trees, Avenue de la Pomme De Pin : BP 20619, ARDON – 45166 OLIVET, CEDEX, France.

★Corresponding author : E-mail : rkrajender@rediffmail.com

1. INTRODUCTION

Sesame (*Sesamum indicum* L.) is an ancient oil seed crop generally known as til or gingelly cultivated in India and many other parts of world, mainly for edible oil and for culinary purposes. The oil is nearly colourless, odourless and has long shelf life and does not become rancid. The oil is also used in Ayurvedic system of medicine. The oil cake is rich in protein and used as cattle feed. The yield potential of this crop is very low when compared to major oil seed crops. Some of the factors that limit the productivity of sesame are : early senescence, extreme susceptibility to biotic (mainly phyllody) and abiotic stress factors and photo-sensitivity.

1.1. Origin and history

Sesame grown in India from antiquity . The centre of origin is often contested. Many authors (Bedigian and Harlon, 1986; Mazzani, 1983; Brar and Ahuja, 1979; Nayar and Mehra, 1970 and Joshi, 1961) based on archeological evidence opined that sesame was originated 400 years ago in India. Recently, Bedigian (1984) reviewed various types of evidences and concluded that sesame originated in India. It was an important crop in the Persian region during 2130 – 2000 BC, the medicinal papyrus of Theophrastus (370 – 286 B.C.) in his Historia Plantarum, in Autharvana veda, the Aurthasastra of Koutilya (3000 B.C.) and Vishnupurana also mentioned about the sesame and its uses.

1.2. Genetics and breeding

The primary aim of sesame breeding is to improve yield and oil contents, especially to meet the demands of sesame oil and the country economy in terms of precious foreign exchange (Reddy, 1995). Generally, the oil content of sesame ranges from 45 – 55 % (Lee and Choi, 1985) and has 25% protein. Improvement of any crop plant needs a through knowledge of gene pool to provide genetic focus especially in transferring useful genes from wild to cultigens. Sesame belongs to Pedaliaceae family and the genus *Sesamum* has 37 species of which *S. indicum* (2n=26) is a dominant cultivated species. The genetic improvement of sesame is in its infancy stage. Biotic and abiotic stresses hampering the realization of the full yield potential of sesame. Wild species of sesame (Table 1) possess genes resistance to biotic and abiotic stresses (Joshi, 1961; Weiss, 1971; Brar and Ahuja, 1979 and Kolte, 1985). The Food and Agricultural Organisation (FAO) of United Nations (Anonymous, 1981 and 1985) has recommended the collection, conservation and

Table 1. Wild species of sesame and useful genes they contain

Group	Species	Chromosome Number (2n)	Special desirable Traits
I	*S. indicum*	26	Cultivated species
	S. malabaricum	26	Resistance to water logging, abundance of flowering
	S. alatum Thonn	26	Drought tolerance, resistant to antigastra pest
	S. capense	26	-
	S. schenkii	26	-
	S. grandiflorum	26	-
II	*S. lanciniatum* Klein	32	Tolerance to drought and Phyllody, resistance to Antigastra
	S. prostratum Retz	32	Tolerance to phyllody, Antigastra drought and salinity
	S. angolense	32	-
III	*S. occidentale* (Heer & Regel)	64	Thermosensitive, tolerant to srought and water logging
	S. radiatu (Schum & Thom)	64	Tolerant to drought, water logging and sensitive to temperature fluctuations

maintenance of the wild species of sesame for enrichment of sesame biodiversity.

This is a self pollinated oil seed crop. The interspecific crosses between *S. indicum* and its wild relatives are limited due to post fertilization barriers mainly embryo abortion. To over come the problems faced by conventional breeding methods to manipulate the architecture of sesame the molecular techniques seems to be ideal.

2. BIOTECHNOLOGY OF SESAME

Advances in plant biotechnology offer immense possibilities for solving a number of basic problems which might help in the crop improvement. The principle component of plant molecular biology is to manipulate the genes by applying the methods of DNA manipulation leading to transgenic varieties capable of stabilizing the yield through increased resistance to biotic and abiotic factors, besides changing the quality traits. There is a paucity of information on the biotechnology of sesame. The major biotechnological strategies for the improvement of sesame are :

(1) Standardization of *in vitro* plant regeneration techniques

(2) Molecular manipulation for improving yield and oil levels

(3) The production of engineered oil and fatty acids.

The current phase of biotechnology focuses around the *in vitro* regeneration of plants, embryo rescue and wide hybridization and genetic engineering.

2.1. Shoot organogenesis

Plant tissue culture and plant regeneration technology has been available to the plant breeders since middle of the twentieth century and has been extensively employed for improvement of several oil seed crops. Somaclonal variation often occurs sponteneously in tissue system, sometimes yielding plants with considerable variation phenotypically and/ or genotypically. If useful genetic change occurs, variants can be added to breeding programmes. However, in view of the highly recalcitrant nature of sesame, very little information is available. The primary step for genetic manipulation of any crop is an efficient protocol for establishment of callus/cell suspension cultures, morphogenesis and plant regeneration, either directly or through somatic embryogenesis. Lee *et al.* (1985) is the first to study *in vitro* culture of sesame. George *et al.* (1987) have established tissue culture from different explants of *Sesamum indicum* L. cv. JLT - 26. The effect of phytohormones and explants on callus induction was reported by Kim *et al.* (1987). Chae and Park (1987) have established herbicide tolerant cell lines but failed to obtain whole plant regeneration. Lee *et al.* (1988) and Kim and Byeon (1991) have investigated the effect of growth regulators on callus induction and organogenesis from different explants of *S. indicum*. Rajender Rao and Vaidyanath (1997) have reported the callus induction potential from various explants of ten different cultivars of *S. indicum* and its two wild species on five different media with various phytohormone combinations. Among the media, genotypes and explants used, MS medium, genotype Rajeswari and hypocotyl explant were found to be the most responsive. The shoot formation with a low frequency (2 - 3%) was obtained from the callus culture on half-strength MS basal medium containing NAA +

BAP + KN after 8 - 10 weeks of culture. The shoots were rooted on MS /L6 medium fortified with NAA/IBA (0.5 mg/l). As there are a few reports on regeneration of sesame via callus phase, it is desirable that more intensive work has to be taken up on this aspect (Rajender Rao and Vaidyanath, 1999).

2.2. Multiple shoot induction

In vitro propagation by meristem culture is a standard prevelent practice in high premium species requiring uniformity. Major attraction of this approach is that it ensures rapid and large scale multiplication, free from pathogens either of bacterial or viral origin. The fact that pathogens are eliminated by culturing the apical meristem and the resultant plant depicts improved vigor and quality, the same can be put to use for the improvement of sesame. The major barriers for sesame crop improvement is phyllody disease caused by mycoplasma and majority of released varieties are prone to this disease, virtually there is no resistance. Besides, several varietal crosses depict appropriate quality of heterosis, which can be tapped to develop hybrid varieties (Rajender Rao and Vaidyanath, 1999). There is a distinct possibility of developing phyllody disease free lines and the propagation and maintenance of valuable genetic strains by the meristem culture and micropropagation methods. There is a paucity of information on micropropagation and the work done in this direction in sesame is presented in Table 2.

Lee *et al.* (1985) developed a method for propagation of *Sesamum indicum* through shoot tip culture on MS medium fortified with NAA +

Table 2. Multiple shoot induction and micropropagation in sesame

Explant	Media and hormones (mg/l)	Result	Reference
Shoot tip	MS + KN (2.0)	Shoot differentiation	Lee *et al.*, 1985
Shoot tip	MS + BA (8.0)	Shoot multiplication	George *et al.*, 1989
Shoot tip	MS/LS/B5/L6 + BA/GA(8.0)	Shoot induction	Rajender Rao and Vaidyanath, 1987
Axillary shoot bud	MS + BA (8.0) + NAA (0.5)	Micropro-pagation	Gangopadhyay *et al.*, 1998

IAA and KN. George *et al.* (1989) have studied the effect of various cytokinins on shoot bud formation from shoot tips when excised with cotyledons and cultured on MS medium containing cytokinin. Rajender Rao and Vaidyanath (1997) have studied the induction of multiple shoots in ten cultivars of *S. indicum* (2n = 26) and its wild relative *S. Occidentale* (2n = 64). The sesame seeds pre- soaked in the aquous solution of BAP or GA_3 (8.0 mg/l) and seedlings were raised on four types of media (MS, LS, B5, L6) for 8 - 10 days, the shoot apical meristems were excised and cultured on all the media. Results were statistically analysed and obtained the significant results (Table 3).

Selection can be performed in cell and tissue cultures, if the desired trait is expressed *in vitro* and out comes are directly correlated with field performance. Somaclonal variation represents potentially a useful

Table 3. ANOVA for multiple shoot induction in sesame

Source of variation	d .f	MSS
Anova between Media and Variety		
Meida	3	26.23*
Variety	9	41.46*
Interaction	27	7.23
Error	40	6.57
Anova between Variety and Treatment		
Variety	9	41.45*
Treatment	1	20.20*
Interaction	9	52.46*
Error	60	0.74
Anova between Media and Treatment		
Meida	3	26.23*
Treatment	1	20.00*
Interaction	3	6.63
Error	72	10.90

*Significant at 5% level

technique for increasing variation. By actual mutation, gene and chromosome rearrangements or activation of repressed genes, variation has been identified under a variety of cell and tissue culture conditions. Somaclonal variation is viewed as essentially random, thus it may not occur reliably for traits of interest. Useful genes may be linked to undesirable traits and epigenetic may accrue. Eliminating such unwanted effects could require generations of selection and breeding. In addition, considerable time and expense may be required to confirm utility, stability and heritability (Stack, 1987). Expressing cells and/or tissues from desirable individuals to directed and stringent *in vitro* selection to induce variation is potentially more efficient than large and lengthy field trials.

2.3. Protoplast culture

The *in vitro* technology of protoplast culture and fusion can provide a mean to enhance genetic variability by creating somatic hybrids among sexually incompatible species and combining nuclei and cytoplasmic organelles in crop plants. Various symmetrical and asymmetrical combinations can be produced that may have valve to breeders and geneticists. The great impetus in this area is due to the possibilities for parasexual manipulation and also genetic transformation by various methods. This approach is of particular help in transferring phyllody disease resistance from wild to cultivated species in sesame. With the advant of this technique (Carlson *et al.*, 1992) several oil seed crops have been tried for the protoplast culture (Bohorova *et al.*, 1986). In sesame there are very few reports in this area of research. Bapat *et al.* (1989) isolated protoplasts and cultured them for callus induction. A high yield of viable protoplasts was obtained from hypocotyl explants of sesame (Murta and Batra, 1990). There is no report on reliable plant regeneration from sesame protoplast so for. Much more intensive studies have to be taken up involving wild species.

2.4. Haploids

The production of haploids by culturing anthers or ovules from elite individuals *in vitro* has generated a considerable interest among geneticists and plant breeders. The breeders have access to many genes in manageble number of genotypes. Creation of haploid plants with one copy of gene at a locus will be ideal for the study of gene expression, besides providing an opportunity for direct selection of desirable traits, which may hasten the inbreed line generation for hybrid seed production. There is an immense importance for the double haploid populations in genetic linkage mapping programmes. Lee *et al.* (1988) and Ryu *et al.* (1993) established callus growth from anthers of sesame on MS medium and studied the effect of cold pretreatment. Ranaweera and Pathirana (1992) and Ryu *et al.* (1992) have reported the *in vitro* anther culture and callus growth on MS medium supplemented with various concentrations and combination of phytohormones. Rajender Rao and Vaidyanath (1999) have studied the anther culture *in vitro* using ten different *S. indicum* cultivars and its wild relative *S. occidentale* on four type of media supplimented with various auxins and cytokinins either individually or in combination. In the case of *in vitro* ovule culture studies in sesame, Alarmelu *et al.* (1992) have reported the callus growth from ovule culture on MS medium containing 2,4-D (2.5 mg/l) + BAP (5.0 mg/l) and glutamine (5-250 mg/l).

2.5. Embryo rescue (wide hybridisation)

Interspecific hybridization in sesame is an important technique mainly to over come the phyllody disease problem which hampers the production levels. Wild relatives of sesame often possess genes for economic traits like quality, resistance to biotic and abiotic factors and male sterility (Table 1). The introgression of useful genes in to cultigens through wide hybridization often do not succeed in all crops, may be due to post fertilization barriers. Some of them may be due to incompatibility between the embryo and endosperm during embryonic growth. *In vitro* culture methods to rescue embryo and production of interspecific/inter generic hybrid plants on a defined media is an ideal approach for sesame crop improvement. One can find paucity of information on embryo culture in sesame. Alarmelu *et al.* (1992) made crosses between diploid *S. indicum* and its wild relatives, diploid *S. alatum* and tetraploid *S. occidentale,* but failed to obtain the hybrid sesame seed due to embryo abortion. They established embryo rescue technique to get the intraspecific hybrid plant. There is a necessity to take up intensive studies on development of this technique .

2.6. Somatic embryogenesis

An efficient regeneration system with cell and tissue cultures is a pre-requisite for the application of biotechnology to plant improvement programmes. Somatic embryogenesis is one of the most powerful morphogenetic schemes among the *in vitro* plant regeneration techniques. Ever since the discovery of Steward *et al.* (1958) that the callus and cell suspension cultures of carrot differentiate and grow into embryo like structures, called as somatic embryos and was later defined by Haccius (1978) as a non - sexual developmental process leading to the differentiation of zygotic embryo like structures from somatic cells. There has been a spate of research in this area and more than 300 species of both mono and dicotyledons have been studied for induction of somatic embryogenesis (Narayanswamy, 1997). However, the regeneration of plant is limited to only 51 species (Rangaswamy, 1986). The advantages of somatic embryos are: 1) Genetically identified propagules provided by avoiding genetic recombination that occurs during meiosis in sexual reproduction. 2) It provides a fast reliable, reproducible method for the mass production. 3) This can be employed for the transfer of genes from secondary and tertiary gene pools through genetic transformation bypassing reproductive barriers. In contrast to organogenesis, somatic embryos exhibit a bipolar structure with a closed vascular system and differentiated shoot and root systems.

Somatic embryogenesis can be initiated either directly without going through the callus phase, i.e. from pre-determined embryonic cells and differentiation into embryonic cells within callus tissue. In sesame, there is a paucity of information of somatic embryogenesis when compared to other oil seed crops. Mary and Jayabalan (1998) have reported the influence of growth regulators on somatic embryogenesis from hypocotyl callus on MS medium. Rajender Rao and Vaidyanath (1998) have standardized a method for obtaining somatic embryos directly from different explants and/or from their calluses. The callus induced on MS medium supplemented with 2,4-D (1-3 mg/l) when transfered on to half-strength MS basal medium fortified with 2,4-D (0.2) + BAP (0.5) and KN (1.0) mg/l resulted in somatic embryogenesis. These embryos were dessicated, matured and regenerated on MS medium. There is no report on reliable plant regeneration from somatic embryos so for .

2.7. Synthetic seed

The concept of synthetic seed or artificial seed (syn seed) in crop improvement was developed by the encapsulation of *in vitro* generated propagating material, *viz.*, somatic embryos, axillary buds, shoot apices etc., in calcium alginate gel, which facilitates easy transportation and can be delivered directly into the soil or by fluid drilling, offers a great promise for the use in cloning of elite germplasm and propagation of transgenic plants. Since the first report of synseed in carrot (Kitto and Janik, 1985) several laboratories reported the production of synseed in various crop species (Redenbaugh, 1983; Fuji *et al.*, 1987). There are no published reports on synseed in sesame. Rajender Rao and Vaidyanath (1998) have demonstrated for the first time such possibilities in sesame. The seed pre treated with BAP were germinated on MS medium containing BAP for the induction of shoot buds. These buds were encapsulated in calcium alginate solution supplemented with growth regulators and were subsequently germinated on MS medium after several days of storage.

2.8. Genetic transformation

Genetic transformation or non sexual addition of specific genes, permits aquization of traits which otherwise are impossible by conventional breeding. This technique can also be used to incorporate additional copies of indigenous genes that are closely linked to undesirable genes circumventing the need for generations of back crossing. Bypassing the sexual process offers time and cost difficulties posed by long generation intervals. Transformation also adapted to reduce expression of existing genes that encode undesirable traits. This antisense technology involves inserting a reverse copy of the gene in question. Successful development of antisense technology will require a better understanding of the extent, stability and understanding mechanisms of suppressed expression. This novel concept of rDNA technology is unique in that it enables the improvement of crop plants in desired direction by altering the properties. The gene transfer techniques and gene manipulation can gain control on altering the quantitative improvement of oil and protein (Downey and Keller, 1993). However, basic to this new technology is the tissue culture much is to be done in developing the biotechnology for sesame. Ahn (1993) and Ogaswara *et al.* (1993) have established hairy root cultures with *Agrobacterium rhizogenes*. A high yield of quinones were produced from hairy root cultures of sesame. These compounds were found to inhibit skin cancer in mice and also have protective activity against the ultra voilet light in inducing peroxidation in membrane lipids (Mimura *et al.*, 1994). These studies indicate that the introduction of align gene through *Agrobacterium* mediated genetic tranformation can be gainfully employed for the improvement of oil seed crop sesame for various agronomical traits.

Introduction of desirable gene into a host organism is a big task, whereas ensuring proper expression is a major technical challenge for biotechnologists/genetic engineers.

3. CONCLUSIONS AND FUTURE PROSPECTS

Rapid technological evolution in plant biotechnology field holds promise to the improvement of oil seed crop, sesame. Conventional breeding methods have not yielded any significant breakthrough in pushing yield and oil percent levels. The integration of classical and molecular techniques holds a promising future for sesame improvement. Tissue culture provides technique for genetic manipulations and offer an effective avenue for developing improved characters in plants. Somatic embryogenesis can be induced directly from explants or their calluses. It has provided opportunities for transferring genes in sesame. The *in vitro* technology further provides an opportunity for transferring wild genes containing the resistance to biotic and abiotic stress factors prevelent in secondary and tertiary gene pools via somatic hybridisation and embryo rescue techniques. As the recent emphasis is on the exploitation of hybrid vigor by producing hybrid varieties, there is an urgent need to develop protocols for obtaining haploids through anther culture.

With proper foresight and balanced support, useful molecular techniques can be devised and drawn in to classical breeding just as other disciplines (*e.g.*, quantitative genetics) were incorporated earlier. Wise investiments coupled with collaboration within and among organisations are necessary to foster productive integration of genetic technologies. The integration of classical genetics and molecular technology is sure to revolutionize the sesame production and oil quality and will help in augumenting world's demand for oil and its products.

REFERENCES

Ahn CS (1993) Establishment of hairy root lines of several economic plant species by inoculum with *Agrobacterium rhizogenes. Korean J. Breed.,* **25** : 235 - 247.

Alarmelu S, Sethupathi R, Sree Rangasamy SR and Narayanan A (1992) *In vitro* callus studies in sesame (*Sesamum indicum*). *Oil Crops News letter,* **9** : 34 - 37.

Anonymous (1981) Descriptors for sesame. AGP : IBPGR/80/71. Inter. Bd. Plant Genet. Res, Secretariat, Rome.

Anonymous (1985) Conclusions and recommendations. In : *Sesame and Safflower : Status and potentials* (Ed Ashri A), FAO plant production and protection paper 66, Rome, pp. 218 – 220.

Bapat VA, George L and Rao PS (1989) Isolation, culture and callus formation of sesame (*Sesamum indicum* L. PT) protoplasts. *Indian J. Exp. Biol,* **27** : 182 - 184.

Bedigian D (1984) *Sesamum indicum* L. : Crop origin, diversity, chemistry and ethnobotany. *Ph.D. Thesis,* University of Illinois, Urbanachampaign, 111.

Bedigian D and Harlan JR (1986) Evidence for cultivation of sesame in the ancient world. *Econ. Bot.,* **40** : 137 – 154.

Bohorova NE, Cocking EC and Power JB (1986) Isolation, culture and callus regeneration of protoplasts of wild and cultivated *Helianthus* species. *Plant Cell Rep.,* **5** : 256 - 258.

Brar GS and Ahuja KL (1979) Sesame : Its culture, Genetics, Breeding and Biochemistry. *Annu. Rev. Plant Sci.,* **1** : 245 – 313.

Carlson PS, Smith HH and Dearing RD (1972) Parasexual plant hybridization. *Proc. Natl. Acad. Sci.,* **69** : 2229 – 2294.

Chae YA, Park SK and Anand IJ (1987) Selection *in vitro* for herbicide tolerant cell lines of *Sesamum indicum*, 2: selection of herbicide tolerant calli and plant regeneration. *Korean J. Breed.,* **19** : 75 - 80.

Commandeur P (1995) Impact of biotechnology on world trade in vegetable oils. *Moniter,* **23** : 73 - 76.

Downey RK and Keller WA (1993) Modifying oil and protein crops : New concepts and approaches. In : International Crop Science Society of America Inc (Madison, USA), pp. 655 - 663.

Fuji JA, Slade DT, Radenbaugh K and Walker KA (1987) Artificial seeds for plant propagation. *TIBTECH,* **5** : 335 - 338.

George L, Bapat VA and Rao PS (1989) Plant regeneration *in vitro* in different cultivars of Sesame (*S. indicum* L.). *Proc. Indian Acad Sci.,* **99** : 135 - 137.

George L, Bapat VA and Rao PS (1987) *In vitro* multiplication of sesame (*S. indicum* L.). through tissue culture. *Ann. Bot.,* **60** : 17 - 21.

George L and Rao PS (1983) *In vitro* multiplication of Safflower through tissue culture. *Proc. Indian Natl. Acad. Sci.,* **48** : 791-794.

Haccius B (1978) Question of unicellular origin on non - zygotic embryos in callus cultures. *Phytomorphology,* **28** : 74 - 81.

Joshi AB (1961) *Sesamum* : Indian central oil seeds committee. Hyderabad-1, India.

Kim HY and Byeon GH (1991) Effect of growth regulators on organ cultures of sesame. *Sub. Tropical Agric.,* **8** : 93 - 103.

Kim MK, Park SK and Chae YA (1987) Selection *in vitro* for herbicide to tolerent cell lines of *Sesame indicum* : Effects of explants and hormone combinations on callus induction. *Korean J. Plant Breed.,* **19** : 70 – 94.

Kitto SL and Janik J (1985) Production of Synthetic seeds by encapsulating asexual embryos of carrot. *J. Amer. Soc. Hort. Sci.,* **110** : 227 - 282.

Kolte SJ (1985) Diseases of Annual Edible Oil Seed Crops. Vol.II : Raton, fla. Rape seed - Mustard and Sesame Diseases. CRC Press, Boca.

Lee JI and Choi BH (1985) Basic studies on sesame plant growth in Korea. In : *Sesame and Safflower : Status and Potentials* (Eds Ashri A) FAO plant production and protection paper 66, Rome.

Lee JI, Park YH, Park YS and Im BG (1985) Propagation of Sesame (*S. indicum* L.) through shoot tip culture. *Korean J. Breed.,* **17** : 367 - 372.

Lee SY, Kim HS and Lee YT (1988) Effect of growth regulators on callus induction and organ regeneration from seedling explant sources of Sesame (*Sesamum indicum* L.) cultivars. *Research Reports of Rural Development Admi. Biotech.* **30** : 69 - 73.

Lee SY, Kim HS, Lee YT and Park CH (1988) Effect of growth regulators, cold pre - treatment and genotype in anther culture of Sesame (*Sesame indicum* L.). *Research Reports of Rural Development Admi. Biotech.,* **30** : 74 - 79.

Mary RJ and Jayabalan (1987) Influence of growth regulators on somatic embyogenesis in Sesame. *Plant Cell Tiss. Org. Cult.*, **49** : 67 - 70.

Mazzani B (1983) Ajonjoli. *In : Cultivoy Mejoramieno de Plants Oleagnosas.* FONAIAP and CENIAP, Ceracas, Venezuela, pp. 169 – 224.

Mimura A, Takebayashi K, Niwano M, Takahara Y, Osawa T and Tokuda H (1994) Antioxidative and anticancer components produced by cell culture of Sesame. *ACS Symposium series,* **547** : 281 - 284.

Murta D and Batra A (1990) High yielding preparation of viable protoplasts from hypocotyls of *Sesamum indicum* L. *Curr. Sci.,* **59** : 325.

Narayana Swamy S (1977) Regeneration of plants from tissue cultures. *In: Applied and Fundamental Aspects of Plant Cell Tissue and Organ Culture* (Eds Reinert J and Bajaj YPS), Berlin, Springer Verlag, 179.

Nayar NM and Mehra KL (1970) Sesame : Its uses, botany, cytogenetics and origin. *Econ. Bot.,* **24** : 20 – 31.

Ogasawara T, Chiba K and Tada M (1993) Production in high yield of napthoquinone by a hairy root culture of *Sesamum indicum. Phytochemistry,* **33** : 1095 - 1098.

Rajender Rao K and Vaidyanath K (1997) Induction of multiple shoots from seedling shoot tips of different varieties of *Sesamum. Indian J. Plant Physiology,* **10** : 257 - 261.

Rajender Rao K and Vaidyanath K (1997) Callus induction and morphogenesis in Sesame (*Sesamum indicum* L.). *Adv. Plant Sci.,* **10** : 21 - 26.

Rajender Rao K and Vaidyanath K (1998) Somatic embryogenesis and regeneration from different explants of *Sesamum indicum* L. National sym. Commercial aspects of Plant Tissue Culture Mol. Biol and Medicinal Plant Biotechnology, Jamia Hamhard, New Delhi.

Rajender Rao K and Vaidyanath K (1999) Biotechnology and *Sesamum* Improvement. In : *Proceedings of National Symposium on "Plant Tissue Culture and Molecular Biology: Emerging Trends"* (Ed Kavi Kishor PB). Orient Longman, Universities Press (India) Limited, pp. 37 - 46.

Rajender Rao K and Vaidyanath K (1998) Synseed and micropropagation in *Sesamum. National Symposium on "Perspectives in Biotechnology"* held at Department of Botany, Kakatiya University, Warangal - 506 009, A.P., India.

Rajender Rao K and Vaidyanath K (1999) Genetic divergence , stability analysis and tissue culture studies on *Sesamum indicum* L. *Ph.D. Thesis,* Osmania University, Hyderabad.

Rajender Rao K and Vaidyanath K (2000) Biotechnology of *Sesamum* : Problems and prospects. *In: Recent Advances in Biotechnology* (Ed Trivedi PC) Panima Publishing Corporation, New Delhi – India, pp. 106 – 120.

Ranaweera KKDS and Pathirana R (1992) Optimization of media conditions for callus induction from anthurs of Sesame cultivars M13. *J. National Science Council of Srilanka,* **20** : 309 - 316.

Rangaswamy NS (1986) Somatic embryogenesis in Sesame. *Plant Cell Tiss. Org. Cult.,* **49** : 67 - 70.

Redenbaugh K (1993) Synseeds : application of synseeds to crop improvement. CRC Press, Boca raton, FL, 1- 481.

Reddy PS (1995) Scope for by production rise. The Hindu survey of Indian Agriculture.

Ryu JH, Doo HS and Kwon TH (1992) Induction of haploid plants by anther culture in Sesame (*S. indicum* L.) (1) Effects of growth regulators and differentiation between genotypes on callus induction. *Korean J. Plant Tiss. Cult.,* **19** : 171 - 177.

Ryu JH, Doo HS and Kwon TH (1993) Induction of haploid plants by anther culture in Sesame (*S. indicum* L.) (2) Effects of growth regulators and differentiation between genotypes on callus induction. *Korean J. Plant Tiss. Cult.,* **19** : 177 - 180.

Stack RW (1987) Long generation times in breeding trees - a pest manegement blessing in disguise. *In: Procedings of the 1987 North Central Tree Improvement Conference.* Fargo, ND, USA, : 72 - 81.

Steward FC, Mapes MO and Mears K (1958) Growth and organized development of cultured cells, II . Organization in cultures, grown from freely suspension cells. *Am. J. Bot.,* **45** : 705 - 708.

Weiss EA (1971) *Castor, Sesame and Safflower.* Leonard Hill London, pp. 311 – 525.

Chapter 9

GENETIC TRANSFORMATION OF *LINUM* SPECIES

Hong-Qing Ling★ and Beat Keller

Institute of Plant Biology, University Zurich, Zollikerstrasse 107, CH-8008, Zurich, Switzerland

Summary

Stable transgenic plants in the genus Linum have been generated by Agrobacterium-mediated transformation, particle bombardment and direct gene transformation with PEG. Agrobacterium-mediated transformation is the most evaluated and used approach for introduction of novel genes into Linum. Transformation efficiency is generally low and varied strongly from laboratory to laboratory. The highest reported transformation frequency is 25%. Some factors influencing transformation efficiency are summarized and discussed. Most transgenic plants showed normal morphology and a segregation pattern of one or two genes in the progeny. Genetic transformation in Linum was successfully applied in cultivar improvement by introduction of herbicide resistance genes and by tagging of the rust resistance genes L^6 *and M after the introduction of the maize Ac element into the flax genome. The next challenge is to develop more efficient transformation systems for flax and to apply the transformation technology for tagging and isolation of agronomically important genes as well as for improvement of flax cultivars for disease resistance, quantities and qualities of oil and fibre.*

Keywords : *Agrobacterium*, flax, genetic transformation, *Linum*, linseed, PEG, particle bombardment

★Correspondence author : E-mail : hqling@genetics.ac.cn

Abbreviations : NPT II : neomycin phosphotransferase II, HPT : hygromycin phosphotransferase, SPEC : spectinomycin resistance gene, GUS : β-glucuronidase, EPSP : 5-enolpyruvyl shikimate-3-phosphate synthase, ALS : acetolactate synthase, PAT : phosphinothricin acetyltransferase, PEG : polyethylene glycol

1. INTRODUCTION

The genus *Linum* is one of the largest genus of flowering plants, comprising more than 200 species (Gill and Yermanos, 1967). The sole crop flax (*Linum usitatissimum* L.) in the genus is one of the oldest and wide-spread crop plants in the world (Hoffmann, 1961). It is cultivated as oil plant (linseed flax) as well as fibre plant (fibre flax). The oil of linseed has very high dry quality and is an important raw materials for paint and lacquer industries in the production of paint, varnish, linoleum; in addition, it has a high value for humann nutrition. The residues of oil production are used as protein source for animal rations because flax seeds have a high content of protein. The fibre of flax is one of the main natural fibres used for textiles as well as for technical applications in other industries. The woody part of stem and roots after yielding of fibres is also a valuable raw material for fibreboard and paper industries. According to FAO data for 1993, 336,000 ha of fibre flax and 3,138,000 ha of linseed flax were cultivated in the world in 1992. Main cultivation countries of flax are China, Canada, India, the former Soviet Union and the United Kingdom. In Canada, linseed flax is the sixth largest crop (McHughen and Holm, 1995a). Apart from its economic value, *L. usitatissimum* is an excellent plant for tagging and isolation of agronomically important genes in crop plants because it has a small genome size and 44% of the genome are comprised by single copy sequences (Cullis, 1981).

Since the first transgenic plants were regenerated in tobacco fifteen years ago (Horsch *et al.*, 1984; Block *et al.*, 1984), plant transformation has become a core reseach tool in plant biology and a practical tool for cultivar improvement. It is used to understand gene function and the molecular regulation of physiological and developmental processes as well as to increase quality and quantity of plant products. Additionally, genetic transformation of plants opens a new opportunity to design new plants for production of pharmaceuticals, nutriceuticals and other beneficial chemicals and for environmental cleanup of pollutants. In the last fifteen years, different transformation techniques were developed and used for introduction of novel genes in plants. *Agrobacterium*-mediated transformation and transformation by particle bombardment as well as direct uptake of naked DNA by isolated protoplasts are widely used techniqes (Hansen and Wright, 1999).

In the genus *Linum*, the first transformation experiment was done by Hepburn and coworkers (1983) with *A. tumefaciens*. They cocultured epicotyl explants of *L. usitatissimum* with a wild strain T37 of *A. tumefaciens* and obtained transformed calli. Transgenic plants were first reported by Basiran *et al* (1987). Since then, a number of transformation experiments in *Linum* have been done in different laboratories of the world. In Table 1, a summary of the transformation work published in *Linum* is presented. *Agrobacterium*-mediated transformation is the most evaluated and used technique in *Linum*. Recently, transformation by particle bombardment (Wijayanto and McHughen, 1999) and direct gene transformation with PEG (Ling and Binding, 1997) were also reported in *Linum*. In this chapter, we will highlight the successes of genetic transformation achieved in *Linum* and discuss factors influencing transformation efficiency as well as the future prospects of transformation in *Linum*.

2. TRANSFORMATION TECHNIQUES

2.1. *Agrobacterium*-mediated transformation

Natural gene transfer mechanisms of soil bacteria (*A. tumefaciens* and *A. rhizogenes*) are widely exploited for introduction of foreign genes into plants. The genes of interest are first inserted into T-DNA (transfer DNA flanked by 25 base pair border sequences) of a disarmed Ti or Ri-plasmid of Agrobacteria, then transferred with the whole T-DNA from *Agrobacterium* to plant cells after inoculation and finally integrated in the genome of plants. The molecular mechanism of *Agrobacterium*-mediated transformation was reviewed by Zupan and Zambryski (1997). Thus, transfer of a gene into plant cells mediated by *A. tumefaciens* or by *A. rhizogenes* is a biologically controlled process. The host range of Agrobacteria also restricts the range of this transformation technique. For the plants which are not susceptible to *Agrobacterium* infection, for example some of monocotyledonous plants, *Agrobacterium*-mediated transformation is still not possible or needs specific artificial conditions.

Linum species are dicotyledonous plants and are susceptible to

Agrobacterium infection. Feasibility of *Agrobacterium*-mediated transformation was demonstrated in 1983 (Hepburn *et al.*, 1983). In the last fifteen years, transformation systems with *Agrobacterium* in *Linum* have been established in different laboratories of the world and some parameters affecting transformation efficiency were evaluated. *A. tumefaciens*-mediated transformation is the most frequently used approach in *Linum* (See Table 1). *Agrobacterium* strains, vectors, plant species and genotypes applied for transformation studies of *Linum* as well as the results of these studies are summarized in Table 1. Explants of hypocotyls and cotyledons as well as isolated protoplasts were used for transformation. Hypocotyl explants from 5-10 days old seedlings are the best source of tissues for transformation because they have much higher capability of shoot regeneration than other tissues (Gamborg and Shyluk, 1976; Mathews and Narayanaswamy, 1976; Murray *et al.*, 1977). Transformation using isolated protoplasts is possible in *L. usitatissimum* by *A. tumefaciens* (Ling and Binding, 1997), but the efficiency is low because shoot regeneration from isolated protoplasts is very inefficient (Ling and Binding, 1987). Therefore, transformation of protoplasts is not recommended unless an efficient system for shoot regeneration from isolated protoplasts is established.

Stable trangenic plants in *Linum* using *A. tumefaciencs* were obtained in several laboratories. Most transgenic plants showed normal morphology. Abnormal flower morphology (Ling and Binding, 1997), color changes of flower and pollen sterility (Rakouský *et al.*, 1999) were also observed in transgenic plants. Phenotypic analysis of the transformed plants in the T_1 generation revealed inheritance of one or two genes in most lines (Dong and McHughen, 1993b; Ling, 1992; McSceffrey *et al.*, 1992; Bretagne-Sagnard and Chupeau, 1996). In some transgenic plants, an irregular segregation rate of the transgene was found. It was suggested that these transgenic plants with irregular segregation patterns might be chimeras consisting of transformed and non-transformed cells (McScheffrey *et al.*, 1992; Dong and McHughen, 1993b). Southern hybridization analysis demonstrated that some transgenic lines containing multiple copies of T-DNA (3-7 copies) in the genome displayed a single gene inheritance in their progeny (Ling, 1992; Bretagne-Sagnard and Chupeau, 1996). A similar phenomenon was described in transgenic petunia (Deroles and Gardner, 1988).

Flax has also been successfully transformed by *A. rhizogenes* (Zhan *et al.*, 1988). Cotyledon explants of 7-days-old seedlings of four flax cultivars (Abyssinian, Akmolinsk, Bombay and Precederia) were

Table 1. Genetic transformation in the genus *Linum*

Transformation methods	Bacterium strain	Vector and plasmid	Cultivars/ species	Selection	Transformation		TF	Reference
					Source	Recovery		
A. tumefaciens	GV2260	p35SGUSINT	NorLin	Km	H	TP	13%	Dong and McHughen, 1991, 1993a, 1993b
			Antarés	Km	H	n	-	Bretagne-Sagnard and Chupeau, 1996
			Torzhokskii 4	Km	H	n		-Polyakov *et al.*, 1998
			Slavnyi 82	Km	H	n	-	
			Dashkovskii 2	Km	H	n	-	
			Berlinka	Km	H	n	-	
	GV2260	pBin19 35SGUS	Bionaa	Km	P	TP	26.3×10^{-6}★	Ling and Binding, 1997
			C.A.N.2612-A	Km	P	TC	13.1×10^{-6}★	
			IR00610	Km	P	nc	-	
			Langeland B	Km	P	TP	5.4×10^{-6}★	
			Otofte 15/47	Km	P	TC	2.5×10^{-6}★	

Table 1. Continued

Transformation methods	Bacterium strain	Vector and plasmid	Cultivars/ species	Selection	Transformation		TF	Reference
					Source	Recovery		
			Ottawa	Km	P	TC	1.6×10^{-6}★	
	C58C1	pGV3850	Antares	Km	H	TP	nd	Basiran *et al.*, 1987
			Sharpes	Km	H	TP	nd	
		pGV3850	Akmolinsk	Km	H, C	n	-	Zhan *et al.*, 1988
			Bisont	Km	H	n	-	
			Stewart	Km	H	n	-	
			Precederia	Km	C	n	-	
	C58	pGV3850*ALS*	NorLin	Km	H	TP	nd	McHughen, 1989
			McGregor	Km	H	TP	nd	
	C58	pGH6	McGregor	Km	H	TP	nd	McHughen and Jordan, 1989
	GV3850	pGV3850	Forge	Km	C	TP	nd	Ellis *et al.*, 1992
	A208	pMON570	McGregor	Km	H	TP	nd	Jordan and McHughen, 1988a
			NorLin	Km	H	TP	nd	

Table 1. Continued

Transformation methods	Bacterium strain	Vector and plasmid	Cultivars/ species	Selection	Transformation		TF	Reference
					Source	Recovery		
		pTiT37:: pMon570	McGregor	Km	H	TP	nd	McHughen and Jordan, 1989
	A281	pGA482	Torzhokskii 4	Km	H	n	-	Polyakov *et al.*, 1998
			Slavnyi 82	Km	H	n	-	
			Dashkovskii 2	Km	H	n	-	
			Berlinka	Km	H	n	-	
	LBA4404	pBI121	Torzhokskii 4	Km	H	TP	nd	Polyakov *et al.*, 1998
			Slavnyi 82	Km	H	TP	nd	
			Dashkovskii 2	Km	H	TP	nd	
			Berlinka	Km	H	TP	nd	
		PBI131	Glenelg	Km	H	TP	25%	Mlynárová *et al.*, 1994
	nd	pTiB6S3-SE	McGregor	Km	H	TP	nd	Jordan and McHughen, 1988b

Table 1. Continued

Transformation methods	Bacterium strain	Vector and plasmid	Cultivars/ species	Selection	Transformation		TF	Reference
					Source	Recovery		
			STS-II	Km	H	TP	nd	
	EHA 18Ac5	nd	NorLin	Km	H	TP	nd	McHughen and Holm, 1995a
	GV3101	pPM90RK:: pPCVRN4	Jitka	Hm	H	TP	nd	Rakouský *et al.*, 1999
			Areco	Hm	H	TP	nd	
			NLN245	Hm	H	TP	nd	
	C58C1Rif^R	pGV3850HPT	*L. suffruticosum*	Hm	P	TP	4.5×10^{-5}★	Ling and Binding, 1997
	A208	pTiT37:: pMON894	NorLin	Km	H	TP	0%-3.7%	McHughen *et al.*, 1989
			Vimy	Km	H	TP	0%-3.7%	
			Andro	Km	H	TP	0%-3-7%	
			McGregor	Km	H	TP	0%-3.7%	
	AGL1	pRECSPEC	Antarés	Sp	H	TP	15%	Bretagne-Sagnard and Chupeau, 1996

Table 1. Continued

Transformation methods	Bacterium strain	Vector and plasmid	Cultivars/ species	Selection	Transformation		TF	Reference
					Source	Recovery		
	nd	pTAB-Epspec	Cass	Sp	C	TP	nd	Anderson *et al.*, 1997
			Hoshangabad	Sp	C	TP	nd	
			m-x32	Sp	C	TP	nd	
	T37	pTiT37	nd	-	E	TC	-	Hepburn *et al.*, 1983
A. rhizogenes	A4	pRiA4	Abyssinian	nd	C	R	-	Zhan *et al.*, 1988
			Akmolinsk	nd	C	R	-	
			Bombay	nd	C	TP	nd	
			Precederia	nd	C	R	-	
	1855	pRi1855	Abyssinian	nd	C	R	-	Zhan *et al.*, 1988
			Akmolinsk	nd	C	TP	nd	
			Bombay	nd	C	TP	nd	
			Precederia	nd	C	TP	nd	
	TR7	pRiTR7	Abyssinian	nd	C	R	-	Zhan *et al.*, 1988

Table 1. Continued

Transformation methods	Bacterium strain	Vector and plasmid	Cultivars/ species	Selection	Transformation		TF	Reference
					Source	Recovery		
			Akmolinsk	nd	C	R	-	
			Bombay	nd	C	TP	nd	
			Precederia	nd	C	R	-	
PEG		pGL2	*L. suffruticosum*	Hm	P	TP	2×10^{-3}#	Ling and Binding, 1997
Bombardment		p35SGUS/INT	Somme	Km	H	TP	nd	Wijayanto and McHughen, 1999

H : hypocotyl explants, C : cotyledon explants, P : protoplasts, TP : transgenic plant, TC : transgenic callus, R : roots, Km : Kanamycin sulfate, nc : no callus formed, nd : not described, TF : transformation frequency, Sp : spectinomycin sulfate, Hm : Hygromycin B, E : epicotyl explants, n : no trangenic plants recovered

★ : absolute transformation frequency (formed calli/total protoplasts), # : relative transformation frequency (transformed calli/total calli regenerated)

inoculated with three bacterial strains (A4, 1855 and TR7) of *A. rhizogenes*. Adventitious roots were regenerated at the cut end of cotyledon explants and some plants were regenerated from the roots cultured subsequently on a shoot-inducing medium. Integration of the T-DNA in chromosomes of transgenic plants was demonstrated by Southern hybridization analysis. The transformed plants showed abnormal morphology (curled leaves, short internodes and more roots in some plants). The authors suggested that transformation by *A. rhizogenes* was an effective alternative to transformation by disarmed *A. tumefaciens*, especially for cultivars for which shoot regeneration is not possible from callus but is possible from roots.

2.2. Transformation by particle bombardment

Particle bombardment is the second widely used technique for introduction of foreign genes into plants and represents a rapid and simple approach (Klein *et al.*, 1987). Bombardment of plant tissues or cells with gold particles coated with DNA delivers foreign DNA into plant cells where it is integrated in the genome. This transformation technique can be used for all tissues and all plants as there is no biological limitation as described for *Agrobacterium*-mediated transformation. The delivery of foreign DNA into plant cells is a purely physical process. Compared with *Agrobacterium*-mediated transformation, the copy number of DNA is normally much higher and the integration pattern is more complex. Particle bombardment is widely used for introduction of foreign genes in plants for which *Agrobacterium* transformation is not efficient, e.g. maize, wheat and barley. Transformation by particle bombardment has two advantages compared to *Agrobacterium*: First, no special transformation vector such as Ti- or Ri-vectors has to be constructed and no residual bacteria from transformation cultures have to be removed. Antibiotics used for this removal are frequently toxic and reduce or inhibit callus formation and shoot differentiation for some plants; Second, several genes in separate plasmids can be introduced together into plant cells (Hadi *et al.*, 1996; Bower *et al.*, 1996 and Wakita *et al.*, 1998). Thus, particle bombardment is also an attractive alternative for gene transfer even in plants which are amenable to *Agrobacterium*-mediated transformation.

Successful transformation by particle bombardment in *Linum* was reported recently by Wijayanto and McHughen (1999). They have sucessfully introduced the reporter gene *uidA* and the marker gene *nptII* in a linseed cultivar Somme by bombardment of hypocotyl segments using gold particles coated with the plasmid p35SGUS/INT. Stable integration of the *nptII* and *uidA* genes in the genome was demonstrated by molecular analysis of putative transgenic plants. Both genes were transmitted into the progeny. Most of the transgenic lines obtained (7 of 10) gave a simple segregation pattern of one gene (3:1) in the T_1 generation. The authors reported that transformation by particle bombardment in flax is more efficient than transformation with *Agrobacterium*. About 25% (10 of 41) of shoots regenerated from bombarded hypocotyl segments on the selection media were confirmed to be transformants. Only one of 10 transgenic lines showed a chimeric phenotype when assayed for *uidA* expression. The frequency of chimeric shoots in transgenic plants obtained by particle bombardment transformation is much lower than by transformation with *Agrobacterium*.

Delivery of foreign DNA into intact cells is the first step of gene transformation. Efficiency of DNA delivery by particle bombardment is affected by many factors, *e.g.* bombardment distance, quantity of DNA, size of gold particle and so on. Wijayanto and McHughen (1999) evaluated delivery efficiency of genes by bombardment in hypocotyl segments under varying conditions including duration of hypocotyl preculture, bombardment distances, quantities of DNA, level of chamber vacuum and size of gold particle. Transient expression of the *uidA* reporter gene was used as indicator. These studies suggested that the best bombardment conditions for gene delivery to hypocotyl segments are 6210 kPa rupture disc pressure, 4-days hypocotyl preculture period, 2 cm gap distance in combination with 6 cm target distance, 66-cm Hg chamber vacuum level, 7.5 μg DNA per DNA-particle mixture and 1.0 μm gold particle size.

2.3. Direct gene transformation with PEG

Direct gene transformation with PEG represents a third method for the introduction of foreign genes into plants (Potrykus, 1991). Isolated protoplasts are used for uptake of foreign DNA in this transformation method. Uptake of naked DNA in protoplasts mediated by PEG is a chemical process. Theoretically, this method is available for all plants if an effective regeneration system has been established for isolated protoplasts. However, shoot regeneration of isolated protoplasts in most plants is still inefficient or impossible and the use of this transformation method is strongly restricted. Because direct gene transformation is a single-cell transformation system and transgenic shoots have to originate from single transformed cells, chimeras obsevered in transformation using multicellular explants by *Agrobacterium* (McHughen and Jordan, 1989) and by particle bombardment (Wijayanto and McHughen, 1999) can not occur.

Direct gene transformation with PEG in *Linum* was reported by Ling and Binding (1997) using the wild species *Linum suffruticosum* spp. *salsoloides* which has a very high regeneration capability of shoots from isolated protoplasts (Ling and Binding, 1987). Stable transformants were obtained by introducing the marker gene *hpt* (conferring resistance to hygromycin B) into isolated protoplasts of shoot tips. The relative transformation rate was about 2.2×10^{-3}. All regenerated calli and shoots were resistant to hygromycin B and contained DNA sequence of the *hpt* gene. As expected, no chimeras were observed in transgenic plants. The length of the integrated fragment of plasmid DNA in plant chromosomes is generally random after direct gene transformation (Potrykus *et al.*, 1985) and the copy number of the introduced gene is normally higher than by transformation with *Agrobacterium*. In the protoplast derived transgenic plants, one to three copies of the *hpt* gene were found in the genome.

3. FACTORS INFLUENCING TRANSFORMATION EFFICIENCY

A. tumefaciens-mediated transformation is the most evaluated and used approach for the introduction of foreign genes into *Linum* because of the ease of the protocol and minimal equipment costs. Therefore, the following discussion of factors influencing transformation efficiency will be focused on this method. As described above, transgenic plants were regenerated in several laboratories using *A. tumefaciens*-mediated transformation. But the efficiency was low and varied strongly from laboratory to laboratory. Obvously, efficiency of transformation is based on the competence of Agrobacteria as donor of the genes as well as on the competence of plants as receptors. The competence of plants includes the competence of the host cells to *Agrobacterium* transformation and the competence of the transformed cells to regenerate into whole plants. Many factors and culture conditions can increase or decrease these competences. Some of them will be discussed in following.

3.1. *Agrobacterium* strains

Agrobacterium strains and plasmid vectors are important factors determining transformation efficiency (Roekel *et al.*, 1993). In *Linum*, there are two reports describing effects of *Agrobacterium* strains on recovery of transgenic plants (Bretagne-Sagnard and Chupeau, 1996; Polyakov *et al.*, 1998). Bretagne-Sagnard and Chupeau (1996) transformed linseed flax cultivar Antarés using two *Agrobacterium* strains GV2260 (p35SGUSINT) and AGL1 (pRECSPEC). No transgenic plants were regenerated from 25,000 hypocotyl explants inoculated with GV2260 (p35SGUSINT) while a 15% transformation rate was achieved by transformation with AGL1 (pRECSPEC) under the same culture and transformation conditions, except for the selection agents (kanamycin sulfate for GV2260 (p35SGUSINT) and spectinomycin sulfate for AGL1 (pRECSPEC)). Different efficiencies of *Agrobacterium* strains were also reported by Polyakov *et al.* (1998) in transformation of four cultivars of fibre flax using three *Agrobacterium* strains A281 (pTiBo542, pGA482), GV2260 (p35SGUSINT) and LBA4404 (pBI121). Transgenic plants were recovered only by transformation with *Agrobacterium* strain LBA4404 (pBI121). The *Agrobacterium* strain GV2260 (p35SGUSINT) did not result in transgenic plants described in the two transformation experiments. However, transgenic plants were obtained using this strain in transformation of the linseed cultivar NorLin (Dong and McHughen, 1991; 1993a and 1993b) and the transformation frequency reached up to 13%. It seems that the effect of *Agrobacterium* strains on transformation efficiency is genotype-dependent. For the transfer of T-DNA from the bacterium to plant cells and subsequent integration into the plant chromosome, three genetic elements of *Agrobacterium* are required: the transferred segment (T-DNA), the virulence region of the Ti-plasmid and a set of at least 11 chromosomal genes (Zupan and Zambryski, 1997). It is not known why different *Agrobacterium* strains give different efficiencies and which genetic elements of *Agrobacterium* play a key role in this process. There is a lack of directly comparable experiments such as comparison of the efficiency of one *Agrobacterium* strain containing different vectors and of different *Agrobacterium* strains containing a same vector.

3.2. *Linum* species and genotypes

In the genus *Linum*, genetic transformation was investigated in the two species *L. usitatissimum* and *L. suffruticosum*. The wild species *L. suffruticosum* was transformed by PEG and by *Agrobacterium* using isolated protoplasts (Ling and Binding, 1997). The transformation frequency of *L. suffruticosum* was generally higher than of *L. usitatissimum* due to the high regeneration capability of calli and shoots from isolated protoplasts.

In *L. usitatissimum*, more than 30 cultivars or genotypes (See Table 1) have been investigated in different laboratories of the world. Transformation frequencies were from 0% to 25%. The differences could be explained by using different *Agrobacterium* strains and transformation protocols. However, it is also likely to that genotypes show different transformation frequencies. Differences of tranformation efficiency between genotypes was observed in transformation of isolated protoplasts of six genotypes of flax (Ling and Binding, 1997). This variation of transformation efficiencies might be due to a different genetic competence for transformation and to different regeneration capabilities of cells and tissues. A genetic competence factor of transformation was identified in potato (El-Kharbotly *et al.*, 1995). Sonti and coworkers (1995) found *Arabidopsis thaliana* mutants which are deficient in T-DNA integration into plant chromosome. They suggested that some plant genes must be involved in the integration process of foreign DNA into the plant genome.

3.3. Preculture

McHughen and coworkers (1989) reported on the necessity of preculture of hypocotyls prior to inoculation with Agrobacteria for recovery of transgenic plants. They investigated the effects of preculture time from 0 to 12 days on production of transgenic plants in transformation of four cultivars (NorLin, Vimy, Andro and McGregor). They used *A. tumefaciens* strain A208 carrying a cointegrate plasmid pTiT37::pMON894 and found that the longer the hypocotyl explants were precultured, the more transgenic plants were recovered. Transformation frequencies (transformed/total inoculated hypocotyls) increased from 0% at 0 and 3 days preculture to 3.75%, at 9 days preculture and the 12 days preculture gave the highest proportion (4.73%) of transgenic to non-transgenic shoots. They recommended 9-12 days preculture time of hypocotyls prior to inoculation with Agrobacteria for transformation. Dong and McHughen (1991) studied the effects of preculture on cellular transformation intensity using transient expression of the *uidA* gene in inoculated hypocotyls. These studies demonstrated that the cellular transformation intensity decreased with increase of preculture time and cells on the upper cut of hypocotyls were very competent to transformation without preculture period. They became progressively more difficult to transform with an increase of preculture time. A significant increase in the number of regenerated shoots per inoculated hypocotyl explant was observed by a preculture period of upto 12 days (McHughen *et al.*, 1989). Thus, effects of preculture on transformation efficiency might be caused by the improvement of the shoot regeneration of inoculated hypocotyls. During preculture of hypocotyls on culture media containing phytohormones, plant cells will redifferentiate and more and more cells will become competent for regeneration of shoots. The proportion of transformed shoot-regenerable cells to total transformed cells with preculture might be higher than without preculture and the chance to recover transgenic shoots will also increase. However, no positive effect of the preculture period on transformation efficiency was reported by Bretagne-Sagnard and Chupeau (1996) in transformation of cultivar Antarés with *Agrobacterium* strain GV2260(p35SGUSINT). Mlynárova and coworkers (1994) described that preculture of hypocotyls prior to inoculation with *Agrobacterium* was not essential for high transformation efficiency because they obtained the highest transformation frequency (25%) reported in *Linum* without any preculture period in transformation of the cultivar Glenelg using *A. tumefaciens* strain LBA4404 (pBI131). It is possible that the preculture period is essential for cultivars which have a poor capacity of shoot regeneration and is not essential for cultivars with a high capacity of shoot regeneration.

3.4. Stripping of epidermis

Hypocotyl explants of *L. usitatissimum* were used in most transformation studies because they possess a much better shoot regeneration capacity than other tissues. However, the adventitious shoots formed by hypocotyls originate mostly from single epidermal cells (Link and Eggers, 1946) which are less competent to *Agrobacterium*-mediated transformation (Dong and McHughen, 1993a). Selection of transformed cells using antibiotics, in inoculated hypocotyls is not efficient to suppress regeneration of non-transformed cells because the transformed cells protect the non-transformed cells from selection agent (Jordan and McHughen, 1988b; McHughen and Jordan, 1989). Many non-transgenic shoots are regenerated from non-transformed epidermal cells and make it difficult to recover transgenic shoots. The regeneration of a number of non-transformed shoots in a hypocotyl explant might also inhibit or reduce differentiation of transformed cells. Jordan and McHughen (1988a) removed strips of epidermis from hypocotyl segments before inoculation with Agrobacteria and found that stripped hypocotyls produced more transgenic plants (10%) than non-stripped (2.1%). They suggested that this increase of transformation rate was due to an increase of wounding by stripping of the epidermis. Polyakov *et al.* (1998) compared the effects of different wounding procedures (peeling of 50% epidermis or by puncturing with thin needles) on the transformation efficiency and found that peeled hypocotyls produced more calli and transgenic plants than the punctured hypocotyls and the hypocotyls without damage. Dong and McHughen (1991) reported that cellular transformation strongly increased by post-peeling of epidermis-stripped hypocotyls after a culture period. The observed improvement of transformation efficiency caused by stripping of the epidermis might have two reasons: First, peeling of epidermis increases the wounded area on hypocotyl segments, increasing cellular transformation intensity. Second, it reduced the regeneration of non-transgenic shoots by removing the non-competent epidermal cells. However, in at least one study, no positive effect by peeling of epidermis on transformation efficiency was observed (Bretagne-Sagnard and Chupeau, 1996). In transformation of the cultivar Glenelg with LBA 4404 containing pBI131 (Mlynárová *et al.*, 1994), a very high transformation rate (25%) was achieved without removing of epidermal cells. The authors concluded that stripping of epidermis was not essential for high transformation efficiency. Apart from effects of the plant genotypes, *Agrobacterium* strains and transformation protocols used in this transformation study, the length of hypocotyl explants might also play a role. Mlynárová and coworkers (1994) used very short segments of hypocotyls (2-3 mm) for transformation while 5 – 10 mm long hypocotyl segments were utilized in the laboratory of McHughen (McHughen *et al.*, 1989). In a more recent study, long hypocotyl segments strongly reduced the regeneration rate of shoots per explant (Bretagne *et al.*, 1994). Transgenic shoots were recovered almost from calli formed at the both cut sides of hypocotyl explants. Cutting hypocotyls into short segments increases the wounded area per segment. Thus, the chance for recovery of transgenic shoots from short hypocotyl explants might be higher than from long segments without peeling of epidermis. It seems that similar results are reached by cutting hypocotyls in short segments or by using longer hypocotyl segments with stripping of epidermal cells.

3.5. Cocultivation duration and concentration of bacteria

After inoculation, a cocultivation period of plant tissues with Agrobacteria is necessary for transfer of T-DNA from *Agrobacterium* cells into plant cells and subsequent integration into the genome. Usually, two days of cocultivation were used in the transformation studies. Dong and McHughen (1991, 1993a) reported that cellular transformation efficiency and recovery of transgenic plants were significantly increased when the cocultivation duration of hypocotyl explants with Agrobacteria was prolonged from 2 days to 5 or 7 days. This increase was possibly due to the proliferation of bacteria, providing a larger population for transformation. The positive effects of a prolonged cocultivation period were also described in *Kalanchoe laciniata* (Jia *et al.*, 1989) and in *melon* (Dong *et al.*, 1991).

Concentration of Agrobacteria used in transformation plays an important role in transformation efficiency. Ling and Binding (1997) found that an increase of the bacterial concentration from 10^6 to 10^7 cells per 1 ml protoplast culture containng 5.0 x 10^5 cells improved transformation frequency by a factor of two. The positive effect of bacterial concentration on the transformation efficiency was genotype-dependent and was more pronounced in the competent genotype.

3.6. Selection agents - Antibiotics

For successful transformation with *Agrobacterium*, antibiotics are necessary in the culture media to eliminate of bacteria as soon as their presence is no longer required and later to select transformed plant cells and transgenic plants.

To eliminate bacteria from transformation cultures, cefotaxime (Basiran *et al.*, 1987; Bretagne-Sagnard and Chupeau, 1996; Ling and Binding, 1997; Mlynárova *et al.*, 1994; Polyakov *et al.*, 1998; Zhan *et al.*, 1988), ticarcillin and timentin (Rakousky *et al.*, 1999) and a combination of cefotaxime with carbenicillin (Dong and McHughen, 1991 and 1993a; Jordan and McHughen, 1988a; McHughen, 1989; McHughen and Jordan, 1989 and McHughen *et al.*, 1989) were applied in transformation of *Linum*. Clearly, different antibiotics have different effects on Agrobacteria and a different toxicity on plant cells and tissues. There are two general criteria for the choice of an antibiotic for elimination of bacteria from transformation cultures: First, the antibiotic has to be effective against the *Agrobacterium* strain used, i. e. bacteria in culture must be completely eliminated at low concentrations and in short time; second, the toxicity for plant cells must be low and it should not inhibit shoot differentiation. Cefotaxime is extensively used in *Linum* transformation. However, an inhibitory effect of cefotaxime on callus growth and shoot regeneration was observed in tomato (Ling *et al.*, 1998), carrot (Okkels and Pedersen, 1988) and *Antirrhinum majus* (Holford and Newbury, 1992). Recently, it was reported that timentin (ticarcillin/potassium clavulanate) was very effective against Agrobacteria even at low concentrations (150 mg/l) and showed a stimulation of callus growth and shoot regeneration in tomato (Ling *et al.*, 1998). The concentration of antibiotic needed in media is possibly dependent on

Agrobacterium strains. Basically, it should not be higher than necessary because most antibiotics have effects on plant cells and tissues and show inhibitory effect on callus and shoot regeneration at a high concentration (Okkels and Pedersen, 1988).

After inoculation and cocultivation with Agrobacteria, only a small proportion of cells are transformed. Chemical agents (antibiotics and herbicides) are used to inhibit multiplication and regeneration of non-transformed cells. In *Linum* transformation, three selection marker genes (*nptII* conferring resistance to neomycins, *hpt* conferring resistance to hygromycin B and *spec* conferring resistance to spectinomycin) were used for selection of transformed shoots. The *nptII* gene is a widely used selection marker, presents in many transformation vectors. Sensitivity of flax hypocotyls to different neomycins (kanamycin, geniticin and paromomycin) was tested (Dong and McHughen, 1993a; Bretagne-Sagnard and Chupeau, 1996). Geniticin was more effective than kanamycin in inhibition of callus formation and shoot regeneration in non-inoculated hypocotyls. Concentrations higher than 200 mg/l of kanamycin sulfate or 100 mg/l of geniticin strongly reduced callus formation and prevented shoot regeneration. Paromomycin was less effective than kanamycin, but it showed a cytokinin effect stimulating shoot regeneration (Bretagne-Sagnard and Chupeau, 1996). However, geniticin might be too toxic for flax cells because much fewer transgenic plants could be recovered using geniticin in comparison with kanamycin sulfate (Dong and McHughen, 1993a). Kanamycin sulfate was used in most transformation studies of *Linum*. The selection efficiency of kanamycin in inoculated hypocotyls was lower than in non-inoculated hypocotyls (Jordan and McHughen, 1988a; 1988b). A number of non-transformed shoots were regenerated from inoculated hypocotyls. Hygromycin B was described as an effective selection agent against untransformed cells (Ling and Binding, 1997; Rakouský, 1999). Bretagne-Sagnard and Chupeau (1996) reported that spectinomycin was more effective than kanamycin in selection of transgenic shoots

After regeneration of shoots, the rooting (Basiran *et al.*, 1987) and the leaf callus assay (McHughen and Jordan, 1989) are two effective methods to eliminate most false transformants. Nevertheless, Southern hybridization analysis of genomic DNA is essential for final identification of transgenic plants. A nopalin assay of transformed plants was not recommended because nopalin can be transported from transformed cells to non-transformed shoots (Jordan and McHughen, 1988b).

4. APPLICATION OF PLANT TRANSFORMATION IN *LINUM*

4.1. Improvement of commercial cultivars by introduction of herbicide resistance genes

Three herbicide resistance genes *als* (acetolactate synthase), *epsp* (5-enolpyruvylshikimate-3-phosphate synthase) and *pat* (phosphinothricin acetyltransferase) were introduced into major commercial linseed cultivars of Canada and one herbicide resistant cultivar has been developed (McHughen *et al.*, 1997).

A mutant *als* gene isolated from *Arabidopsis* confers resistance to the selective herbicide sulfonylurea (Haughn *et al.*, 1988). The *als* gene was introduced into the commercial cultivars NorLin and McGregor of Canada using *A.tumefaciens*-mediated transformation (McHughen, 1989). Stable transgenic plants were isolated from both cultivars. McSheffrey and coworkers (1992) characterized transgenic plants and showed that most transgenic lines (10 of 14) displayed the segregation pattern of a single gene. ALS-protein from transgenic lines was more resistant to chlorsulfuron than the enzyme from the parents (2.5 to over 60-fold). The transgenic lines were 25- to 260-fold more resistant to chlorsulfuron in a root growth assay. The variability of the herbicide resistance among various transgenic lines might be attributed to position effect. Field test showed that the resistance of transgenic plants to chlorsulfuron was stable and that there was no significant difference between the transgenic lines and the parent for agronomic performance such as oil content, yield, height, maturity in soil untreated with the herbicide (McHughen and Holm, 1991; 1995b; McHughen and Rowland, 1991). One transgenic line selected from the cultivar NorLin was fully resistant to field doses of the herbicide (McHughen and Holm, 1995b) and this transgenic line has been developed to the new sulfonylurea herbicide resistant cultivar CDC Triffid in Canada (McHughen *et al.*, 1997).

McHughen and Holm (1995a) tranferred the *pat* gene, conferring tolerance to the non-selective herbicide glufosinate into the genome of the cultivar NorLin by transformation with *A. tumefaciens*. Five transgenic lines were selected for field tests. Only one line survived in soil sprayed with glufosinate at 600 g/ha in the first year. The surviving line showed high tolerance to the non-selective herbicide glufosinate in subsequent field tests and suffered minimal damage from the herbicide at a concentration which killed non-transformed flax plants. No significant difference of agronomic traits such as yield and maturity was observed between the transgenic line and the parent under untreated conditions. The yield of the transgenic line increased significantly under treated conditions, possibly due to weed control.

The modified *epsp* gene from *Petunia* confers enhanced resistance to the herbicide glyphosate (Shah *et al.*, 1986). This gene was transferred into two commercial cultivars of linseed flax in Canada using *A.tumefaciens*-mediated transformation (Jordan and McHughen, 1988a). Trangenic plants were regenerated and showed resistance to glyphosate *in vitro* and in the greenhouse. Unfortunately, some damage of the meristematic tissues occurred, leading to the death of the apical meristem and subsequent regrowth from axillary meristems. As a consequence, the transgenic plants had delayed maturity. Field test of transgenic plants revealed a similar result as test in greenhous (Jordan and McHughen, 1993).

4.2. Transposon tagging and isolation of the rust resistance genes

Transposon tagging is a powerful technique for the isolation of agronomically important genes in plants. Ellis and coworkers (1992) developed a transposon tagging system in flax for isolation of rust resistance genes. They introduced the maize transposable elements *Ac* (*Activator*) and *Ds* (*Dissociation*) in genome of flax cultivar Forge carrying four different rust-resistance genes (L^6, *M*, *N* and P^2) by inoculation of cotyledon explants with *A. tumefaciens*. Transgenic plants containing the maize transposable element *Ac* or *Ds* were regenerated. Progeny examination of three transformed plants with *Ac* and 15 plants transformed with *Ds* and the *Ac* transposase gene showed that the transposon was active only in one transgenic plant having a high *Ac* copy number (10 copies) and jumped to new locations in its progeny. 25-30% of the progeny of some members of this family contained newly transposed *Ac* elements, indicating that gene tagging with this line was feasible. One active *Ac* element was located at about 29 map units from the L^6 rust resistance gene. The descendants of this plant were crossed to the susceptible cultivar Hoshangabad. Rust-susceptible mutants were screened in progeny. 29 susceptible mutants were identified (Lawrence *et al.*, 1993). One of the mutants contained the resistance gene L^6 tagged by *Ac*. From this mutant line, the L^6 gene was successfully isolated (Lawrence *et al.*, 1995). The *M* rust resistance gene from flax was also cloned recently using the same strategy (Anderson *et al.*, 1997).

5. CONCLUSIONS AND FUTURE PROSPECTS

Genetic transformation in *Linum* has made great progress during the last fifteen years. Transformation systems have been established in several laboratories of the world. Transgenic plants could be generated *via Agrobacterium*-mediated transformation, particle bombardment and direct gene transfomation with PEG. *Agrobacterium*-mediated transformation is the most widely evaluated and used technique because of its effectivity and the easy protocol coupled with minimal equipment costs. Transformation by particle bombardment is an attractive alternative for introduction of novel genes in *Linum* plants while direct uptake of naked DNA using isolated protoplasts medaited by PEG is not recommended unless an efficient shoot regeneration system from isolated protoplasts can be established. Generally, transformation efficiency was not satisfactory in most transformation studies although transgenic plants were obtained. The development of a more efficient transformation system is still necessary. Factors and culture conditions such as plant genotypes, bacterial strains, preculture, peeling of epidermis, cocultivation time and bacterial concentration as well as selection agents have to be optimized.

L. usitatissimum is the sole versatile crop yielding both valuable oil and fibre. Isolation of the genes involved in fibre development and oil synthesis is the first step to understand the molecular regulation of oil and fibre synthesis *in vivo* and their subsequent manipulation. Isolation of the rust resistance genes L^6 and *M* demonstrates that transposon tagging in flax is possible. Furthermore, gene tagging using T-DNA in *Linum* should be an effective method for isolation of interesting genes because *L. usitatissimum* has a small genome and more than 40% of the genome are comprised by single copy sequences (Cullis, 1981). Insertional mutants of T-DNA have a big advantage in that once a mutant has been identified it is easy to isolate the tagged gene. For screening of insertional mutants, a large number of transgenic plants has to be produced. Therefore, an efficient transformation system is necessary.

The successful development of the herbicide resistant cultivar CDC Triffid by introducing the modified *als* gene from *Arabidopsis* (McHughen *et al*., 1997) demonstrates that plant transformation in *Linum* is a powerful tool for improvement of cultivars. The improvement of rust disease resistance is another major breeding goal. Thirty-one different genes conferring resistance to different races of the fungal pathogen *Melampsora lini* have been identified in flax and mapped to five loci K, L, M, N and P (Ellis *et al*., 1997). The resistance genes *L* and *M* have been isolated recently (Lawrence *et al.*, 1995; Anderson *et al.*, 1997). The L locus is a single gene with 13 alleles expressing different rust resistance specificities whereas the M locus is complex containing about 15 related genes in tandem, conferring resistance to seven specific races (Anderson *et al.*, 1997). It is not possible to pyramid different L specificities in one flax cultivar by traditional breeding methods because the L locus is one gene with multiple alleles. Genetic transformation provides the possibility to combine different L alleles into one cultivar to control rust disease. Additionally, rust resistance genes from wild *Linum* species which are sexually incompatible with flax such as *L. marginale* can be introduced into flax by gene transformation.

Considerable progress in improvement of yield and quality of oil and fibre in flax has been made using conventional breeding methods. However, conventional breeding suffers from the limited gene pool available in *Linum*. Gene transformation allows to use genes from unrelated plant species and even from animal and microbial species to improve flax cultivars, expanding the breeder´s repertoire for the necessary improvement of flax in the future.

REFERENCES

Anderson PA, Lawrence GJ, Morrish BC, Ayliffe MA, Finnegan EJ and Ellis JG (1997) Inactivation of the flax rust resistance gene *M* associated with loss of a repeated unit within the leucine-rich repeat coding region. *Plant Cell,* **9** : 641-651.

Basiran N, Armitage P, Scott RJ and Draper J (1987) Genetic transformation of flax (*Linum usitatissimum*) by *Agrobacterium tumefaciens*: Regeneration of transformed shoots *via* a callus phase. *Plant Cell Rep.,* **6** : 396-399.

Block DM, Herrera-Esterella L, Montagu MV, Schell J and Zambryski P (1984) Expression of foreign genes in regenerated plants and their progeny. *EMBO J.,* **3** : 1681-1689.

Bower R, Elliott AR, Potier BAM and Birch RG (1996) High-efficiency, microprojectile-mediated cotransformation of sugarcane, using visible or selectable markers. *Mol. Breed.,* **2** : 239-249.

Bretagne B, Chupeau MC, Chupeau Y and Fouilloux G (1994) Improved flax regeneration from hypocotyls using thidiazuron as a cytokinin source. *Plant Cell Rep.,* **14** : 120-124

Bretagne-Sagnard B and Chupeau Y (1996) Selection of transgenic flax plants is facilitated by spectinomycin. *Transgenic Res.,* **5** : 131-137.

Cullis CA (1981) DNA sequence organisation in the flax genome. *Biochim. et Biophys. Acta,* **652** : 1-15.

Deroles SC and Gardner RC (1988) Expression and inheritance of kanamycin resistance in a large number of transgenic petunias generated by *Agrobacterium*-mediated transformation. *Plant Mol. Biol.,* **11** : 355-364.

Dong JZ and McHughen A (1991) Patterns of transformation intensity on flax hypocotyls inoculated with *Agrobacterium tumefaciens*. *Plant Cell Rep.,* **10** : 555-560.

Dong JZ and McHughen A (1993a) An improved procedure for production of transgenic flax plants using *Agrobacterium tumefaciens*. *Plant Sci.,* **88** : 61-71.

Dong JZ and McHughen A (1993b) Transgenic flax plants from *Agrobacterium* mediated transformation: incidence of chimeric regenerants and inheritance of transgenic plants. *Plant Sci.,* **91** : 139-148.

Dong JZ, Yang MZ, Jia SR and Chua NH (1991) Transformation of melon (*Cucumis melo* L.) and expression from the cauliflower mosaic virus 35S promoter in transgenic melon plants. *Bio/Technology,* **9** : 858-863.

El-Kharbotly A, Jacobsen E, Stiekema WJ and Pereira A (1995) Genetic localisation of transformation competence in diploid potato. *Theor. Appl. Genet.,* **91** : 557-562.

Ellis JG, Finnegan EJ and Lawrence GJ (1992) Developing a transposon tagging system to isolate rust-resistance genes from flax. *Theor. Appl. Genet.,* **85** : 46-54.

Ellis J, Lawrence G, Ayliffe M, Anderson P, Collins N, Finnegan J, Frost D, Luck J and Pryor T (1997) Advances in the molecular genetic analysis of the flax-flax rust interaction. *Annu. Rev. Phytopathol.,* **35** : 271-291.

Gamborg OL and Shyluk JP (1976) Tissue culture, protoplasts, and morphogenesis in flax. *Bot. Gaz.,* **137** : 301-306.

Gill YS and Yermanos DM (1967) Cytogenetic studies on the genus *Linum*: I. Hybrids among taxa with 15 as the haploid chromosome number. *Crop Sci.,* **7** : 623-627.

Hadi MZ, McMullen MD and Finer JJ (1996) Transformation of 12 different plasmids into soybean *via* particle bombardment. *Plant Cell Rep.,* **15** : 500-505.

Hansen G and Wright MS (1999) Recent advances in the transformation of plants. *Trends Plant Sci.,* **4** : 226-231.

Haughn G, Smith J, Mazur B and Somerville C (1988) Transformation with a mutant *Arabidopsis* acetolactate synthase gene renders tobacco resistant to sulfonylurea herbicides. *Mol. Gen. Genet.,* **211** : 266-271.

Hepburn AG, Clarke LE, Blundy KS and White J (1983) Nopaline Ti-plasmid, pTiT37, T-A insertions into a flax genome. *J. Mol. Appl. Genet.,* **2** : 211.

Hoffmann W (1961) Lein, *Linum usitatissimum* L. In: *Handbuch der Pflanzenzüchtung* (Eds Kappert H and Rudorf W), Paul Parey Berlin und Hamburg, Bd **5** : 264-366.

Holford P and Newbury HJ (1992) The effects of antibiotics and their breakdown products on the *in vitro* growth of *Antirrhinum majus*. *Plant Cell Rep.,* **11** : 93-96.

Horsch RB, Fraley RT, Rogers SG, Sanders PR, Lloyd A and Hoffmann N (1984) Inheritance of functional foreign genes in plants. *Science,* **223** : 496-498.

Jia SR, Yang MZ, Ott R and Chua NH (1989) High frequency transformation of *Kalanchoe laciniata*. *Plant Cell Rep.,* **8** : 336-340.

Jordan MC and McHughen A (1988a) Glyphosate tolerant flax plants from *Agrobacterium*-mediated gene transfer. *Plant Cell Rep.,* **7** : 281-284.

Jordan MC and McHughen A (1988b) Transformed callus does not necessarily regenerate transformed shoots. *Plant Cell Rep.,* **7** : 285-287.

Jordan MC and McHughen A (1993) Transformation in *Linum usitatissimum* L. (Flax). In: *Biotechnology in Agriculture and Forestry*, Vol. 22, Plant Protoplasts and Genetic Engineering III (Ed. Bajaj YPS). pp. 244-252.

Klein TM, Wolf ED, Wu R and Sanford JC (1987) High-velocity microprojectiles for delivering nucleic acids into living cells. *Nature* (Lond.) **327** : 70-73.

Lawrence G, Finnegan J and Ellis J (1993) Instability of the L^6 gene for rust resistance in flax is correlated with the presence of a linked *Ac* element. *Plant J.,* **4** : 659-669.

Lawrence G, Finnegan EJ, Ayliffe MA and Ellis JG (1995) The L^6 gene for flax rust resistance is related to the *Arabidopsis* bacterial resistance gene *RPS2* and the tobacco viral resistance gene *N. Plant Cell,* **7** : 1195-1206.

Ling HQ (1992) Versuche zur Entwicklungsbiologie und somatischen Genetik *in vitro* mit Arten der Gattung *Linum*. Dissertation zur Erlangung des Doktorgrades, in der Mathematisch-Naturwissenschaftlichen Fakultät der Christian-Albrechts-Universität zu Kiel, Germany.

Ling HQ and Binding H (1987) Plant regeneration from protoplasts in *Linum. Plant Breed.,* **98** : 312-317.

Ling HQ and Binding H (1997) Transformation in protoplast cultures of *Linum usitatissium* and *L. suffruticosum* mediated with PEG and with *Agrobacterium tumefaciens. J. Plant Physiol.,* **151** : 479-488.

Ling HQ, Kriseleit D and Ganal MW (1998) Effect of ticarcillin/potassium clavulanate on callus growth and shoot regeneration in *Agrobacterium*-mediated transformation of tomato (*Lycopersicon esculentum* Mill.). *Plant Cell Rep.,* **17** : 843-847.

Link GKK and Eggers V (1946) Mode, site, and time of initiation of hypocotyledonary bud primordia in *Linum usitatissimum* L. *Bot. Gaz.,* **107** : 441-451.

Mathews VH and Narayanaswamy S (1976) Phytohormone control of regeneration in cultured tissues of flax. *Z. Pflanzenphysiol.,* **80** : 436-442.

McHughen A (1989) *Agrobacterium*-mediated transfer of chlorsulfuron resistance to commercial flax cultivars. *Plant Cell Rep.,* **8** : 445-449.

McHughen A and Holm F (1991) Herbicide resistant transgenic flax field test: Agronomic performance in normal and sulfonylurea-containing soils. *Euphytica,* **55** : 49-56.

McHughen A and Holm FA (1995a) Development and preliminary field testing of a glufosinate-ammonium tolerant transgenic flax. *Can. J. Plant Sci.,* 117-120.

McHughen A and Holm FA (1995b) Transgenic flax with environmentally and agronomically sustainable attributes. *Transgenic Res.,* **4** : 3-11.

McHughen A and Jordan MC (1989) Recovery of transgenic plants from "escape" shoots. *Plant Cell Rep.,* **7** : 611-614.

McHughen A, Jordan M and Feist G (1989) A preculture period prior to *Agrobacterium* inoculation increases production of transgenic plants. *J. Plant Physiol.,* **135** : 245-248.

McHughen A and Rowland GG (1991) The effect of T-DNA on the agronomic performance of transgenic flax plants. *Euphytica,* **55** : 269-275.

McHughen A, Rowland GG, Holm FA, Bhatty RS and Kenaschuk EO (1997) CDC triffid transgenic flax. *Can. J. Plant Sci.,* **77** : 641-643

McSheffrey SA, McHughen A and Devine MD (1992) Characterization of transgenic sulfonylurea-resistant flax (*Linum usitatissimum*). *Theor. Appl. Genet.,* **84** : 480-486.

Mlynárová L, Bauer M, Nap JP and Pretová (1994) High efficiency *Agrobacterium*-mediated gene transfer to flax. *Plant Cell Rep.,* **13** : 282-285.

Murray BE, Handyside RJ and Keller WA (1977) *In vitro* regeneration of shoots on stem explants haploid and diploid flax (*Linum usitatissimum*). *Can. J. Genet. Cytol.,* **19** : 177-186.

Okkels FT and Pedersen MG (1988) The toxicity to plant tissue and to *Agrobacterium tumefaciens* of some antibiotics. *Acta Hort.,* **225** : 199-207.

Polyakov AV, Chikrizova OF, Kalyaeva MA, Zakharchenko NS, Balokhina NV and Bur´yanov YI (1998) The transformation of fiber flax plants. *Rus. J. of Plant Physiol.,* **45** : 764-769.

Potrykus I (1991) Gene transfer to plants: Assessment of published approaches and results. *Annu. Rev. Plant Physiol. & Plant Mol. Biol.,* **42** : 205-225.

Potrykus I, Shillito RD, Saul MW and Paszkowski J (1985) Direct gene transfer state of the art and future potential. *Plant Mol. Biol. Rep.* **3** : 117-128.

Rakouský S, Tejklová E, Wiesner I, Wiesnerová D, Kocábek T and Ondrej M (1999) Hygromycin B-an alternative in flax transformant selection. *Biol. Plant.,* **42** : 361-369.

Roekel JSC, Damm B, Melchers LS and Hoekema A (1993) Factors influencing transformation frequency of tomato (*Lycopersicon esculentum*). *Plant Cell Rep.,* **12** : 644-647.

Shah DM, Horsch RB, Klee HJ, Kishore GM, Winter JA, Tumer NE, Hironaka CM, Sanders PR, Gasser CS, Aykent S, Siegel NR, Rogers SR and Fraley RT (1986) Engineering herbicide tolerance in transgenic plants. *Science* **233** : 478-481.

Sonti RV, Chiurazzi M, Wang D, Davies CS, Harlow GR, Mount DW and Signer ER (1995) *Arabidopsis* mutants deficient in T-DNA integration. *Proc. Natl. Acad. Sci. U.S.A.,* **92** : 11786-11790.

Wakita Y, Otani M, Iba K and Shimada T (1998) Co-integration, co-expression and co-segregation of an unlinked selectable marker gene and *NtFAD3* gene in transgenic rice plants produced by particle bombardment. *Genes & Genetic Systems,* **74** : 219-226.

Wijayanto T and McHughen A (1999) Genetic transformation of *Linum* by particle bombardment. *In Vitro Cell. Dev. Biol. Plant,* **35** : 456-465.

Zhan XC, Jones DA and Kerr A (1988) Regeneration of flax plants transformed by *Agrobacterium rhizogenes. Plant Mol. Biol.,* **11** : 551-559

Zupan J and Zambryski P (1997) The *Agrobacterium* DNA transfer complex. *Critical Rev. Plant Sci.,* **16** : 279-295.

Chapter 10

BIOTECHNOLOGY OF THE OIL PALM (*ELAEIS GUINEENSIS* JACQ)

Alain Rival*[1], Tregear James[2], Jaligot Estelle[2], Morcillo Fabienne[2], Aberlence Frédérique[2], Billotte Norbert[1], Richaud Frédérique[2], Beule Thierry[2], Borgel Alain[2] and Duval Yves[2]

[1]*CIRAD-CP. TA80/PSIII. Bd de la Lironde. F-34398 Montpellier Cedex 05, France*
[2]*CIRAD-CP/IPD. GeneTrop. Centre IRD. BP 5045. F-34032 Montpellier Cedex 01, France*

Summary

Biotechnological approaches play an increasing role in breeding strategies for oil palm. They are fully integrated in the rapid integration of genetic progresses aimed at broadcasting improved material to the planters.

Clonal micropropagation : *In the mid-seventies, the first results obtained by the strategy of reciprocal recurrent selection (RRS) and the emergence of the methods of cloning by in vitro culture led to the development of a technique of micropropagation through somatic embryogenesis which was tested initially in Côte d'Ivoire, then in Malaysia and Indonesia. This work established the utility of clonal micropropagation which was found to enable the production of high yielding clones. In addition, this development phase highlighted the difficulties related to scaling-up in relation to, on the one hand, mass production required to meet the needs of planters and, on the other hand, the genetic fidelity of the regenerated plant material. These two concerns led the Cirad-IRD team to look further into the underlying mechanisms involved in somatic embyogenesis and the somaclonal variation events induced by the regeneration techniques. The development of a regeneration protocol based on the use of*

*Corresponding author : E-mail : alain.rival@cirad.fr

embryogenic suspensions has provided us with a method which allows production on a large scale of single somatic embryos displaying structural similarities with zygotic seed embryos. Studies of the late phases of embryogenesis were carried out through the comparative analysis of the development and maturation of zygotic and somatic embryos. Research work resulted in a better understanding of the physiology of the somatic embryo during its development, the accumulation of various types of storage molecules and of the mechanisms of acquisition of tolerance to desiccation. These data are prerequisites for the production of embryos which can be stored, i.e. the production of artificial seeds and/or cryopreserved embryos.

Structural and functional genomics : *The Cirad/IRD laboratory is currently establishing a collection of systematically sequenced EST (Expressed Sequence Tag) cDNA clones representing genes expressed in specific tissues, developmental stages or environmental conditions of interest. Our aim is to assemble an extensive catalogue of oil palm genes which can be screened either on the basis of their sequence affinities (similarity to know genes of interest) or by using high throughput macro- or microarray screening to examine their expression patterns. In this connection, we are establishing a collection of EST clones as part of a French government-funded Genopole project. A key part of our EST project is centred on studying the functioning of the oil palm shoot apical meristem.*

Marker-assisted breeding is being implemented in a long-term multi-stage project in order to make optimum use of the existing network of field experiments by capitalising upon recent advances in molecular maker technology. Areas of interest include: molecular analysis of genetic diversity in both E. guineensis and E. oleifera germplasms; large scale development of PCR-based microsatellite marker; parallel development of three genome mapping and QTL detection projects studying key agronomic characters: resistance to Fusarium wilt in Africa; increased and stable oil palm production; and detection and introgression of E. oleifera genetic factors conferring resistance to Bud Rot in Latin America. A network of specifically designed field trials is used to validate QTL markers, to implement marker-assisted breeding strategies in oil palm, and to pursue physical mapping towards the characterisation, cloning and tagging of useful genes.

Post-genomics : *In order to tackle the problem of the mantled* *flowering abnormality, which is induced during the oil palm micropropagation process, the Cirad-IRD group has carried out studies of gene expression in tissue cultures as a means of establishing an early clonal conformity testing procedure. Today, a number of genes whose expression appears to be modulated in tissue cultures according to their mantled status have been identified and characterised. Efforts will now be directed towards the exploitation of this new knowledge in the development of clonal conformity tests. In particular, we seek to address two key questions. Firstly, we wish to ascertain when in the tissue culture process testing should best be carried out. This aspect can be tackled by monitoring the expression of the genes of interest throughout the regeneration process. Secondly, we wish to identify the best methodology for clonal conformity testing by comparing RNA, protein and DNA (PCR) based approaches. Studies carried out by the Cirad/IRD group on genomic DNA methylation changes induced by tissue culture suggest that the latter may play a key role in the determination of the mantled abnormality. We recently demonstrated, by the use of two complementary methods for evaluating methylation rates at the genome-wide level, that there is a highly significant DNA hypomethylation in leaves of abnormal regenerants and calli, compared to their normal counterparts. New investigations are now aimed at elucidating the mechanisms and/or sequences through which epigenetic misregulation could provoke the onset of the mantled phenotype. We are currently establishing protocols combining "regional" studies of chromatin conformation and the quantification of corresponding DNA methylation rates. Moreover, we hope that by isolating oil palm relatives of the Arabidopsis thaliana METI DNA-methyltransferase gene, we will obtain formation to help explain how the mantled abnormality is generated, dysfunctions of genes of this family having been found to be linked to a number of developmental abnormalities.*

Genetic engineering : *We are now at only the beginning of research on genetic engineerng in oil palm even if the regeneration of somaclones through somatic embryogenesis has been obtained since the mid-70s. Several projects are under way, with the aim of introducing foreign genes into oil palm in order, for example, to modify oil yield and quality or resistance to pests and diseases.*

Keywords : Clonal propagation, genetic engineering, structural and functional genomics, oil palm.

1. INTRODUCTION

Oil palm (*Elaeis guineenesis* Jacq) belongs to the family of the *Aracaceae* in the order of the Spadiciflores, sub-order of Palmales. The size of oil palm genome, as recently estimated by flow cytometric analysis (Rival *et al.*, 1997a) is $3.4.10^9$ bp with a 2n = 32 caryotype.

At the adult age, this arborescent monocotyledon shows the typical aspect of palms with a crown consisting of 40-50 opened palmate leaves (Fig. 1).

Growth is ensured by a single terminal meristem which produces an average of two leaves per month. Vertical growth rates may vary from 30 to 75 cm per year, depending on the genetic origin of the palm. The root system is composed of a large number of fasciculate adventive roots which secure the anchorage of the plant. Oil palm growths preferentially in rich in temporarily damp alluvial soils (Hartley, 1988). It can reach 30 m in height, with a crown of leaf of 10-16 m in diameter, and so requires a large soil area to properly ensure correct growth. The standard plantation density in industrial plantations is generally 143 palms per ha. The oil palm is a temporal dioecious species (Cruden, 1988) which displays alternate male and female flowering cycles throughout the life of the plant.

Oil palm is the second source after soybean of edible vegetable oil (Fig. 2). Industrial exploitation of oil palm started at the beginning of the 20th century, firstly in South East Asia (Malaysia, Indonesia), then in Africa on the edges of the Gulf of Guinea and then in Brazil in the 1920's. Since the 60's, its cultivation area has been extended to South America, mainly in the Amazon basin (Equator, Colombia, Peru) and to the Pacific coast (Costa Rica). During this period, intensive planting programmes were developed in South East Asia and to a lesser extent in West Africa. World palm oil production has undergone a tenfold increase since 1948 and it was near 22 million tonnes per year in 2000 (Table 1). Oil production is still growing due to a continuous increase in the area planted. Malaysia and Indonesia together account for more than 80% of world production, with an average increase of 7% over the last five years (1995-2000), while production remained static in Latin America, particularly in Colombia. The consumption of oleaginous products continues to rise (Table 2), due to demographic pressure, the increasing domestic use of oil palm in producing countries and to the demand from highly populated countries such as India, China and Pakistan, which are

Figure 1 : Oil palms planted at CNRA La Mé Research Station in Cote d'Ivoire (West Africa)

Table 1. Palm oil production statistics (10^3 tons). Source : Oil World (2001).

	1995	1996	1997	1998	1999	2000
World	15 477	16 285	17 487	16 839	20 525	21 728
Malaysia	7 811	8 386	9 057	8 315	10 553	10 840
Indonesia	4 480	4 540	5 380	5 006	6 250	6 900*
Nigeria	660*	670*	680*	690*	720	740
Colombia	387	410	441	422	500	524

*estimated

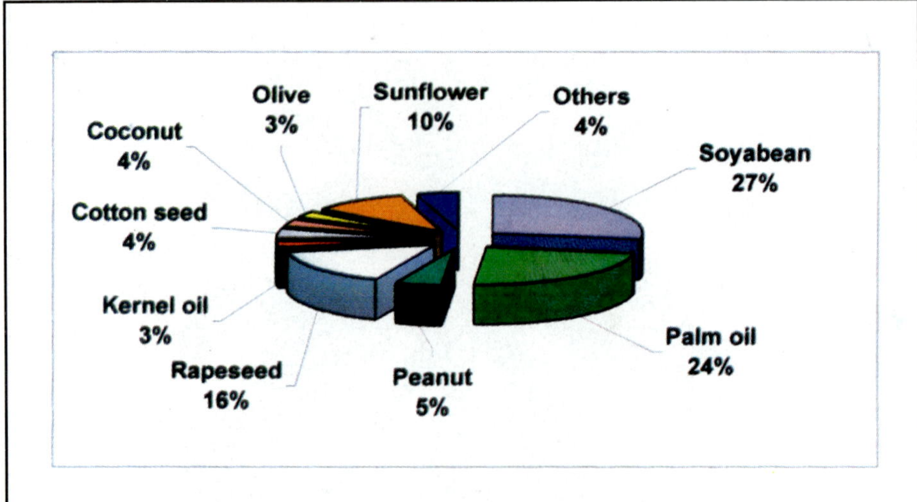

Figure 2 : World production in vegetable oils (Oil World, PORLA, 2000)

at present relatively small consumers of oleaginous products (10, 12 and 16.2 kg per capita per year respectively in 1998). The price of palm oil, after a continuous growth (from 300 to 650 USD per ton of CPO – Crude Palm Oil) during the 1990-95 period, stabilised at an average price of 550 USD/Ton during the 1996-99 period, before it dropped dramatically to 'each 310 USD/Ton on average during year 2000.

Table 2. Palm oil consumption statistics (10^3 tons). Source : Oil World (2001).

	1995	1996	1997	1998	1999	2000
World	14 627	16 042	17 737	17323	19429	21760
Indonesia	2 115	2 503	2 841	2784	2885	2947
E.U.	1 698	1 803	1 923	2000	2136	2348
China	1 305	1 211	1 726	1594	1406	1643
India	768	1 222	1 388	1772	2953	3750
Malaysia	1 098	1 236	1 191	985	1231	1388
Pakistan	1 177	1 148	1 124	1119	1062	1083
Nigeria	698	732	792	779	776	849

2. CHALLENGES IN OIL PALM BIOTECHNOLOGY

Due to the increasing demand for palm oil, lack of plantable land and the foreseeable increases in cultivation costs, it is necessary to make available to the farmers planting material with high genetic potential.

Nowadays, planting material consists solely of *tenera* hybrids (fruits with shell of intermediate thickness), originating from crosses between *dura* (thick shell) and *pisifera* (thin shell) types, the thickness character being controlled by a monofactorial gene (Beirnaert and Vanderweyen, 1941).

The breeding strategies developed by seed companies are aimed at producing *dura* × *pisifera* hybrids with a high productivity of oil containing a high proportion of unsaturated fatty acids, a low growth rate and, in certain cases, resistance/tolerance to diseases such as vascular wilt caused by *Fusarium oxysporum elaeidis* in Africa (Hardon *et al.*, 1976). Breeding schemes now incorporate the exploitation of genetic resources able to provide tolerance/resistance to diseases which are confined to specific parts of the oil palm cultivation area *Ganoderma* disease in South East Asia; and also bud rot in Latin America.

Since each selection cycle lasts for around 10 years, genetic improvement is very slow, even if much progress has been achieved over the last 50 years (Jacquemard *et al.*, 1997). A very high heterogeneity is still observed among hybrids, some palms producing 60% more oil than the average of the progeny of a given cross (Noiret, 1981). These characteristics must be considered along with the low planting density (generally *ca* 143 palms per hectare) and the necessity of establishing seed orchards for the production of commercial planting material; thus it can be seen that oil palm improvement is labour intensive, time consuming and therefore expensive.

Biotechnological approaches applied to oil palm breeding have considerably increased in their importance, not only in micropropagation, but also in marker-assisted breeding, and more generally, in physiological and molecular studies of the expression of genes of paramount agronomic value (flowering, abscision, etc.).

Our group in Cirad has capitalised on developments in this area by regularly incorporating new biotechnological approaches into its overall strategy, beginning with classical tissue culture-based approaches and progressing to molecular markers, differential gene expression analysis and genetic transformation.

Our final aim, in accordance with Cirad's mandate for development of tropical agriculture, is to obtain, protect, multiply and disseminate genetically improved oil palm material to the planters.

3. CLONAL PROPAGATION

The constraints which exist in oil palm breeding make desirable the development of a vegetative propagation technique which would allow:

- the exploitation of the variability existing among the *tenera* hybrids by cloning elite individuals (Noiret, 1981).
- an increase of the production of high quality seeds by cloning the best male parents (*pisifera*), since pollen production can be a limiting factor (Hartley, 1988; Krikorian, 1989);
- the exploitation of (*E. guineensis* × *E. oleifera*) interspecific hybrids, a limited number of which are fertile, but which can show a good tolerance to pests and diseases, notably in South America (Meunier, 1975).
- the production of biclonal seeds from somatic embryogenesis-derived parents (Hartley, 1988).
- the true-to-type regeneration of future genetically-engineered material bearing useful agronomic traits (Rival, 2000).

The biological characteristics of the oil palm do not allow its vegetative propagation by conventional horticultural means, in spite of work carried out on the reversion of floral buds or the study of viviparous palms (Chevalier, 1910; Henry, 1948; Davis, 1980).

The *in vitro* culture of apices was attempted with young palms (Staritsky, 1970; Abdullah, 1990) but they did not produce satisfactory results. Moreover, the excision of the apex leads to the death of the mother tree.

Therefore, the only possibly way to clonally propagate elite oil palms is by means of somatic embryogenesis. Cloning of oil palm (*Elaeis guineensis* Jacq) is performed by inducing somatic embryogenesis on calli derived from various tissue sources, using tissue culture protocols recently reviewed in detail by Rival (2000).

3.1. Current protocols

Approximately 25 years ago, two major groups initiated research programmes for oil palm micropropagation: in the UK and in Malaysia, the Unilever Plantations and Harrison and Crossfields Plantations group (Smith and Jones, 1970; Corley *et al.*, 1977) and in France and Côte d'Ivoire, the IRHO/ORSTOM group (which became Cirad-CP/Ird in the 80's) (Rabéchault *et al.*, 1970; Noiret, 1981). These programmes were initiated to complement in-house breeding strategies and were aimed at multiplying the elite germplasm available in the plantations for commercial use. Given the potentially important commercial applications of the results, only limited information was published concerning the techniques set up, as underlined by various authors (Krikorian, 1989; Blake, 1990). During the eighties, several teams from producing countries started their own research programmes. Naturally, due to the dynamism of oil palm industry in Malaysia, several local teams developed large scale research programmes, notably at PORIM (Palm Oil Research Institute of Malaysia), now renamed as MPOB (Malaysia Palm Oil Board) (Paranjothy and Othman, 1982). Ten years ago, Wooi (1990) estimated that in Malaysia, around 10 commercial laboratories were carrying out research on *in vitro* propagation of the oil palm through somatic embryogenesis. Most of them are still conducting commercial or semi-commercial activities in oil palm clonal micropropagation, sometimes with a diversification of their micropropagation activities towards more remunerative planting material, such as banana.

The performance of oil palm micropropagation by somatic embryogenesis suggests that cloning is possible for most of the planting material currently available, including *E. guineensis* × *E. oleifera* interspecific hybrids (Duval *et al.*, 1997).

In a recent review (Rival, 2000) an overview was given of regeneration protocols applied to oil palms from various different genetic origins.

With overall success rates of over 90%, the first stages of the procedure can be considered to be fully mastered, albeit extremely slow. Indeed, it takes at least two years to regenerate a significant number of plants. However, the frequency at which proliferating embryos are obtained *in vitro*, enabling mass production, is just 40%. This stage is currently the major stumbling block for large-scale ramet production to multiply all selected ortets. Wong *et al.* (1996) recently reported 53% proliferating embryo lines, which is similar to the values obtained by the

French Cirad-CP/Ird group and its partners. The phenomenon of somatic embryo proliferation is certainly worth studying in detail in order to increase the success rate for this stage. It corresponds to the secondary embryogenesis described for other species such as woody plants (Merkle, 1995).

3.2. Large scale propagation through embryogenic suspension cultures

Progress has been made in plant production through somatic embryogenesis, particularly by developing systems based on the artificial seed concept (Redenbaugh, 1993), in which the somatic embryogenesis process is used to produce individual embryos with relatively synchronous development (Redenbaugh *et al.*, 1986). Embryo development can be halted at a given stage, either by mimicking the natural procedure that occurs during zygotic embryo quiescence or by using artificial methods, such as low temperature storage. The somatic embryos, which are often encapsulated with antifungal/nutritive coating are then used as seeds, either *in vitro* or sown directly in the field (Fujii *et al.*, 1992). The initiation of embryogenic oil palm cell suspensions has been reported by several authors (de Touchet *et al.*, 1991; Teixeira *et al.*, 1995) thus demonstrating the feasibility of this technique for a rather "recalcitrant" model.

Research has been carried out in our group (de Touchet *et al.*, 1991; Duval *et al.*, 1995a and b; Aberlenc-Bertossi *et al.*, 2000) in order to develop new methods of large scale micropropagation for oil palm (see also Bajaj, 1991 for a review).

In oil palm, embryogenic suspensions are established from friable, nodular calli, which are isolated from nodular compact calli (Duval *et al.*, 1995a). The embryogenic suspension was prepared by dissociating the initial callus in a stirred liquid medium, then by selecting the meristematic nodules that multiply in the presence of 2,4-D during subsequent subcultures. Between 10 and 12 $mg.l^{-1}$ of biomass, which represents more than 10^5 aggregates, were inoculated into a medium containing 100 $mg.l^{-1}$ of 2,4-D and 2 $mg.l^{-1}$ of activated charcoal. A series of histological observations revealed the considerable homogeneity of the tissues and how the suspension proliferates. The suspension growth rate was about 2 to 3 per monthly culture cycle. Embryogenesis expression was then triggered by the transfer of cell aggregates to a growth regulator-free liquid medium for 4 weeks. The cell aggregates then differentiated into proembryos consisting of meristematic cells. This stage, which amounts to weaning from auxin, is essential to stop proliferation and to induce the expression of embryogenesis. At the end of this stage, the suspension was filtered (1 mm mesh) so as to keep only the smallest aggregates, which are the only ones capable of developing into individualised plantlets. Embryo maturation was then achieved by spreading the filtrate onto a solid medium, at a rate of 0.05 ml PCV (Packed Cell Colume) per Petri dish, leading to the differentiation of ca 300 embryos per dish under these conditions. The early phases of this step were characterised by the formation of an epidermis, giving a shiny white appearance to the embryos, followed by the differentiation of both shoot and root poles. In many cases, only root development could be observed. Modification of the auxin/cytokinin ratio through the addition of 5μM BAP (benzylaminopurine) in the maturation medium led to a significant increase in the number of individualised shoots, with a concomitant halt in root development. The production scheme currently used in our laboratory involves 4 distinct stages, as shown in Figure 3.

To date, embryogenic suspensions have been successfully isolated for more than 20 different clonal lines. The average concentration was found to be *ca* 10^5 cell clusters per litre with a multiplication factor reaching 4x per month. These characteristics allow mass propagation. Sondahl (1991) reported on the successful culture of oil palm cell suspension in bioreactors. For oil palm, Sondahl (personal communication) estimated the price of an encapsulated embryo produced by bioreactor technology in the USA at ca 0.20 USD, to which has to be added the cost of *in vitro* culture for germination, rooting and acclimatisation. The direct transfer into greenhouse cultivation of coffee (*Coffea arabica*) somatic embryos produced by temporary immersion has been reported recently (Barry-Etenne *et al.*, 1998). Such improvements, aimed at improving the quality of microplants while reducing production costs, have a direct impact on the future economic viability of the industrial production of elite plant material.

In the case of oil palm, it is clear that research efforts on the implementation of production strategies based on the large scale use of embryogenic cell suspensions have been severely hampered by the critical problem of somaclonal variation.

Field trials are under way for the assessment of the clonal fidelity of plantlets originating from cell suspension cultures. The first results obtained from Cirad suspension culture-derived clones are quite encouraging, even though obtained from a small number of samples.

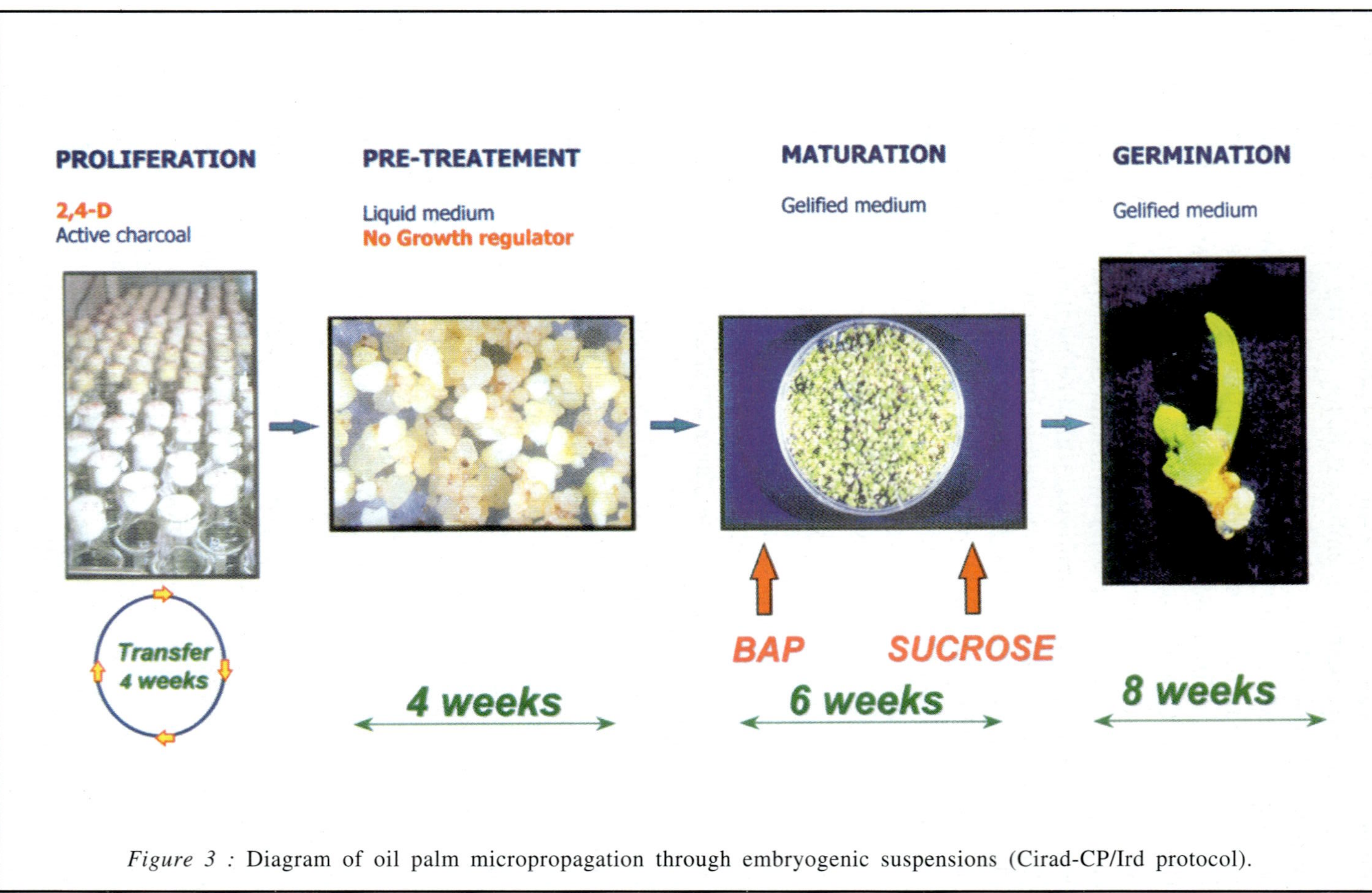

Figure 3 : Diagram of oil palm micropropagation through embryogenic suspensions (Cirad-CP/Ird protocol).

Various research groups from the oil palm industry in Malaysia have now developed embryogenic suspension cultures growing on 2,4-D free or even auxin-free media. These protocols are presently under assessment and initial observations of trueness-to-type have revealed a generally very low percentage of abnormal palms.

A management strategy, based on the sequential cryopreservation of embryogenic lines which is now feasible in our Lab (Chabrillange *et al.*, 2000), could be followed in order to lower the risk of losing embryogenic capacity in suspensions during multiplication. It is necessary to plan the simultaneous production of a limited number of embryos from a large number (10-20) of clonal lines with the aim of maintaining the widest possible genetic basis for clonal production.

3.3. Embryo maturation

Oil palm somatic embryos produced from embryogenic suspension cultures, exhibit an incomplete maturation and develop directly towards germination without passing through a quiescent phase (Aberlenc-Bertossi *et al.*, 1999). We have carried out studies to improve our knowledge of the biology of embryonic development and to identify factors involved in maturation, a critical step in the determination of somatic embryo quality for several species (Merkle *et al.*, 1995). The acquired knowledge will allow us to monitor somatic embryo desiccation tolerance and the vigour of regenerated plants and to propose novel strategies for storage, thus facilitating the large-scale production of oil palm clones. Using zygotic embryo development as a reference, physiological and biochemical characteristics of maturity were studied. As a first step we analysed changes in several key molecules between mid-embryogenesis and fruit abscision. Changes in the abundance of compounds involved in the determination of desiccation tolerance (such as oligosaccharides and abscisic acid [ABA]), or in the vigour of regenerated plantlets (such as storage proteins) were investigated. On the basis of the results obtained, we were then able to modify culture conditions in order to enhance somatic embryo maturation.

3.3.1. Acquisition of desiccation tolerance

Zygotic embryos of oil palm can withstand complete desiccation and dehydrated seeds can be stored for two to three years. Dry matter, water content, sugar and ABA contents were investigated throughout zygotic embryo development in relation to the acquisition of desiccation tolerance.

Embryo dry weight was found to increase between the 3rd and 4th month post anthesis (mpa) and was then stable until shedding (*ca* 6 mpa). Embryos underwent dehydration but the water content remained high at maturity (*ca* 1.5 g H_2O g^{-1} DW). Throughout embryo development, desiccation tolerance was significantly correlated with embryo dry weight and water content and tolerance was acquired around 3.5 mpa. Sucrose, the main sugar present throughout embryo development, accounted for approximately 24% of the dry weight. Glucose and fructose contents decreased to less than 1 $mg.g^{-1}$ dry weight in embryos at maturity. Raffinose was detected at 4 months as embryos became tolerant. Stachyose appeared later, in 5-month-old embryos. ABA accumulation peaked in embryos between 3.5 and 4 mpa. The acquisition of desiccation tolerance between the 3rd and the 4th month was associated with sugars and ABA synthesis, underlying the role of these compounds in the phenomenon (Aberlenc-Bertossi *et al.*, 1995 and 2001).

The main changes in biochemical and physiological parameters occurring *in planta* during oil palm embryogenesis are summarised in Fig. 4. Taking these results in account, the effects of sugars and ABA on desiccation tolerance, soluble sugar content and germination rate of oil palm somatic embryos were investigated. The enrichment of media

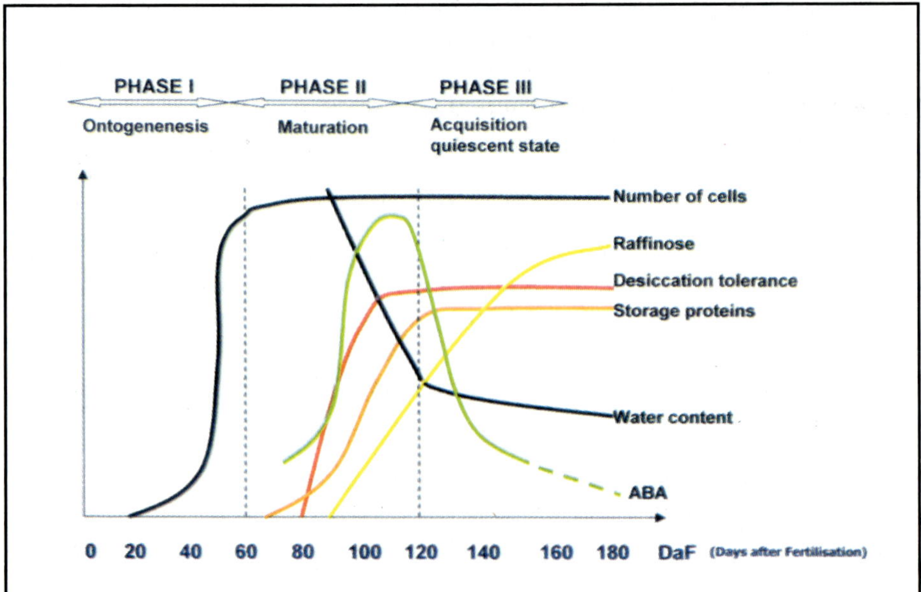

Figure 4 : Changes in biochemical and physiological parameters occurring *in planta* throughout oil palm embryogenesis.

with sucrose induced a dehydration of embryos and an increase in dry weight. Sucrose accumulated in embryos but monosaccharides content increased and oligosaccharides were not detected. Sucrose was found to significantly enhance embryo desiccation tolerance. Supplementation of the medium with exogenous ABA had no effect on embryo dry weight and water content but produced a change in sugar content (Aberlenc-Bertossi *et al.*, 2001). Monosaccharide levels decreased in ABA-treated embryos. Sucrose contents increased and raffinose was detected in embryos when extended culture on ABA containing medium was performed. ABA significantly improved embryo tolerance to rapid desiccation. In addition, the presence of ABA in germination media induced inhibition of embryo shoot emission.

The enrichment of the medium with sucrose and ABA improved somatic embryo maturation and desiccation tolerance which remains, however, incomplete as compared to zygotic embryos.

3.3.2. Accumulation of storage proteins

Storage proteins accumulated during oil palm embryo development were extracted, purified and characterised (Morcillo *et al.*, 1997). Only water- and low salt-soluble proteins, with respective sedimentation coefficients of 2S and 7S, were detected in mature embryos. After purification by gel filtration, the various classes of protein were identified and characterised by electrophoresis and amino acid composition analysis. The 2S proteins comprise polypeptides of 22 kD and 19 kD, which are acidic (pI<6) and basic (pI>9) respectively. The 7S proteins predominate and are heterogeneous oligomers (Mr of 156 and 201 kD), comprising a polypeptide triplet of Mr between 45 and 65 kD with no disulphide bonds. Their amino acid composition is broadly similar to those of the 7S proteins of other monocotyledonous embryos, but differs from those of the legume 7S vicilins. Histological examinations and electrophoresis showed that the 2S and 7S proteins appeared at the 3rd month after fertilisation, no qualitative changes being detected up to the 6th month of embryo development.

The accumulation of 7S globulins was studied in zygotic and somatic embryos (Morcillo *et al.*, 1998). Antibodies raised against these proteins were used for their detection by Western blotting and quantification by ELISA (Enzyme-Linked Immunosorbent Assay). In zygotic embryos, the 7S globulins were found to accumulate mainly between the 3rd and the 4th months after anthesis, corresponding to the end of the embryo growth. They constitute 10% of dry weight and 50% of soluble protein content. The amounts of soluble protein and 7S globulins in somatic embryos were found to increase rapidly during the early stages of development, but were almost 80 times lower than in zygotic embryos. In somatic embryos, 7S globulins represented 0.3% of dry weight and 4% of soluble proteins. After 22 days of development, protein content declined slowly. The *in vitro* production of 7S glubulins (and more generally salt-soluble proteins) was improved by the addition to the culture medium of glutamine, arginine, sucrose and ABA, the effects of these components being additive (Morcillo *et al.*, 1999).

In order to improve embryonic maturation and the vigour of regenerated plantlets, the effects of modifying *in vitro* culture conditions were investigated with respect to the accumulation of the major oil palm storage proteins, the 7S globulins (Morcillo *et al.*, 1999). In this study, the effect of arginine and glutamine on globulin accumulation was studied using somatic embryos of two different genotypes by means of ELISA. Arginine and glutamine were both found to enhance protein accumulation but in different ways which were best illustrated by measurements of soluble proteins per embryo and 7S globulin content per dry weight. Arginine increased soluble proteins by 46% irrespective of the clone, and glutamine by 19% and 63% depending on the clone. The clone which accumulated the least protein in the presence of glutamine was that which contained most of the proteins initially. Only arginine favoured an increase in 7S globulin content on a per dry weight basis, irrespective of the clone considered (+26%).

We have recently undertaken (Morcillo *et al.*, 2001) the characterisation and expression analysis of the oil palm *GLO7A* gene encoding a 7S globulin protein. To investigate further the regulation of 7S globulin gene expression in both zygotic and somatic embryos of oil palm, we isolated a cDNA clone, *GLO7A*, for use as a probe in northern hybridisation studies. The nucleotide sequence of the *GLO7A* cDNA reveals that it encodes a polypeptide of 572 amino acids (66 kD) sharing significant sequence similarities with various vicilin-like proteins of both dicotyledonous and monocotyledonous plants. Northern hybridisation analysis shows that 7S globulin mRNA accumulation in zygoitc embryos is temporally regulated with a profile essentially the same as that observed at the protein level. In somatic embryos, 7S globulin proteins were found to occur in amounts approximately 80 times lower than those in zygotic embryos. This lack of 7S globulin protein accumulation in somatic embryos is mirrored by a low accumulation of the *GLO7A* mRNA. The *in vtiro* production of 7S globulins (and more generally salt-soluble proteins) is improved by the addition to the culture medium of arginine, sucrose and ABA, the effects of these three components being additive. To investigate further the action of the 3 molecules of interest, we performed parallel studies on mRNA and protein abundance. Our studies of transcript accumulation suggest that ABA and sucrose act directly on mRNA synthesis or stability; however, it appears that there are also translational or post-translational regulatory factors which act to limit protein accumulation in somatic embryos. To assess whether *GLO7A* gene expression might be modulated by *cis*-acting promoter elements related to those found in other plants, the *GLO7A* gene promoter was cloned and sequenced. Two motifs resembling ABREs (ABA-responsive elements) and one motif resemblig a seed-specific promoter element were identified within the 5' flanking sequence.

In summary, our group succeeded in identifying markers of maturity (sugars, ABA and storage proteins) in zygotic embryo which can be used to evaluate somatic embryo quality. The results achieved on somatic embryo allow us to define *in vitro* culture conditions for improved tolerance to desiccation and accumulation of storage proteins. Further investigations will be undertaken to improve maturation protocols.

These results open the way for the use of strategies based on the "artificial seeds" concept for oil palm (Fig. 5). The latter could be obtained from embryos produced at a high rate from embryogenic suspensions, which would then be correctly treated to achieve maturation (enrichment in storage proteins). Single somatic embryos would then be encapsulated and stored (either at room temperature or at 0-4°C) before delivery at the appropriate time, according to the planting season, to tissue culture laboratories or nurseries situated close to the plantations, where germination either *in vitro* or *ex vitro* could be carried out.

3.4. Physiology of vitroplants

3.4.1. Physiology of *in vitro* rooting

Following induction by auxin treatment, the *in vitro* rooting performance of oil palm shoots varies considerably. Study implemented by Rival *et al.* (1997) has described a preliminary evaluation of gaiacol-peroxidase activity as a marker of *in vitro* rooting. It was carried out on 17 different clones obtained through the standard micropropagation procedure; no rooting occurred in the absence of exogenous auxins. Changes in

peroxidase activity were found to similar to those described previously for other species, attaining peak between 10 and 14 days under optimum initiation conditions (see review by Gaspar *et al.*, 1992). The peak was found to shift in time when the rooting success rate was low. Peroxidase activity measurements, carried out prior to induction on 24 production batches belonging to 13 different clonal lines, revealed a significantly greater heterogeneity in batches with a low success rate (>50%). Thus it was found that root initiation was applied to material with a highly variable physiological status. This study has enabled major improvements in the standard rooting protocol, which initially involved two separate induction/expression phases. A single rooting step is now used, with an auxin treatment applied over a much longer time (8 weeks) using lower NAA (Naphthalene Acetic Acid) concentrations (0.5 - 1.0 mg.1^{-1}). This quite long induction/expression phase probably acts as a buffer stage, which is able to stabilise the physiological status of shoots (peroxidase activity, levels of endogenous auxins) before the inductive treatment by auxin may act.

Studies on changes in peroxidase activity during the *in vitro* rooting of oil palm clonal plantlets have led to the implementation of an improved rooting protocol, which has been successfully developed on a large scale in production units.

3.4.2. *In vitro* photosynthesis and acclimatisation

Several studies have been conducted on oil palm in order to reduce acclimatisation losses (Rival *et al.*, 1994, 1996, 1997b,c, 1988a). This physiological problem has a very severe impact on production costs, because it occurs at the ultimate stage of the tissue culture process.

The *in vitro* photosynthetic parameters of somatic seedlings have been measured throughout the Cirad-CP/Ird somatic embryogenesis-based cloning procedure, with the aim of characterising the physiological status of the *in vitro* regenerated plants and thus optimising success rates during acclimatisation. A similar strategy has been followed in our group for the study of microplant physiology in several species of palms: coconut (Triques *et al.*, 1997 a and b) and date palm (Masmoudi *et al.*, 1999).

Various photosynthetic parameters (photochemical activities, CO_2 exchange and carboxylase enzymatic activities) were studied during four characteristic stages of *in vitro* micropropagation: proliferating somatic embryos, shoots (1st and 2nd caulogenesis cycles) and rooted plantlets (Rival *et al.*, 1997c).

In vivo chlorophyll fluorescence measurements indicated that the maximum photochemical activity of photosystem II (PSII) was very low in proliferating embryos and strongly increased in later development stages, finally reaching an activity very close to that measured in acclimatised plants. The quantum yield of photosynthetic electron transport followed the same trend except that a marked depression of electron transport activity was observed in the rooted plantlets. CO_2 exchange measurements showed that absolute levels of *in vitro* photosynthesis were low, but measurable.

Photosynthetic activity was also investigated by focusing on the primary steps of carbon metabolism in plants. the activities of two of the primary enzymes of CO_2 fixation, namely PEPC (Phosphoenol pyruvate carboxylase, EC 4.1.1.31) and Rubisco (Ribulose 1,5-bisphosphate carboxylase, EC 4.1.1.39) were measured throughout the micropropagation process. The PEPC:Rubisco ratio progressively decreased, due to a substantial depletion of PEPC activity. Specific RubisCO activity did not show any significant alteration, except a transient increase during the first caulogenesis stage on gelified medium. Quantitation of Rubisco was carried out via rocket immuno-electrophoresis, using a polyclonal antibody raised against Rubisco from green tobacco levees. The relative amount of Rubisco increased during somatic embryo development (from 3.2% in proliferating embryos to 38.8% in 2nd-cycle-shootlets), then it decreased during the rooting treatment (26.4%). The impact of sucrose enriched media (60 $g.l^{-1}$) on photosynthetic activity is most probably involved in the decrease observed during the rooting phase.

In order to characterise the physiological phenomena which occur during the acclimatisation of *in vitro*-grown oil palms, a comparison of the growth and carboxylase activities of *in vitro* propagated plants and seedlings was carried out over a 100 day period (Rival *et al.*, 1998a). Growth parameters (total FW, relative foliar FW and the number of expanded leaves) and biochemical characteristics (total soluble protein and chlorophyll content, specific PEPC, Rubisco activities and relative Rubisco content) were studied. Oil palm *in vitro* propagated plants were found to undergo an original pattern of acclimatisation, as their PEPC/ Rubisco ratio was not affected during transplanting to the greenhouse environment and remained at the same level (ca. 0.05) as was measured *in vitro* growing leaves. At about D_{60} after sowing (or *ex vitro* transplanting) the main physiological characteristics (chlorophyll and soluble protein contents, and PEPC/Rubisco ratio) were similar in both seedlings and *in vitro* propagated plants, but growth characteristics were markedly different. Rocket immuno-electrophoresis (Rival *et al.*, 1996) revealed that relative Rubisco amounts were in a comparable range (ca. 230 $mg.g^{-1}_{prot.}$) in leaves from *in vitro* grown and already acclimatised *in vitro*-propagated plants and were found to be lower than in greenhouse-cultivated adult oil palms (350 $mg.g^{-1}_{prot.}$).

All the studied photosynthetic parameters (carboxylase activities, photochemical activity, CO_2 exchanges) indicate that, in oil palm, photosynthetic activity could be measured as early as during the first caulogenesis step, showing a noticeable increase during the second step. Subsequently, photosynthesis decreased root growth.

With respect to Grout's classification (1988), it can be assumed that the oil palm belongs to the class of plants in which *in vitro*-grown leaves can contribute to autotrophy and then play an active part in acclimatisation.

It is therefore highly probable that acclimatisation losses could be due principally to poor or incomplete rooting and/or to a poor management of the environment of vitroplants (especially concerning RH) during this very critical step.

4. MOLECULAR ANALYSIS OF SOMACLONAL VARIATION

In oil palm, approximately 5% of somatic embryo-derived palms show abnormalities in their floral development, involving an apparent feminisation of male parts in flowers of both sexes, called the "*mantled*" phenotype (Corley *et al.*, 1986; Rival, 2000). This somaclonal variation phenomenon may result in partial or complete flower sterility, thus directly affecting oil production, depending on the severity of the abnormality. Interestingly, reversions to the normal phenotype over time have been found to occur, leading to a complete recovery of the normal phenotype for 100% of the slightly *mantled* individuals, and for 50% of the severely *mantled* ones after 9 years in the field (Rival *et al.*, 1998b).

Several potential biochemical markers of the *mantled* abnormality in oil palm have been investigated by studing the polypeptide patterns (Marmey *et al.*, 1991) and endogenous cytokinins (Maldiney *et al.*, 1986; Besse *et al.*, 1992). Nevertheless, it has been very difficult to assess the validity of such markers on a large number of samples, because of the lack of reproducibility (in the case of proteins) or the high cost (in the case of endogenous cytokinins) of such estimations.

Flow cytometric analysis performed on NCC and FGC, and on plantlets originating either from seeds or from somatic embryos, demonstrated a uniform 2C ploidy level (Rival *et al.*, 1997c). Extensive Random Amplified Polymorphic DNA (RAPD) experiments, involving the examination of 8900 markers, also failed to show banding patterns discriminating either the mother palm genome from its clonal offspring, or true-to-type regenerants from somaclonal variants (Rival *et al.*, 1998c).

Previous studies have shown that the *mantled* abnormality is epigenetic in nature. Firstly, it has been observed that reversion to a normal floral phenotype may occur in the field (Rival *et al.*, 2000); secondly, although the *mantled* abnormality is strongly transmitted through tissue culture, only a weak non-Mendelian transmission occurs via seeds (Rao and Donough, 1990). Thirdly, our previous studies died not allow us to identify any major alternations in genomic DNA structure that could be linked with the *mantled* phenotype.

4.1. The DNA methylation hypothesis

In recent years, evidence has been accumulating that DNA methylation at the genome wide level plays a key role in regulating plant development (Finnegan *et al.*, 2000).

Changes in DNA methylation on deoxycytidine (dC) residues have been shown to be involved in the regulation of gene expression at the transcriptional level, particularly during the differentiation/dedifferentiation processes and as a response to a variety of environmental stresses (Brown, 1989; Phillips *et al.*, 1994; Oakeley *et al.*, 1997; Finnegan *et al.*, 1998; Sheldon *et al.*, 1999; Demeulemeester *et al.*, 1999; Steward *et al.*, 2000). Furthermore, micropropagation protocols often involve growth regulator treatments, which might affect the level of DNA methylation. Phenotypic alterations have been observed in plants regenerated from cultured cells and tissues. Genetic and phenotypic variation found among regenerated plants has been termed somaclonal variation by Larkin and Scowcroft (1981). It is now clear that a diversity of genetic and epigenetic changes are underlying this instability, which may not be the result of a single causal mechanism (Finnegán *et al.*, 1993; Kaeppler *et al.*, 2000).

We firstly chose to use a global approach for the investigation of DNA methylation rate, aimed at revealing differences between normal and variant plant material. Using two genome-wide quantification methods, (HPLC estimation of total 5mdC concentrations, and *in vitro* saturation of CG sites with methyl groups by the *Sss*I Methylase-Accepting Assay), we demonstrated the occurrence of a significant genomic hypomethylation in abnormal calli (-4.5%; $p < 10^{-5}$) and leaves (-1.2%; $p < 10^{-5}$) from "*mantled*" regenerants (Fig. 6), compared with their normal counterparts. The same patterns were observed (Fig. 7) in immature inflorescences (Jaligot *et al.*, 2000).

This study on global methylation rates provides us with a first glimpse of the molecular changes associated with the *mantled* abnormality and it is consistent with the epigenetic characters observed, including reversion. Despite the highly significant decrease in methylation rate observed in *mantled* palms, it is likely that very few of the corresponding cytosines play a direct role in the triggering of the somaclonal variant phenotype.

Global DNA hypomethylation associated with local genetic or epigenetic defects has already been documented in several cases of development abnormalities, in plants (Finnegan *et al.*, 1996 and 1998; Nakano *et al.*, 2000) as well as in animals (Hansen *et al.*, 1999; Baylin and Herman, 2000).

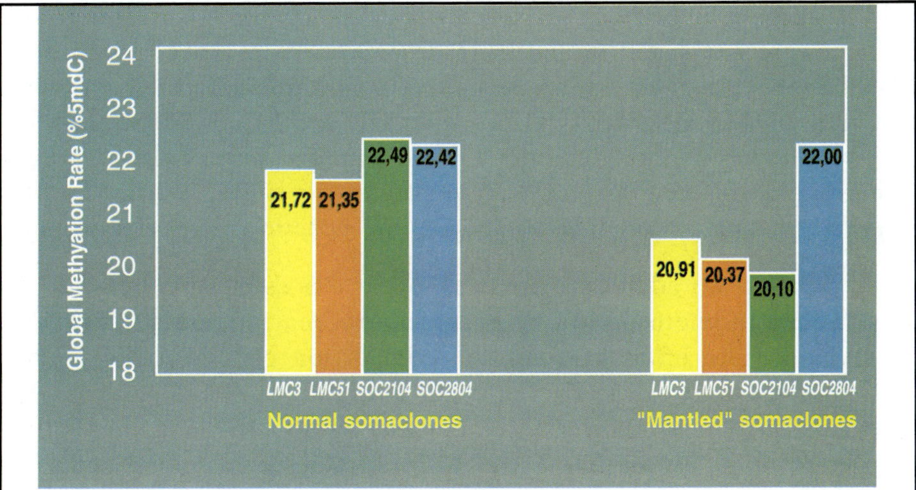

Figure 6 : Comparison of Global Methylation Rates (%) of genomic DNA in mature leaves from normal and *mantled* variants. 7 sampled palms/clone/type. 3 hydrolyses. *Interaction:* $F(3,113) = 35.40;\ p < 0.0000.$

Therefore, we considered it necessary to target more precisely those sequences which, when misregulated, could potentially account for the "*mantled*" phenotype, or which could be used as markers for the early detection of DNA methylation perturbation in the regeneration process. To this end, we have been carrying out methylation-sensitive RFLP and

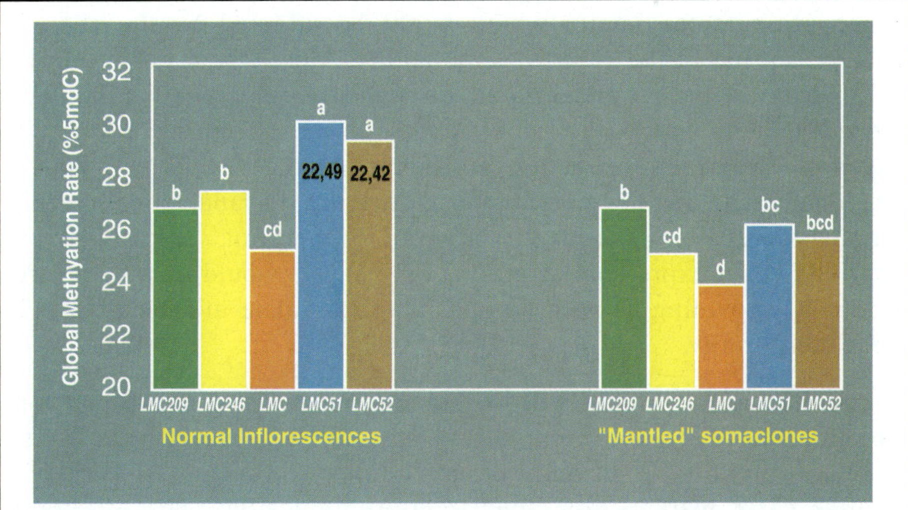

Figure 7 : Comparison of Global Methylation Rates (%) in immature inflorescences from normal and *mantled* variants. *Interaction:* $F(4,23) = 12.20;\ p < 0.05.$

AFLP studies, involving the isoschizomeric enzymes *Msp*I and *Hpa*II. RFLPs were envisaged primarily in order to screen a pool of oil palm cDNA clones for methylation-dependent polymorphism (Table 3), while the aim of MSAP (methylation-sensitive amplified polymorphism) investigations was to generate a large number of relevant markers, exhibiting a differential methylation pattern depending on the normal/*mantled* phenotype but independent of the genetic origin of clones.

In parallel, we addressed the question of a possible link between the observed hypomethylation and chromatin rearrangement, as DNA methylation rate often parallels the compaction of a genomic domain (Van Blockland *et al.*, 1997; Callebaut *et al.*, 1999; Wade *et al.*, 1999). As a continuation of this work, we hope that by isolating oil palm relatives of the *Arabidopsis thaliana METI* DNA-methyltransferase gene, we will obtain useful information to help explain how the mantled abnormality is generated, dysfunctions of genes of this family having been found to be linked to a number of developmental abnormalities.

4.2. Analysis of differential gene expression

The fact that the *mantled* abnormality involved a characteristic homeotic modification of oil palm floral architecture implies that the activity of a specific subset of genes has been altered in somaclonal variant plants, at least within the flower and fruit tissues. We are currently attempting to identify floral homeotic genes of the "MADS box" transcription factor family which might be affected by the chain of events resulting in the *mantled* abnormality. Most of our efforts to date have however been concentrated on investigating whether *mantled*-dependent gene expression patterns may also exist during the *in vitro* stages of the micropropagation procedure. At the vegetative stages of development, no distinct morphological criteria are available for screening out abnormal palms. The identification of "fingerprint" genes displaying *mantled*-related expression at pre-planting stages would thus be of great benefit by providing us with the means to formulate molecular tests for clonal conformity.

In collaboration with MPOB, within the framework of the "Somaclonal Variation in Oil Palm" initiative, we have carried out extensive studies of gene expression in oil palm tissue cultures as a means of identifying putative early markers for clonal conformity testing. The differential display (*ddRT-PCR*) technique (Liang and Pardee, 1992) was used for this purpose. The latter technique exploits the rapidity of the polymerase

Table 3. Characterisation of Southern blot polymorphism revealed by cDNA probes on isoschizomeric methylation-sensitive RFLP digestions of DNA extracted from oil palm embryogenic calli.

	LMC 458				LMC 464			
	Enzyme-dependent polymorphism		Type-dependent polymorphism between *IIpa*II digestions	Type-dependent polymorphism between *Msp*I digestions	Enzyme-dependent polymorphism		Type-dependent polymorphism between *Hpa*II digestions	Type-dependent polymorphism between *Msp*I digestions
	($C^{m}CGG$)		($^{?}C^{m}CGG$)	($^{m}C^{?}CGG$)	($C^{m}CGG$)		($^{?}C^{m}CGG$)	($^{m}C^{?}CGG$)
Probe	NCC	FGC			NCC	FGC		
CPHO 1	+	+	-	-	+	+	-	-
CPHO 5	+	+	-	-	+	+	-	-
CPHO 6	-	-	-	-	-	-	-	-
CPHO 7	+	+	+	-	+	+	-	-
CPHO 12	-	-	-	-	-	-	-	-
CPHO 17	+	+	-	-	+	+	-	-
CPHO 23	+	+	-	-	+	+	-	-
CPHO 27	+	+	-	-	+	+	-	-
CPHO 30	-	-	-	-	-	-	-	-
CPHO 31	+	+	-	-	+	+	-	-

Table 1. Continued

	LMC 458				LMC 464			
	Enzyme-dependent polymorphism		Type-dependent polymorphism between *Hpa*II digestions	Type-dependent polymorphism between *Msp*I digestions	Enzyme-dependent polymorphism		Type-dependent polymorphism between *Hpa*II digestions	Type-dependent polymorphism between *Msp*I digestions
	(C^mCGG)		($^?C^mCGG$)	($^mC^?CGG$)	(C^mCGG)		($^?C^mCGG$)	($^mC^?CGG$)
Probe	NCC	FGC			NCC	FGC		
CPHO 32	+	+	-	-	+	+	-	-
CPHO 34	+	+	-	-	+	+	-	-
CPHO 39	+	+	-	-	+	+	-	-
CPHO 42	+	+	-	-	+	+	-	-
CPHO 45	-	-	-	-	-	-	-	-
CPHO 48	+	+	-	-	+	+	+	-
CPHO 49	+	+	+	-	+	+	+	-
CPHO 53	+	+	+/-	-	+	+	+/-	-
CPHO 54	+	+	-	-	+	+	+	-
CPHO 59	-	-	-	-	-	-	-	-

Table 1. Continued

	LMC 458				LMC 464			
	Enzyme-dependent polymorphism		Type-dependent polymorphism between *IIpa*II digestions	Type-dependent polymorphism between *Msp*I digestions	Enzyme-dependent polymorphism		Type-dependent polymorphism between *Hpa*II digestions	Type-dependent polymorphism between *Msp*I digestions
	(C^mCGG)		($^?C^mCGG$)	($^mC^?CGG$)	(C^mCGG)		($^?C^mCGG$)	($^mC^?CGG$)
Probe	NCC	FGC			NCC	FGC		
CPHO 60	+	+	-	-	+	+	-	-
CPHO 62	+	+	+	+	+	+	+	+
CPHO 63	+	+	-	+	+	+	-	+
CPHO 64	+	+	-	-	+	+	-	-
CPHO 66	+	+	-	-	+	+	-	-
CPHO 69	+	+	-	-	+	+	-	-
CPHO 70	+	+	-	-	+	+	-	-
CPHO 70								

Plus (+) and minus (-) signs refer respectively to the presence or to the absence of detection of a category of differential banding pattern with a given probe. Non-reproducible patterns are marked (+/-).

chain reaction (PCR) to compare mRNA abundance between two or more RNA samples extracted from the plant material of interest. The isolation, validation and characterisation of expression markers of the *mantled* abnormality involves a number of stages summarised in Fig. 8. In order to study epigenetic factors associated with the *mantled* phenotype it is desirable to use material which is homogeneous. Several factors must be taken into account; firstly the physiological state of the material, secondly its genotypic background and thirdly its tissue composition. In order to be able to compare cultures of comparable physiological status, sets of clonal lines are established at the same zero time point.

By initiating a culture from a normally flowering seed-derived palm and, at the same time, an abnormal ramet palm previous cloned from the latter, it is possible to obtain two cultures of the same genetic background differing in their status with respect to the *mantled* abnormality. Additionally, a third culture may be established from a normal ramet palm previously cloned from the same original seed-derived palm. This culture may have a normal status with respect to the mantled abnormality, but the risk of abnormality is likely to be increased owing to the extra phase of tissue culture which it has undergone compared with the ortet-derived culture. We have studied gene expression by differential display at three different stages of the tissue culture process, namely the callus, proliferating embryo and leafy shoot stages. Initially, we focused our attention upon the proliferating somatic embryo stage; however, this material suffers the drawback that tissue composition very variable, leading to problems of homogeneity. This led us to concentrate our efforts mostly on the leafy shoot and callus stages for subsequent work.

As Figure 8 illustrates, the differential display stage provides candidate markers which must be validated and characterised using a range of different molecular techniques. Initially, Northern blotting is used to confirm differential expression using the same material as for the differential display experiment. Next, markers confirmed as being differential are tested on various other cultures to assess whether gene expression is consistently associated with *mantled* status. Any markers considered sufficiently promising after this stage are characterised in detail by the isolation of the cDNA coding sequence, Northern studies of tissue specificity and genomic 5' flanking region isolation. The latter allows us to investigate whether any conserved *cis*-acting elements are present in the promoter region which might shed light on the regulation of the gene in question.

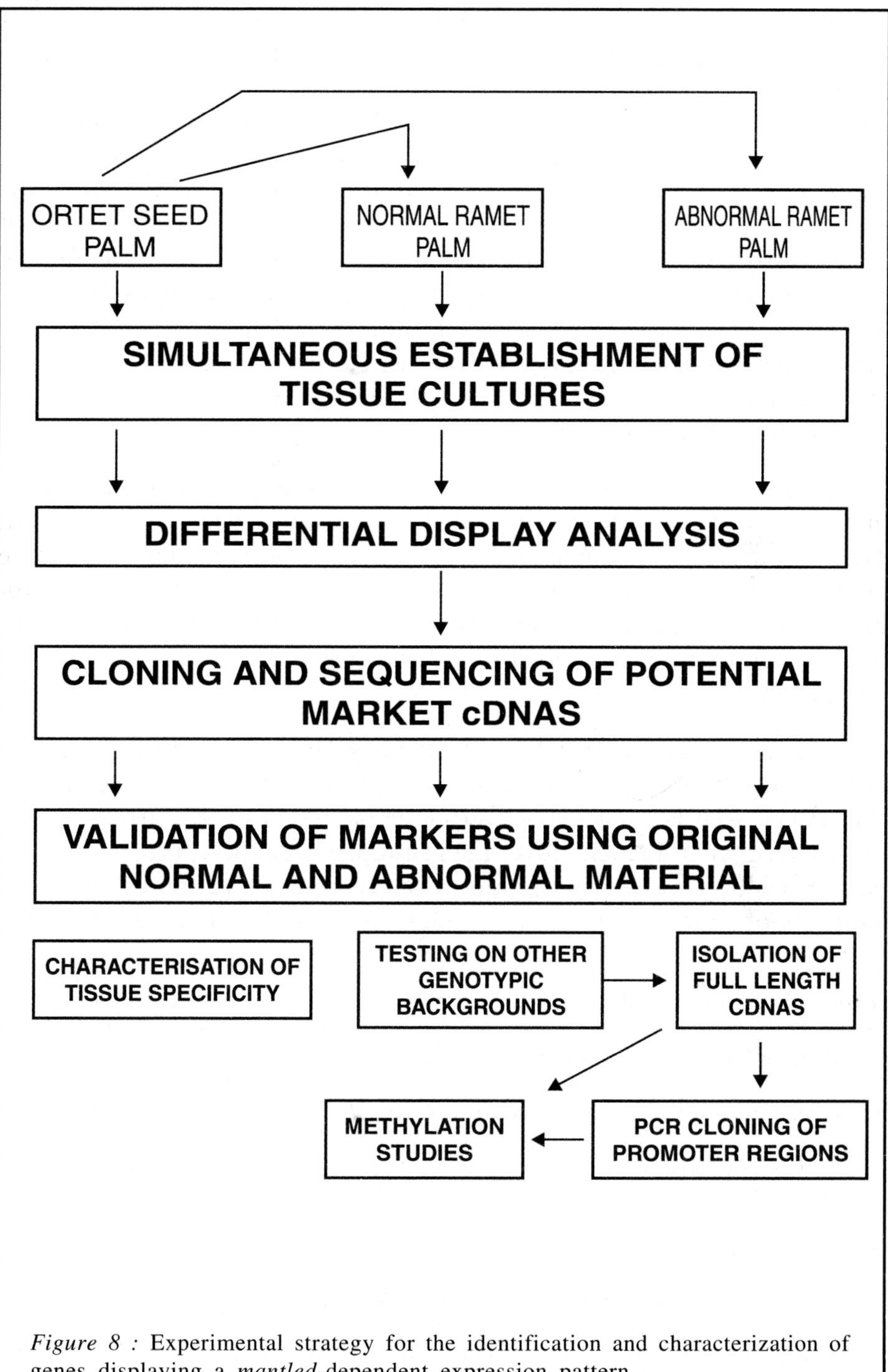

Figure 8 : Experimental strategy for the identification and characterization of genes displaying a *mantled*-dependent expression pattern.

An illustration of the results obtained in the various stages of marker isolation and validation is shown in Table 4. In this case, data obtained for leafy shoot material is presented. We have identified and characterised putative expression markers of the *mantled* abnormality for several different stages of the micropropagation procedure (callus, embryoids, leafy shoots and greenhouse-harvested leaves). One of the most promising markers is currently being jointly patented by Cirad and MPOB. Research papers discussing the data obtained in the course of the project are being prepared for publication (J. Tregear *et al.*, submitted). In the next stage of this work, we will be testing the reliability of the markers on a wider range of genotypes. This will provide a clearer picture of their reliability and enable us to select specific markers of interest for pilot scale clonal conformity testing.

The latter might be achieved by a number of different techniques, including Northern hybridisation, immunodetection (see below) or PCR. The final choice will depend on a number of factors, notably the nature of the marker gene and the developmental stage and tissue chosen for the testing. This choice of maker will also have to integrate the real cost of the technique, in comparison with the production costs of oil palm vitroplants.

In parallel with the work described above, the Cirad-CP/Ird laboratory is also currently establishing a collection of systematically sequenced EST (Expressed Sequence Tag) cDNA clones. This work is being carried out as part of a French government-funded Genopole project. A key part of our EST project is centred on studying the functioning of the oil palm shoot apical maristem. We thus plan to build up a collection

Table 4. Summary of differential display marker data for leafy shoot material

No of potential markers identified by differential display	46
No of individual cloned cDNA fragments obtained	58
No of cDNAs producing differential signals in first stage of northern retesting	13
No of cDNAs producing stronger northern signal in normal tissue	6
No of cDNAs producing stronger northern signal in abnormal tissue	7
No of cDNAs producing consistent signals for several genotypes (confirmed to date)	3

representing genes expressed in response to environmental conditions of interest. Our aim is to assemble an extensive catalogue of oil palm genes in this way, which will be screened either on the basis of their sequence affinities (similarity to known genes of interest) or by using high throughput macro- or microarray screening to examine their expression patterns.

4.3. Protein studies

In order to improve our understanding of the differential gene expression phenomena underlying the appearance of the *mantled* abnormality and to develop strategies for the implementation of simple and inexpensive clonal conformity testing, we are in the process of initiating protein studies as a complementary approach to the transcriptome-based work described above. Initial work will be aimed at characterising variations in the abundance of individual polypeptides in relation to clonal conformity using two dimensional SDS polyacrylamide gel electrophoresis (2D SDS-PAGE).

Although the latter technique has been practised for many years, it originally suffered from problems of reproducibility which have been considerably improved, thanks to recent efforts in this area (Wilkins *et al*, 1995). Thus there has emerged the new field of proteomics which enables global comparisons to be made between mRNA and protein accumulation. We will be using this approach in an attempt to identify early protein markers of normal or variant tissues. One inherent advantage of a protein-based approach is that it allows faster progression towards an antibody test compared with RNA studies.

4. MARKER ASSISTED BREEDING

Oil palm breeding programmes are all based on a major *Sh* gene determining three variety types (Beirnaert and Vanderweyen, 1941): *dura* (homozygous loci Sh+/Sh+) with thick-shelled fruits, *pisifera* (homozygous loci Sh-/Sh) which is generally sterile female and produce shell-less fruits, and *tenera* (heterozygous loci Sh+/Sh-) with medium-thickness shells.

The originality of oil palm lies in its high capacity for multiplication from seed, which allows breeders to build up families of full-sib individuals which enable an effective evaluation of the general and specific combining abilities of their parents, if tested in appropriate structures.

This quality, together with the coexistence of several generations of selected palms and genetic tests set up in a vast experimental network stretching throughout Africa, Southeast Asia and Latin America make oil palm a prime model plant for tropical tree crop quantitative genetics.

5.1. Oil palm breeding strategy and update on molecular markers

The genetic improvement scheme implemented by CIRAD and its partners is a variation on the reciprocal recurrent selection (RRS) scheme applied to maize by Comstock *et al.* (1949). It is based on two groups of genotypes, and exploits their complementary yield components: group A (Deli, Angola) and group B (African populations), tested in combination with respect to each other (Meunier and Gascon, 1972). The advanced stages of the selection scheme are based on a relatively limited number of parents which differ in terms of origin and genetic characteristics. Although the techniques used have yet to be optimised, vegetative propagation by somatic embryogenesis as developed by the Cirad-CP/Ird group has opened the way for clonal selection and mass distribution of the best selected individuals that cannot be reproduced from seed. The genetic progress in terms of oil production is 15-18% for the second selection cycle at Cirad-CP (Cochard *et al.*, 1993), compared to the first cycle, and there should be a further 30% gain by clonal selection of the best individuals obtained from the best progeneis (Baudouin and Durand-Gasselin, 1991).

However, this success has come up against the inherent limitations of several factors, notably:

- The duration of a generation for the crop and the long selection cycles – ten to twelve years – that have to be set up on vast experimental areas,
- The limited knowledge of the genetic diversity and degree of heterozygosity of the material tested,
- The complex phenotypic expression of the main quantitative characters which have been selected, such as:
- oil production – which depends on relatively heritable but negatively correlated characters, or on characters influenced by the environment;
- tolerance of vascular wilt (*Fusarium oxysporum* f.sp. *elaeidis*); a major disease in Africa;

- tolerance to Bud Rot disease in South-America;
- vegetative development: the aim being to reduce vertical growth in order to prolong the economic lifespan of the plantations.
- The impossibility of determining at the nursery stage the variety of individuals to be planted, which results in not testing *tenera* × *tenera* individuals whose 25% *pisifera* individuals are too numerous, and not planting plots of pure male *pisifera* chosen for commercial *dura* × *pisifera* seed production.

In short, in addition to the perennial nature and dimensions of the plant, it is our poor knowledge of the oil palm genome that prevents optimum conventional breeding.

We now have a clear idea of the requirements relating to better knowledge of the genetic structure of populations, checks on the identity and heterozygosity of progenies, more effective studies of the genetic links between selected characters and, above all, the provision of early selection aids (Jones, 1989; Brown, 1993; Jack and Mayes, 1993; Mayes *et al.*, 1996).

The use of oil palm genetic markers was initiated in Malaysia and the U.K. at the start of the 1990s, for clonal identification and genetic mapping (Cheah and Wooi, 1993; Mayes *et al.*, 1997). It developed prospectively, with the appearance of new techniques: genetic diversity studies of *E. guineensis* using RFLP or RAPD markers (Cheah *et al.*, 1991: Shah *et al.*, 1994); RFLP genotyping of *E. guineensis* and *E. oleifera* accessions (Jack *et al.*, 1995; Mayes *et al.*, 1996); search for cDNA markers of genes involved in floral development (Singh and Cheah, 1996 and 2000) or genes coding for fatty acid biosynthesis enzymes (Rashid and Shah, 1996; Shah and Cha, 2000); search for RAPD markers of somaclonal variants (Paranjothy *et al.*, 1993; Chowdhury, 1993; Rival, 1998); and RAPD detection of Bud Rot tolerance (Ochoa *et al.*, 1997). The Plant Breeding Institute published the first RFLP genetic map of oil palm (Mayes *et al.*, 1997). Although it is incomplete, with 860 cM mapped and 24 linkage groups compared to 16 chromosome pairs, the map was a first step towards a rational use of molecular markers.

At first, the work of CIRAD was focused on the molecular analysis of genetic diversity in the *E. guineensis* germplasm used in breeding programmes, with isoenzyme markers (Ghesquière, 1983), of *E. oleifera* germplasm using cDNA or genomic RFLP probes (Barcelos *et al.*, 1999) and of pathogen strains of *Fusarium oxysporum* F. sp. *elaeidis* responsible for the vascular wilt disease, with RFLP markers (Mouyna *et al.*, 1996), the joint use of selection indexes and molecular markers to optimise the breeding scheme (Purba *et al.*, 2000 and 2001).

5.2. Marker-assisted breeding strategy for oil palm

Molecular markers are generally DNA labels used to identify and to assess genes of agronomic interest in a plant. They are neutral with respect to the environment and can be detected at every stage of plant growth. Molecular tools indicate the presence or absence of such genes in a given material and will be used to step up selection pressure on each cycle of the oil palm genetic improvement programme and, for identical experimental areas, to achieve greater genetic progress more rapidly. Generally speaking, the molecular markers development in oil palm will make it possible to assess all the parents used in breeding programmes and to choose only those bearing the favourable alleles of the agronomic genes, right from the nursery stage, for further genetic improvement or for seed production. In addition, molecular markers being sought elsewhere to meet the specific objectives of certain geographical regions (resistance to vascular wilt in Africa, resistance to Bud Rot in Latin America) will, once identified, make it possible to screen the parents available and carry out marker-assisted selection of germplasm more suitable for those regions.

Most agronomic characters are governed by several interacting genes of small effects (called Quantitative Trait Loci or QTL). Little is known about the number and effects of these genes, which leads to imprecise phenotypic selection. The marker-assisted breeding (MAB) concept assumes a strong genetic link between the markers and loci of the genes studied. It necessarily involves prior genetic mapping of species (Mohan *et al.*, 1997). The markers are located on the genome sufficiently densely for each locus of the genome to be strongly linked to at least one of them. In a second stage, the locus detection and the evaluation of the effects of worthwhile genes are based on a study of the relation between the molecular polymorphism observed on markers and the variation in the characters studied (Charcosset, 1996).

At this stage, the applied value of detecting QTL using markers depends on adopting an experimental design enabling an estimate of the type (additive, dominance) and stability of QTL effects (genetic bases, environment). Lastly, it is important to pay particular attention to applying the knowledge and molecular tools already acquired on other plants, to automating procedures (currently based on the PCR technique) and to exploiting and transferring results (from one plant to another, from one topic to another, from one laboratory to another).

Biomolecular research applied to marker-assisted oil palm breeding is a long-term plan. The different stages identified and undertaken by CIRAD for oil palm are presented in detail below (Billotte and Baudouin, 1997).

5.2.1. Mass development of PCR based-markers in oil palm

Simple Sequence Repeats (SSRs), also called microsatellites, are tandem arrays of simple nucleotide motifs that are ubquitous components of eucaryotic genomes (Delseny *et al.*, 1983; Tautz, 1989; Tautz and Renz, 1984). Inherited in a Mendelian fashion (Saghai Maroof *et al.*, 1994; Weissenbach *et al*, 1992), their hypervariable length polymorphism is simply revealed by the Polymerase Chain Reaction (PCR) using flanking primers that generate co-dominant markers (Fig. 9). Smith *et al.* (1997) described SSR technology as presenting the potential advantages of reliability, reproducibility, discrimination, standardisation and cost effectiveness. A mass development of oil palm microsatellite markers has been undertaken by CIRAD since 1998, with the view to developing genetic diversity studies, to variety identification, to pedigree analysis as well as to genome mapping and QTL detection towards marker-assisted selection activities.

A quick and simple technique for building microsatellite-enriched libraries was developed by Billotte *et al.*, (1999), from a hybridization-based capture methodology using biotin-labelled microsatellite oligoprobes and streptavidin-coated magnetic beads (Fig. 10). This technique allowed the construction of several oil palm (GA)n, (GT)n or (CCG)n enriched-libraries from total genomic as well as purified chloroplastic DNA. About 200 functional SSR primer pairs have already been developed by CIRAD in oil palm from microsatellite clones that were sequenced in close collaboration with the National Sequencing Cetnre in Evry France (*Genoscope*).

The characterisation of 21 SSR loci was published by Billotte *et al.* (2001) together with primer sequences, estimates of allele size as well as expected heterozygosity in *E. guineensis* and in the closely related species *E. oleifera* where an optimal utility of the SSR markers was observed. Multivariate data analyses showed the ability of SSR markers

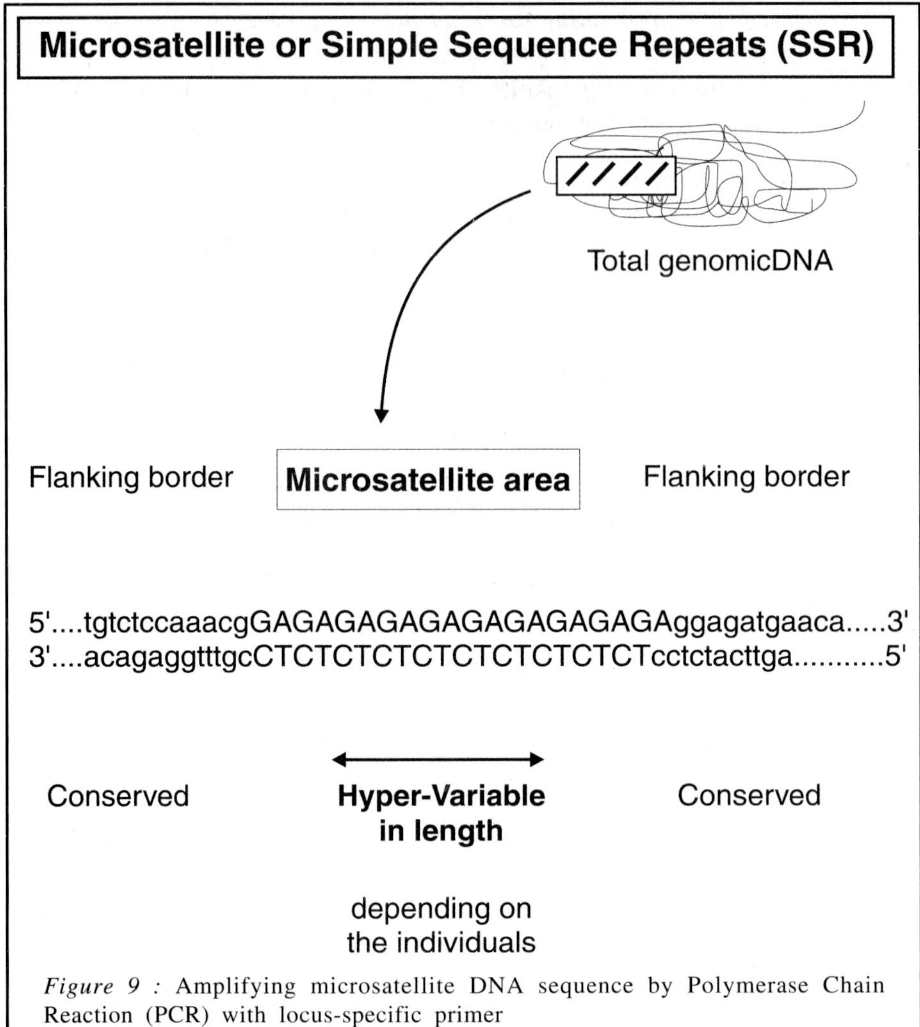

Figure 9 : Amplifying microsatellite DNA sequence by Polymerase Chain Reaction (PCR) with locus-specific primer

to efficiently reveal the genetic diversity structure of the genus *Elaeis* in accordance with known geographical origins and measured genetic relationships based on previous molecular studies (Fig. 11). High levels of allelic variability indicated that *E. guineensis* SSRs will be a powerful tool for genetic studies of the *Elaeis* genus including variety identification and intra- or inter-specific genetic mapping (Table 5). PCR amplification tests, from a subset of 16 other palm species, and allele sequence data showed that *E. guineensis* SSRs are putative transferable markers across palm taxa. In addition, phenetic information based on SSR flanking region sequences makes *E. guineensis* SSR markers a potentially useful

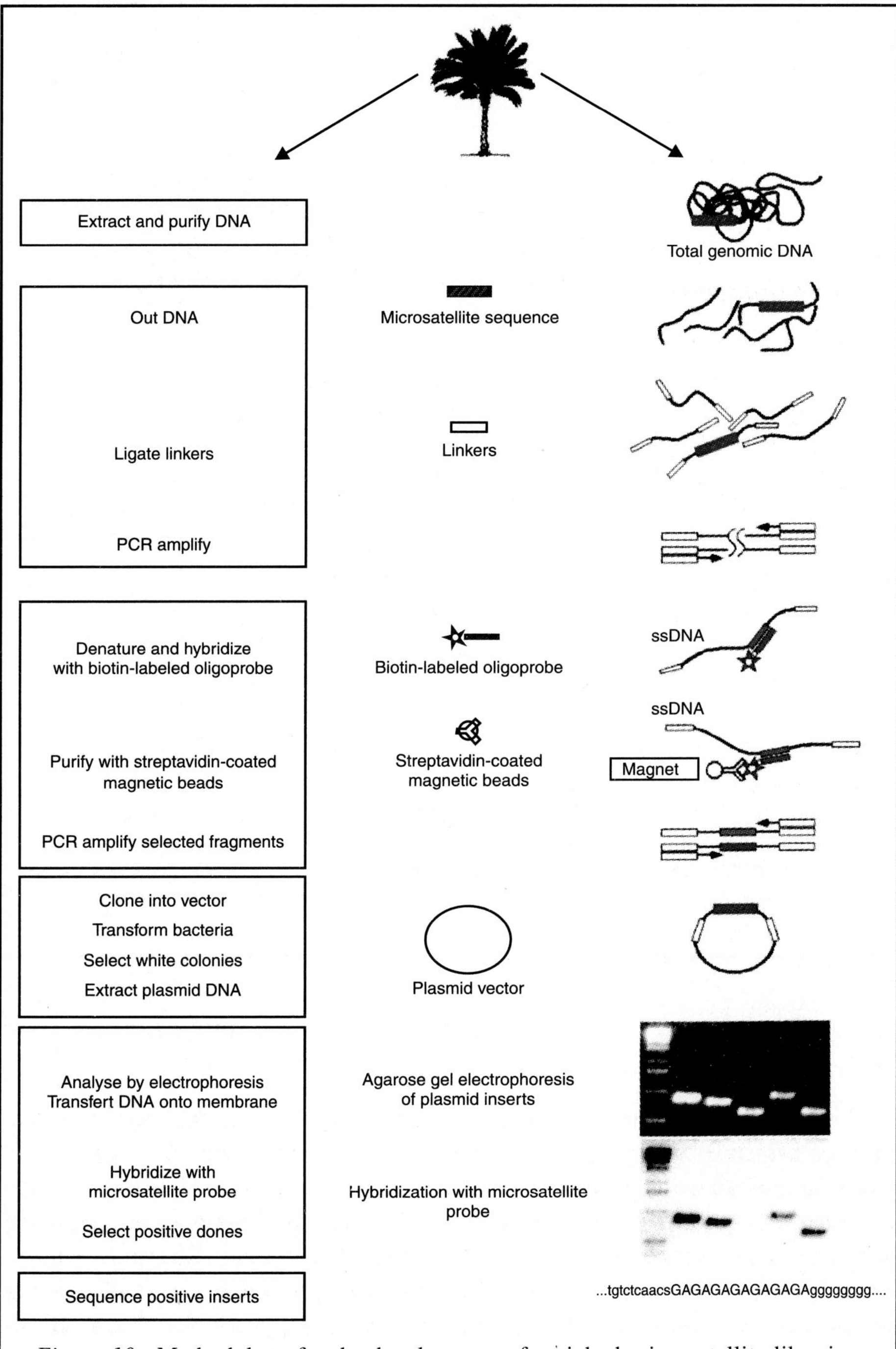

Figure 10 : Methodology for the development of enriched-microsatellite libraries, according to Billotte *et al.* (1999).

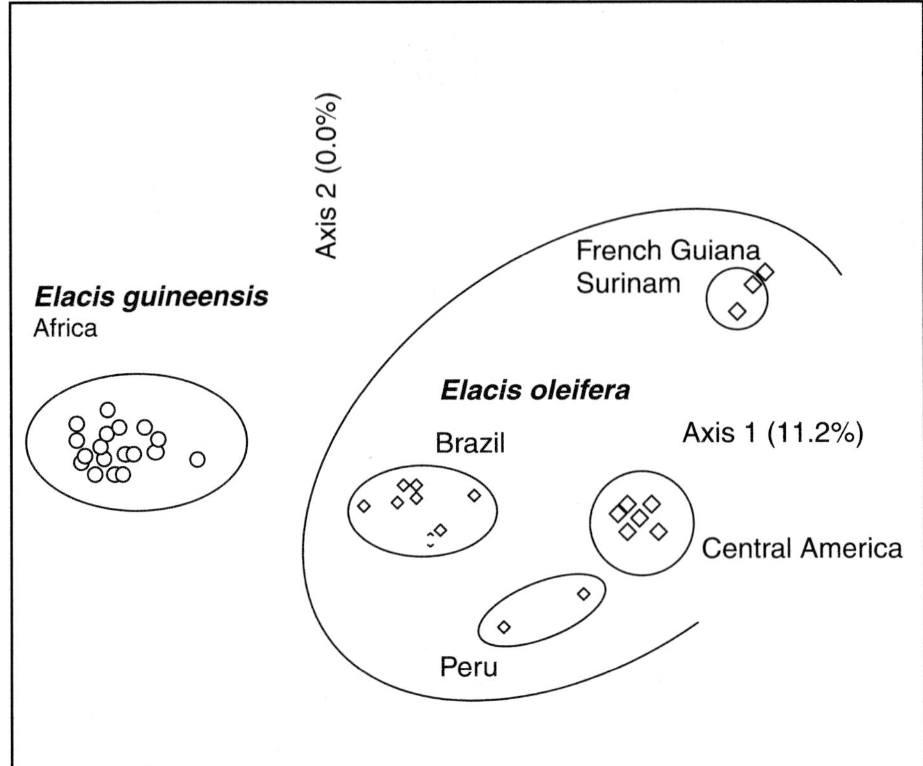

Figure 11 : Factorial analysis of correspondences performed on 20 single-locus microsatellite markers over 18 accessions of *E. guineensis* (Africa), 19 accessions of *E. oleifera* (Brazil, Central America, French Guyana, Peru and Surinam) and 1 accession of *B. odora*. Note: Axis 1 and 2 represent 20.2% of the total molecular variability.

molecular resource for any researcher studying the phylogeny of palm taxa.

These microsatellite markers are the main technical basis of the marker-assisted breeding strategy which CIRAD is developing for oil palm in collaborating with its partners (Fig. 12), partly within a new Coconut-Oil palm research project funded by the European Community and involving, in addition to CIRAD, the Max Planck Institute (project co-ordinator, Germany), PT SOCFIN Indonesia (SOCFINDO Plantation, Indonesia), the NEIKER Cetre (Spain), the Malaysian Oil Palm Board (MPOB, Malaysia) and the Indonesian Palm Oil Research Institute (IOPRI, Indonesia).

Table 5. Type of repeats and average values of SSR allele numbers, expected heterozygostity (He) and probability of identity (PI) of 20 *E. guineensis* SSR loci in *E. guineensis* and *E. oleifera*.

Sample	SSR Motif	Repeats Number	# Alleles *E.g.*	# Alles *E.o.*	#Shared alleles	Total # alleles	He			PI	
							E.g.	*E.o.*	*E.g.+E.o.*	*E.g.*	*E.o.*
9	(GA)n	17	7,1	7,3	3,0	11,4	0,7	0,7	0,8	0,02-0,15	0,02-0,38
7	(GT)n	10	4,3	3,9	1,6	6,6	0,4	0,4	0,6	0,02-1,00	0,09-1,00
4	(CCG)n	6	2,7	3,5	1,7	4,5	0,4	0,6	0,6	0,11-0,40	0,12,-0,23
Average	-	-	5,2	5,3	2,2	8,4	0,5	0,6	0,7	0,3	0,2
Polymorphic markers							80%	95%	100%		

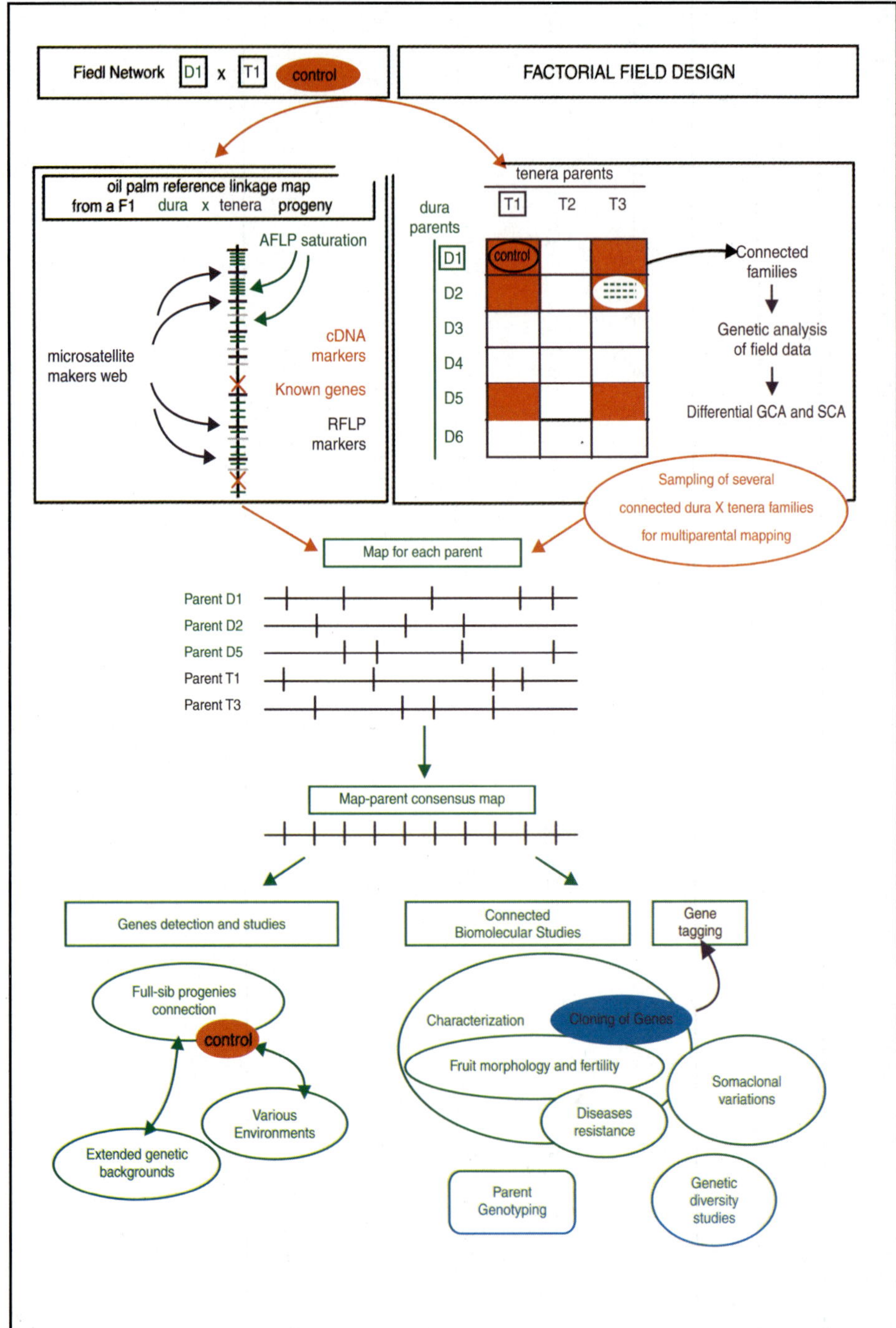

Figure 12 : An example of biomolecular research strategy towards marker-assisted breeding of oil palm developed at CIRAD, France.

5.2.2. Establishing a reference genetic map

In addition to providing further knowledge of the genome, the aim of drawing up reference maps is a rational choice of markers and parents to be used later in the plan. A control *dura* Deli × *tenera* La Mé F_1 progeny widely used in the breeding scheme was chosen (cross DA10D × LM2T). The homologous and heterologous markers mapped will primarily be co-dominant, locus-specific and easily transferable from one population to another (mostly microsatellites but also genomic RFLP (Restricted Fragment Length Polymorphism) or cDNA markers, and EST (Express Sequenced TAG) markers). Highly polymorphic microsatellite markers will make up the web of the reference genetic map; they are effective in genetic diversity studies (Jarne and Lagoda, 1996) notably of oil palm (Shah and Nyuk, 1996), and suitable for genome mapping as they provide good cover (Akagi *et al.*, 1996). These markers will be topped up with RFLP and cDNA "anchor points" AFLP (Amplified Fragment Length Polymorphism) markers will be used to saturate the reference linkage map.

5.2.3. Parallel establishment of a multi-parent consensus map

The theoretical work performed by Muranti (1996 and 1997) showed that a multi-parent consensus map, established with several parents (at least four to six) and full-sib families linked within a diallel or factorial design is the most effective in the search for QTL suitable for use in marker-assisted breeding. In addition to ensuring more accurate detection compared to a single two-parent map, it also enables an evaluation of the effects of QTL depending on their type (additive, dominance) and of difference genetic bases. Such a multi-parent map will be established by CIRAD in collaboration with PT SOCFIN Indonesia, using microsatellite and cDNA markers on an existing factorial genetic design already observed for most vegetative and production characters. The map population consists of about 600 palms and several *dura* × *tenera* full-sib families, including the reference map population, obtained from different La Mé, Deli and Yangambi parents. All individual maps of the system will be connected between themselves by common parents and by co-segregating microsatellite or cDNA/EST markers.

5.2.4. The search for worthwhile QTL or major genes

QTL for vegetative and yield characters, as well as of vascular wilt tolerance (based on nursery tolerance test performed in Côte d'Ivoire in collaboration with the CNRA (Centre National de Recherche

Agronomique) Research Station of La Mé, will be detected by studying the correlation existing between the markers and the phenotypic characters of the individuals chosen for the individual or consensus maps. The variability of QTL effects will be studied depending on the environment, based on the control progeny duplicated under different conditions, in Indonesia and in Côte d'Ivoire.

At the present time, two complementary studies have been conducted to identify AFLP or microsatellite markers linked to this Sh gene: by BSA (Bulk Segregant Analysis) of segregant groups, a methodology published by Michelmore *et al.* (1991) for detecting major genes, and by genetic mapping (Billotte *et al.* 2001). A total of 124 *Eco*RI/*Mse*I AFLP primer pairs and 88 microsatellite primer pairs was used to analyse 3 *dura, tenera* and *pisifera* segregant groups, each consisting of the DNA of 8 individuals obtained by selfing of a *tenera* LM2T. Out of a total of 9,330 loci screened by the AFLP technique, an AFLP-BSA condidate marker AggCAA20 was identified (Fig. 13) then validated by individual analyses of the DNA making up the mixes. Microsatellite screening did not reveal any marker on these mixes.

A F_1–type progeny (*dura* × *tenera*) of 90 individuals obtained from a DA115D x LM2T cross was chosen to carry out genetic mapping of the oil palm and its Sh gene, in the LM2T parent. The genetic map of LM2T, constructed with MAPMAKER at LOD score = 5.0 and r = 0.3 (apart from minor exceptions), distributed a set of 149 AFLP or microsatellite markers in 15 linkage groups, 3 pairs and 6 unlinked markers (Figure 13). The number of linkage groups, close to the number n = 16 pairs of chromosomes of the plant, and a total map size of 1355 cM already indicate relatively good coverage of the genome. The AFLP-BSA marker, which had reveled a co-dominance reading of the Sh+ gene on the *dura* and *tenera* phenotypes, was likewise mapped in LM2T. The gene Sh and its AFLP-BSA marker were mapped onto the longest linkage group of the map (219.5 cM), at 7.2 cM or 12.6 cM according to the respective JOINMAP or MAPMAKER genetic mapping softwares.

The positive results of the study showed the pertinence of BSA analysis in the search for molecular markers of the Sh gene, and the complementarity of the genetic mapping approach. The AFLP-BSA marker and the genetic map of LM2T are an important step towards marker-assisted selection of the *Sh* gene, and of other genes of agronomic interest in oil palm.

5.2.5. The development of five topics linked to genetic mapping

- Topic 1 : The molecular dissection of the chromosomal region surrounding major gene Sh, coding for the existence or lack of a shell, with a view to cloning and (intragenic) labelling of the related genes responsible for fruit morphology and fertility (degree of shell lignification, kernel volume in *dura* and *tenera* varieties, female sterility of *pisifera* genotypes).
- Topic 2 : Characterisation, cloning and tagging of vascular wilt tolerance genes.
- Topic 3 : Overall molecular studies of the genetic diversity of oil palm, which will be based partly on using QTL markers and will determine the potential for applying natural diversity, particularly to predict heterosis.
- Topic 4 : The search for material tolerant to Bud Rot in Latin America, by the genetic mapping of tolerance genes in the American species *E. oleifera* (Le Guen *et al.*, 1991) with a view to their introgression into *E. guineensis* by marker-assisted backcrossing. This latter project is based on the genetic mapping of an inter-specific back-cross of first generation (BC_1) for which the resistance to Bud Rot of each genotype will be evaluated through the field test of clonal descendants planted in an area strongly affected by the disease.
- Topic 5 : Biomolecular studies of somaclonal variation factors. Like the current physiological studies, the final aim will be to contribute towards transferring production of true-to-type clonal material to an industrial scale.

5.2.6. The validation of molecular markers

Field Trial Systems (FTS) consisting of several populations to be used for future marker-assisted breeding programmes will be established using new factorial genetic designs. Trials will be planted at different locations consisting of 5000 genotypes and involving at least 20 parents (full-sib families).

The objectives are to test he accuracy of molecular markers on multi-parent and connected full-sib families issued from extended genetic backgrounds and dedicated to a wider identification of intra- and inter-population QTL diversities. Also, the project results will integrate into current oil palm breeding programmes in the form of an applied multi-characters marker-assisted breeding strategy for worthwhile genes.

Figure 13

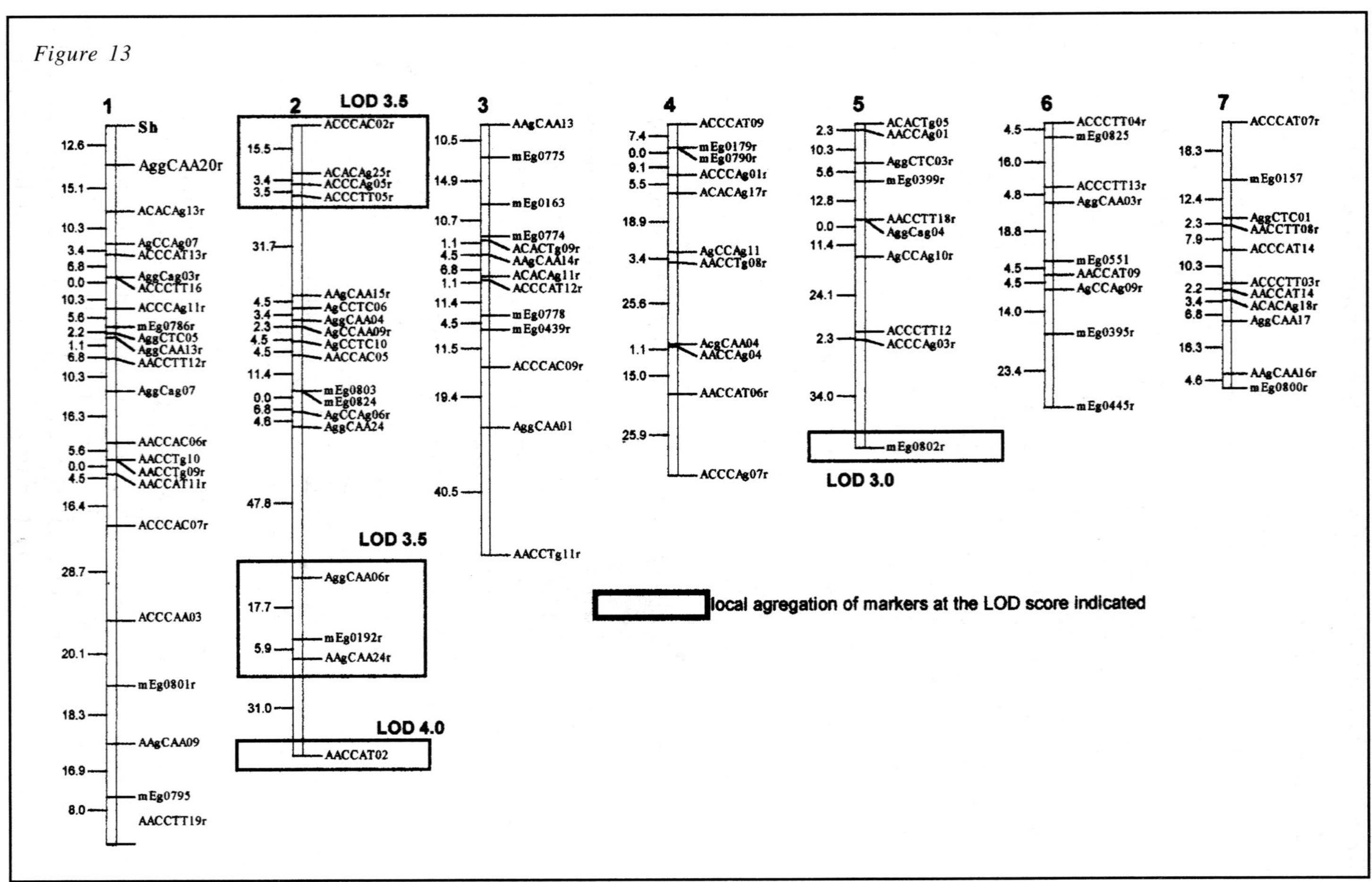
1
2
3
4
5
6
7
LOD 3.5
LOD 3.5
LOD 4.0
LOD 3.0
local agregation of markers at the LOD score indicated

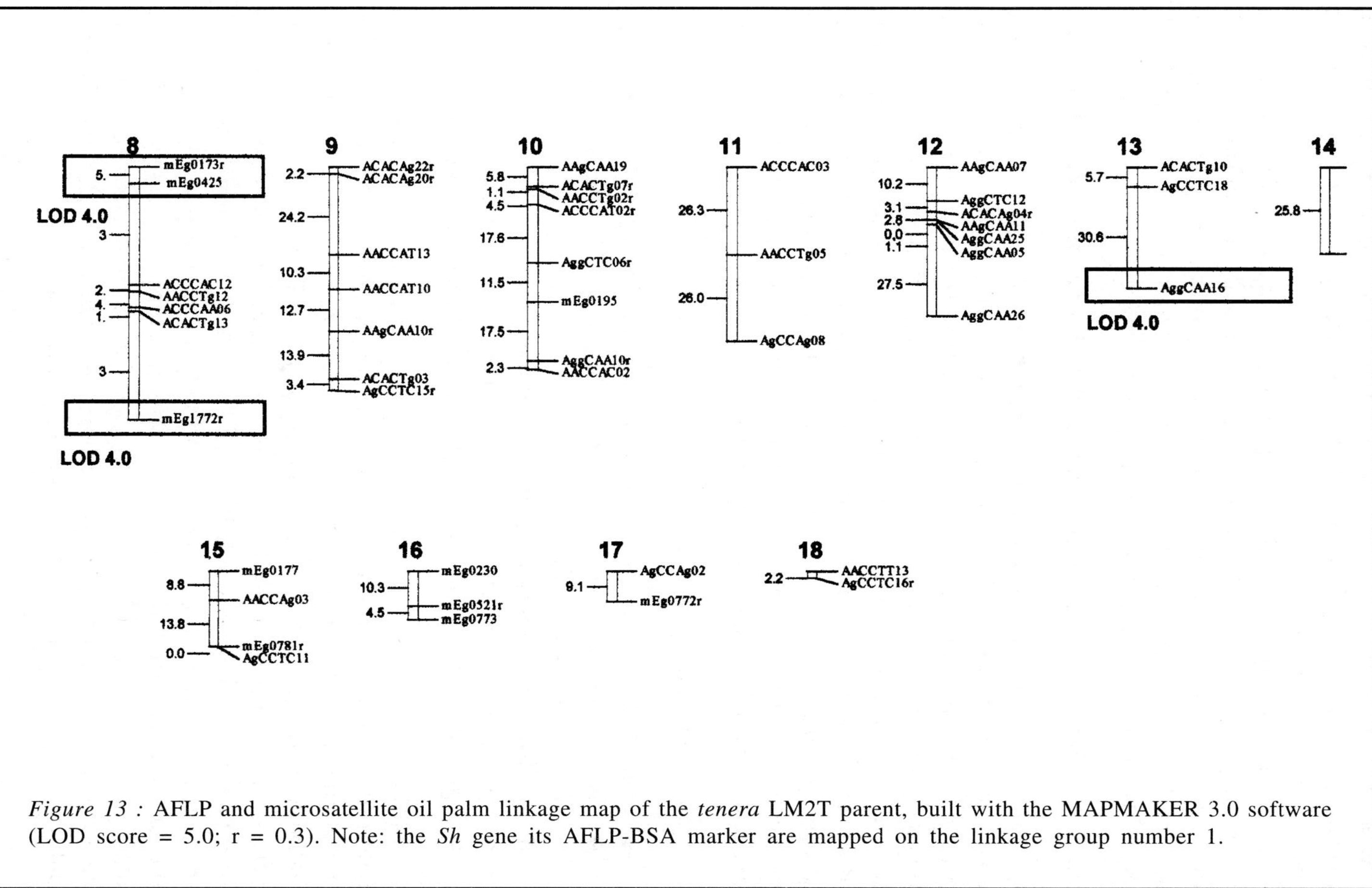

Figure 13 : AFLP and microsatellite oil palm linkage map of the *tenera* LM2T parent, built with the MAPMAKER 3.0 software (LOD score = 5.0; r = 0.3). Note: the *Sh* gene its AFLP-BSA marker are mapped on the linkage group number 1.

Various parents will be used including palms suspected or know to present genetic resistance factors to the lethal diseases caused by *Ganoderma* or *Fusarium oxysporum*.

Some trials will test inter-specific back-crosses of first or second generation for the marker-assisted detection and fast introgression of important *Elaeis oleifera* characters into oil palm: low height increment and high oil quality, but also genetic resistance factors to the lethal diseases *Ganoderma, Fusarium wilt* and Bud Rot.

5.3. Perspectives in marker-assisted breeding of oil palm

Molecular markers data will be associated with selected phenotypic characters to define and apply the marker-assisted breeding scheme. Such use of molecular markers will increase the efficiency of conventional breeding in terms of accuracy and time saving. It is an early selection tool. It will be easier to determine the genotype of material to be tested at an early stage, rather than measuring certain phenotypic characters at each generation, for ten years or more. This will limit the time lapse between selected generations, fasten genetic gain and so increase the expected quality of commercial seeds.

Major applications will concern the following areas:

- variability management (germplasm, combining ability groups),
- genotyping for characters highly influenced by the environment or costly to measure,
- control and monitoring of recombinations,
- predicting genotypic values for complex characters, and
- predicting the value of a cross from information about the parents.

Varietal identification and legitimacy analyses (determination of paternity, reproduction schemes) are one major application. In oil palm, molecular markers will be used to select *dura, tenera* or *pisifera* individuals at the nursery stage. In the same way, it will be possible to detect and then plant *pisifera* individuals in male parent plots for *tenera* seed production. It will be partly possible, using gene markers, to evaluate the genetic value of these *pisifera* individuals, which was impossible in the field until now due to their female sterility. Lastly, effective management of the entire variability of the *E. guineensis* species will be a possibility, with field testing of *tenera* × *tenera* progenies, of which only *tenera* individuals will be retained (Fig. 14).

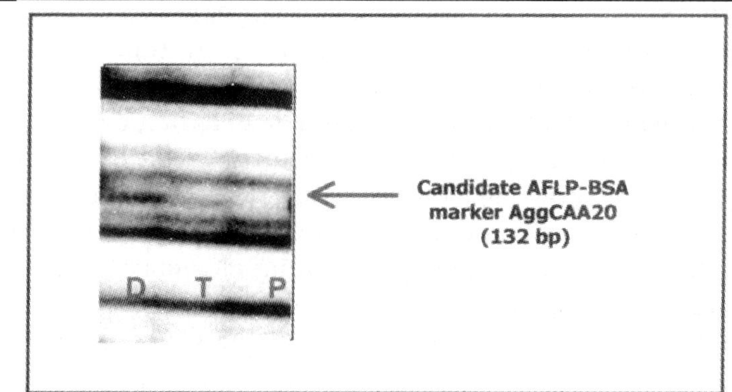

Figure 14 : Candidate AFLP marker of the Sh+ allele evidenced by bulk segregant analysis of groups issued from the selfing of a *tenera* LM2T parent. Note: D = *dura* DNA bulk; T = *tenera* DNA bulk; P = *pisifera* DNA bulk.

Automatic genotyping will concern the main worthwhile parents and the survey populations will be integrated into the oil palm breeding scheme. Here, biomolecular results, in terms of favourable gene detection and evaluation with the molecular tools, will be applied to the whole improvement scheme.

6. GENETIC ENGINEERING

We are now at only the beginning of research on genetic engineering in oil palm even the regeneration of somaclones through somatic embryogenesis has been obtained since the mid-70s. Several projects are under way, with the aim of introducing foreign genes into oil palm in order, to modify oil yield and quality or resistance to pests and diseases.

Preliminary work has recently been published by Chowdhury *et al.*, (1997) on the evaluation of promoters for use in the genetic transformation of oil palm using microprojectile particle bombardment. The transient expression of reporter genes has been achieved. To our knowledge, transgenics in oil palm has not yet been developed.

As a further complementary approach to help us understand the molecular phenomena underlying the *mantled* abnormality, we have recently started work on establishing an experimental oil palm transformation system in our Laboratory. This will enable us to study the *in vivo* activities of the various different gene promoters isolated during the course of the work described previously.

We are currently in the process of preparing promoter::GUS constructs which will be initially tested for their activity by transient expression analysis using particle bombardment. This may provide indications as to whether DNA methylation plays a direct role in determining the differential expression pattern of a given gene, since the activity of the gene promoter carried in an exogenous plasmid-derived DNA will be probably be unaffected by clonal conformity in this case. Conversely, if the differential expression pattern is determined by signalling elements further upstream in a regulatory cascade, the activity of the bombarded promoter would be expected to be influenced by tissue conformity. In the longer term, we will be establishing a system for stable transformation of oil palm, which will enable us to study the developmental regulation of marker gene promoters during the *in vitro* regeneration process. Transient expression and stable transformation of oil palm tissues has already been demonstrated by Chowdhury *et al.* (1997) and Kadir and Parveez (2000), respectively.

For our transformation work, we have chosen to concentrate on the use of embryogenic suspensions as the starting plant material, since this should simplify the process of regeneration from tissue culture : low size, abundant and very homogenous (de Touchet *et al.*, 1991). For initial bombardment optimisation experiments, transient expression was achieved using plasmid constructs containing the reporter gene beta-glucuronidase (GUS) under the control of either the 35S cauliflower mosaic virus or the modified maize *Adh1* promoter (Emu). Cells calibrated at 1-2 mm were transformed by biolistic bombardment using a PDS 1000/He apparatus (Biorad) whilst varying several key parameters, such as:

- *plasmid type* (35S or Emu promoter)
- *hypertonic pretreatment* in 180 g/l sucrose for 5 hours, 1 day or 3 days
- *bombardment* after 5, 8 or 15 day of sub-culture into fresh medium.

The results obtained are shown in Table 6. An overall higher of expression was obtained with the 35S promoter. The plasmid used is pe35SGUS-7S, derived from pUC119 (Viera and Messing, 1987) in which the 35S promoter contains a duplication of the region situated between 343 bp and 90 bp upstream up the transcription start site. The Emu promoter used is contained in the plasmid pEmu-GN (Last *et al.*, 1991). These results are in broad agreement with those of Chowdhury *et al.*,

Table 6. Number of blue spots counted on 500 mg fresh weight of embryogenic cell suspension material, 3 days after biolistic bombardment. For each treatment, 10 replicates were performed. Letters indicate significant differences after Neuman and Keul's multirange test ($p < 0.001$).

Plasmid	Plating (days)			Culture (days)		
	0.2	1	3	5	8	15
Emu :: *gus*	4^{a}	29^{a}	11^{a}	6^{a}	41^{a}	16^{a}
35S :: *gus*	20^{a}	119^{b}	144^{b}	100^{b}	104^{b}	120^{b}

(1997) although the latter authors reported a higher expression from the Emu promoter than from its 35S equivalent. This may be due to the fact that embryogenic callus was used for the bombardment in the latter case rather than embryogenic cells from suspension culture. Concerning the hypertonic pre-treatment of the bombarded material, we observed a positive effect on expression levels; however a 5-hour treatment appears to be insufficient to attain the necessary equilibration of osmotic pressure. The expression potential of the bombarded cells was not observed to vary during the first 15 days after sub-culturing, during which period the cells are dividing actively. We are now investigating whether any changes occur during the second two weeks of the sub-culturing period. In addition to the experiments mentioned above, which were performed with 1-2

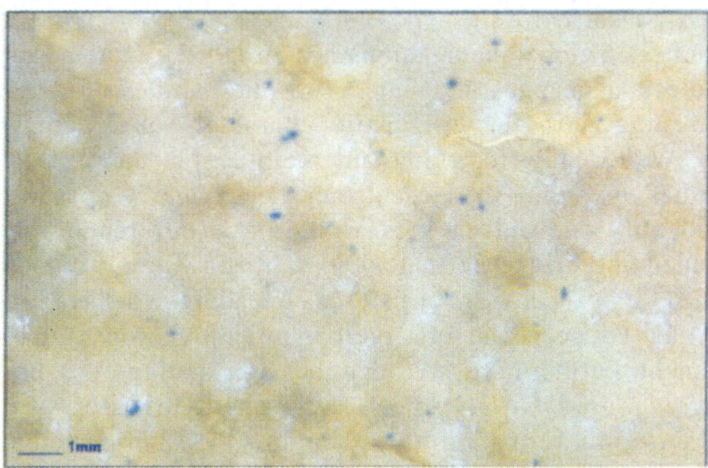

Figure 15 : Cell suspension calibrated at < 1 mm showing transient expression of *35S::gus* three days after boilistic transformation. 172 ± 38 spots were counted on 500 mg FW of plated cells (10 replicates).

mm diameter cells, we have also carried out bombardment of < 1 mm cells under the same conditions. After a 2 days hypertonic pre-treatment, the 35S::GUS construction was found to produce 172 ± 38 spots per 500 mg fresh weight of cells (Fig. 15).

7. CONCLUSIONS AND FUTURE PROSPECTS

Today, a range of biotechnological approaches play an increasingly important role in Cirad's breeding strategies for oil palm improvement. They are fully integrated into our programme aimed at obtaining, multiplying and disseminating genetically improved material to the planters.

Biotechnological approaches applied to oil palm breeding have considerably increased in their importance, not only in micropropagation, but also in marker-assisted breeding, and more generally, in physiological and molecular studies of the expression of genes of paramount-agronomic value (flowering, abscision, disease resistance, etc.)

Recent results in micropropagation using embryogenic suspension open the way for the use of strategies based on the "artificial seeds" concept for oil palm. The latter could be obtained from embryos produced at a high rate from embryogenic suspensions, which would then be correctly treated to achieve maturation (enrichment in storage proteins). Single somatic embryos would then be encapsulated and stored (either at room temperature or at 0 - 4°C) before delivery at the appropriate time, according to the planting season, to tissue culture laboratories or nurseries situated close to the plantations, where germination either *in vitro* or *ex vitro* could be carried out.

Our studies on global methylation rates provided us with a first glimpse of the molecular changes associated with the *mantled* abnormality and it is consistent with the epigenetic characters observed, including reversion. We are now carrying out methylation-sensitive RFLP and AFLP studies, involving the isoschizomeric enzymes *MspI* and *HpaII*, in order to identify relevant markers, exhibiting a differential methylation pattern depending on the normal/*mantled* phenotype but independent of the genetic origin of clones. In parallel, we addressed the question of a possible link between the observed hypomethylation and chromatin rearrangement. Furthermore, we hope that by isolating oil palm relatives of the *Arabidopsis thaliana METI* DNA-methyltransferase gene, we will obtain useful information to explain how the *mantled* abnormality is generated.

Molecular markers data will be associated with selected phenotypic characters to define and apply the marker-assisted breeding scheme. Such use of molecular markers will increase the efficiency of conventional breeding in terms of accuracy and time saving. It is an early selection tool.

Our group in Cirad, while collaborating with several groups in producing countries, has capitalised on developments in this area by regularly incorporating new biotechnological approaches into its overall strategy, beginning with classical tissue culture-based approaches and progressing to molecular markers, differential gene expression analysis and genetic transformation.

ACKNOWLEDGEMENTS

Research work described in this chapter was conducted under a joint research programme between Ird (ex-ORSTOM: Institut de Recherche pour le Dévelopment) and Cirad-CP (Centre de Coopération Internationale en Recherche Agronomique pour le Dévelopment – Départment Cultures Pérennes).

Research on somaclonal variation was partially funded with the aid of a grant from the Malaysian Palm Oil Board (formerly PORIM) through a Collaborative Research Project (n° CRB-96-001).

The authors thank colleagues involved in the Oil Palm Project: Jean-Luc Verdeil at Ird-GeneTrop; Tristan Durand-Gasselin, Philippe Amblard and Francoise Potier at Cirad-CP.

Biotechnology research would not have been possible without the excellent collaboration of our colleagues working in producing countries : CNRA at La Mé in Côte d'Ivoire, INRAB at Pobé in Benin, SOCFINDO and IOPRI in Indonesia, and FELDA and MPOB in Malaysia.

REFERENCES

Abdullah R (1990) Clonal micropropagation of oil palm through apical meristem culture, *VII International Congress of Plant Tissue and Cell Culture,* Amsterdam. Abstract p. 85.

Aberlenc-bertossi F, Morcillo F, Rival A and Duval Y (1995) Factors involved in desiccation tolerance in oil palm embryos. In: Proc. *International Congress on Integrated Studies on Drought Tolerance of Higher Plants.* 31 Aug. – 2 Sept. 1995, Montpellier, France, Abstract, IV-1.

Aberlenc-Bertossi F, Noirot M and Duval Y (1999) BA enhances the germination of oil palm somatic embryos derived from embryogenic suspension cultures. *Plant Cell Tiss. Org. Cult.,* **56** : 53-57.

Aberlenc-Bertossi F, Chabrillange N and Duval Y (2000). Tolérance á la dessiccation des embryons de palmier á huile. Poster, Colloque AUPELF-UREF "*Des modéles biologiques a l'amélioration des plantes*", Montpeller, 3-5 juillet 2000.

Aberlenc-bertossi F, Chabrillange N and Duval Y (2001) Abscisic acid and desiccation tolerance in oil palm (*Elaeis guineensis* Jacq.) somatic embryos. *Genetics Selection Evaluation*, (in press).

Akagi H, Yokozeki Y, Inagaki A and Fujimura T (1996) Microsatellite DNA markers for rice chromosomes. *Theor. Appl. Genet.,* **93** : 1071-1077.

Bajaj YPS (1991) *Biotechnology in Agriculture and Forestry*, vol 17. High Tech and Micropropagation I. Springer, Berlin, Heidelberg, New York.

Barcelos E, Second S, Kahn F, Amblard P, Lebrun P and Seguin M (1999) In *Evolution, variation and classification of palms.* (Eds. Henderson A and Borchsenius F). Memoirs of the New York Botanical Garden, 83. The New York Botanical Garden Press. New York. pp. 191-201.

Barry-Etienne D, Teisson C, Berthouly M, Cote F, Bertrand B and Etienne H (1998) Direct transfer in greenhouse of coffee somatic embryos produced by temporary immersion. In: *Plant biotechnology and in vitro biology for the 21st century*, IX IAPTC Congress. Abstract 113.

Boudouin L and Durand-Gasselin T (1991) Genetic transmission of characters linked to oil yields in oil palm by cloning results for young palms. *Oléagineux,* **46** : 313-319.

Baylin SB and Herman JG (2000) DNA hypermethylation in tumorigenesis. Epigenetics joins genetics. *Trends Genet.,* **16** : 168-174.

Beirnaert A and Vanderweyen R (1941) Contribution á l'étude génétique et biométrique des variétés *d'Elaeis guineensis* Jacq., *Publ. De l'I.N.E.A.C., Série Sci.,* n° 27.

Besse I, Verdeil JL, Duval Y, Sotta B, Maldiney R and Miginiac E (1992) Oil palm (*Elaeis guineensis* Jacq.) clonal fidelity: endogenous cytokinins and indoleacetic acid in embryogenic callus cultures, *J. Exp. Bot.,* **43** : 983-989.

Billotte N and Baudouin L (1997) Strategy in biomolecular research applied to oil palm breeding. ISOPB Seminar. December 1997. Kuala Lumpur. Malaysia. pp. 9.

Billotte N, Frances L, Amblard P, Durand-Gasselin T, Noyer JL and Courtois B (2002) Search for AFLP and microsatellite molecular markers of the *Sh* gene in oil palm (*Elaeis guineensis* Jacq.) by Bulk Segregant Analysis (BSA) and genetic mapping. (submitted).

Billotte N, Lagoda PJL, Risterucci AM and Baurens FC (1999) Microsatellite enriched-libraries: applied methodology for the development of SSR markers in tropical crops. *Fruits,* **54** : 277-288.

Billotte N, Risterucci AM, Barcellos E, Noyer JL, Amblard P and Baurens FC (2001) Development, characterisation, and across-taxa utility of oil palm (*Elaeis guineensis* Jacq.) microsatellite markers. *Genome,* **44** : 1-14.

Blake J (1990) Coconut (*Cocus nucifera* L.) Micropropagation. In: *Biotechnology in Agriculture and Forestry, Legumes and Oil Seed Crops,* Vol. 10, Springer-Verlag, Heidelberg, pp. 538-554.

Brown PTH (1993). The role of biotechnology in oil palm breeding. *Proc. ISOPB Symposium,* Malaysia pp. 158-162.

Brown PTH (1989) DNA methylation in plants and its tissue culture. *Genome,* **31** : 717-729.

Callebaut I, Courvalin JC and Mornon JP (1999) The BAH (bromo-adjacent homology) domain : a link between DNA methylation, replication and transcriptional regulation. *FEBS Lett.,* **446** : 189-193.

Chabrillange N, Aberlenc-Bertossi F, Noirot M, Duva IY and Engelmann F (2000) Cryopreservation of oil palm embryogenics suspensions. In: (Eds. Engelmann F and Takagi H) *Cryopreservation of tropical plant germplasm. Current research progress and application.* JIRCAS, Tsukuba, Japan/IPGRI, Rome, Italy, pp. 172-177.

Charcosset A (1996) L'identification de locus affectant des characters quantitatifs (QTL) a l'aide de marqueurs génétiques est-elle justifiée pour to sélection? *Le Sélectionneur Francais,* **46** : 35-45.

Cheah SC, Siti Nor Akmar A, Leslie CL, Ooi SC, Abdul Rahman A, and Madon M (1991) Detection of DNA variablity in the palm using RFLP probes. *Proc. Int. Palm Oil Conf. Malaysia,* 144-150

Cheah SC and Wooi KC (1993). Application of molecular techniques in oil palm tissue culture. *Proc ISOPB Symposium Malaysia,* 163-170.

Chevalier A (1910) Les végétaux utiles de l'Afrique tropicale française. *Documents sur le palmier à huile,* Vol. **7** : 127.

Chowdhury MKU (1993) Detection of variation in tissue culture-derived normal and abnormal clones of oil palm using RAPD method. *Proc ISOPB Symposium Malaysia,* pp. 190-194.

Chowdhury MKU, Parveez GKA and Saleh NM (1997) Evaluation of five promoters for use in transformation of oil palm (*Elaeis guineensis* Jacq.), *Plant Cell Rep.,* **16** : 277-281.

Cochard B, Noiret JM, Baudouin L, Flori A and Amblard P (1993) Second cycle reciprocal recurrent selection in oil palm, *Elaeis guineensis* Jacq. Results of Deli X La Me' hybrid tests. *Oléagineux,* **48** : 441-451.

Comstock RE, Robinson HF and Harvey PH (1949) A breeding procedure designed to make maximum use of both general and specific combining ability. *Agron. J.,* **41** : 360-367.

Corley RHV, Barrett JN and Jones LH (1977) Vegetative propagation of oil palm via tissue culture, *Oil Palm News,* **22** : 2-8.

Corley RHV, Lee CH, Law LH and Wong CY (1986) Abnormal flower development in oil palm clones, *Planter,* Kuala Lumpur. **62** : 233-240.

Cruden RW (1988) Temporal dioecism: systematic breadth, associated traits, and temporal patterns, *Bot. Gaz.,* **149** : 1-15.

Davis TA (1980) Double reversal of spadices in an African oil palm. *Planter,* Kuala Lumpur, **56** : 212.

Delseny M, Laroche M and Penon P (1983) Detection of sequences with Z-DNA forming potential in ligher plants. *Biochem. Biophys. Res. Commun.,* **116** : 113-120.

Demeulemeester MAC, Van Stallen N and De Proft MP (1999) Degree of DNA methylation in chicory (*Cichorium intybus* L.): influence of plant age and vernalization. *Plant Sci.,* **142** : 101-108.

Duval Y, Aberlenc F and de Touchet B (1995a) Use of embryogenic suspensions for oil palm micropropagation. *ISOPB International symposium on Recent Development in Oil Palm Tissue Culture and Biotechnology,* 24-25th September, 1993, Kuala Limpur, Malaysia.

Duval Y, Engelmann F and Durand-Gasselin T (1995b) Somatic embryogenesis in oil palm (*Elaeis guineensis* Jacq.) In : Somatic embryogenesis and Synthetic Seed I, *Biotechnology in Agriculture and Forestry* (ed Bajaj YPS) Springer Verlag, vol **30** : 335-352.

Duval Y, Amblard P, Rival A, Konan E, Gogor S and Durand-Gasselin T (1997) Progress in oil palm tissue culture and clonal performance in Indonesia and Côte d'Ivoire. *Planter* (Kuala Lumpur) 291-307.

Finnegan EJ, Brettell RIS and Dennis ES (1993) The role of DNA methylation in the regulation of plant gene expression. In: *DNA methylation: Molecular biology and biological significance* (Eds. Jost JP and Saluz HP). Birkhauser, Basel, Switzerland, pp. 218-261.

Finnegan EJ, Genger RK, Peacock WJ and Dennis ES (1998) DNA methylation in plants. *Annu Rev. Plant. Physiol. Plant Mol. Bol.,* **49** : 223-247.

Finnegan EJ, Peacock WJ and Dennis ES (1996) Reduced DNA methylation in *Arabidopsis thaliana* results in abnormal plant development. *Proc. Natl. Acad. Sci. USA,* **93** : 8449-8454.

Finnegan EJ, Brettel RIS and Dennis ES (1993) The role of DNA methylation in the regulation of plant gene expression, In: *DNA methylation : Molecular biology and biological significance,* (Eds. Jost JP and Saluz HP), pp. 218-261.

Finnegan EJ, Peacock WJ and Dennis ES (2000) DNA methylation, a key regulator of plant development and other processes. *Curr. Opin. Genet. Dev.,* **10** : 217-223.

Fujii J, Slade D, Aguirre-Rascon J and Redenbaugh K (1992) Field planting of alfalfa artificial seed. *In Vitro Cell. Dev. Biol.,* **28P** : 73.

Gaspar T, Kevers C, Hausman JF, Berthon JY and Ripetti V (1992) Practical uses of peroxidase activity as a predictive marker of rooting performance of micropropagat shoots, *Agronomie,* **12** : 757-765.

Grout BWW (1988) Photosynthesis of regenerated plantlets *in vitro*, and the stresses of transplanting, *Acta Hort.,* **230** : 129-135.

Guesquiere M (1983) Contribution a létude de la variabilité génétique du palmier á huile (*Elaeis guineensis* Jacq.) Le polymorphisme enzymatique. *Ph.D Thesis*. Université Paris-Sud. Cenre d'Orsay. France, pp. 116.

Hansen RS, Wijmenga C, Luo P, Stanek AM, Canfield TK, Weemaes CMR and Gartler SM (1999) The DNMT3B DNA methyltransferase gene is mutated in the ICF immunodeficiency syndrome. *Proc. Natl. Acad. Sci. USA,* **96** : 14412-14417.

Hardon JJ, Gascon JP, Noiret JM, Meunier J, Tan GY and Tam TK (1976) Major oil palm breeding programmes, In: *Development in Crop Science 1 Oil Palm Research.* (Eds. Corley RHV, Hardon JJ and Wood BJ) Elsevier Sci. Publ. Co. pp. 109-126.

Hartley CWS (1988) The Oil Palm, *Tropical Agriculture Series,* Longman London, 761.

Henry P (1948) Un *Elaeis* remarquable: le palmier á huile vivipare, *Revue Int. Bot. Appl. Agric. Trop.,* **28** : 422.

Jack PL, Dimitrijevic TAF and Mayes S (1995) Assessment of nuclear mitocondrial and chloroplast RFLP markers in oil palm (*Elaeis guineensis* Jacq.). *Theor. Appl. Genet.,* **90** : 643-649.

Jack PL and Mayes S (1993) Use of molecular markers for oil palm breeding. II. Use of DNA markers (RFLPs). *Oléagineux,* **48** : 1-8.

Jacquemard JC, Baudouin L and Noiret JM (1997) Le Palmier a luile, in: Charrier A, Jacquot M, Hamon S and Nicolas D (eds), L'amélioration des Plantes Tropicales, Editions Reperes, ORSTOm/CIRAD. 507-532.

Jaligot E, Rival E, Beule T, Dussert S and Verdeil JL (2000) Somaclonal variation in oil palm (*Elaeis guineensis* Jacq.) : the DNA methylation hypothesis, *Plant Cell Rep.,* **19** : 684-690.

Jones LH (1989) Prospects for biotechnology in oil palm (*Elaeis guineensis* Jacq.) and coconut (*Cocos nucifera*) improvement. *Biotechnology and Genetic Engineering Reviews,* **7** : 281-296.

Kadir G and Parveez A (2000) Production of transgenic oil palm (*Elaeis guineensis* Jacq.) using biolistic techniques, In: *Molecular biology of woody plants* (Eds. Jain SM and Minocha SC), Vol. 2, Kluwer, Dordrecht.

Kaeppler SM, Kaeppler HF and Rhee Y (2000) Epigenetic aspects of somaclonal variation in plants, *Plant Mol. Biol.,* **43** : 179-188.

Krikorian AD (1989) The context and strategies for tissue culture of date, African oil and coconut palms, In : *Applications of Biotechnology in Forestry and Horticulture,* (Ed Vibha Dhawan), Plenum Press Yew-York, 119-144.

Larkin PJ and Scowcroft WR (1981) Somaclonal variation – a novel source of variability from cell cultures for plant improvement. *Theor. Appl. Genet.,* **60** : 197-214.

Le Guen V, Amblard P, Omore A, Koutou A and Meunier J (1991) IRHO *Elaeis oleifera* × *Elaeis guineensis* interspecific hybrid programme. *Oléagineux,* **46** : 479-487.

Liang P and Pardee AB (1992) Differential display of eukaryotic messenger RNA by means of the polymerase chain reaction, *Science,* **257** : 967-971.

Maldiney R, Leroux B, Sabbagh I, Sotta B, Sossountzov L and Miginiac E (1986) A biotin-avidin-based enzyme immunoassay to quantify three phytohormones: auxin, abscisic acid and zeatin riboside, *J. Immunol. Methods.,* **90** : 151-158.

Marmey P, Besse I and Verdeil JL (1991) Mise en évidence d'un marqueur protéique différenciant deux types de cals issue de mème clones chez le palmier á huile (*Elaeis guineensis* Jacq). *C.R. Acad. Sci. Paris,* t 313, série III, 333-338.

Masmoudi R, Rival A, Nato A, Lavergne D, Drira N and Ducreux G (1999) Carbon metabolism in *in vitro* cultures of date palm: *Phoenix dactylifera* L. : the role of carboxylases (PEPC and Rubisco). *Plant Cell, Tiss. Org. Cult.,* **57** : 139-143.

Mayes S, James C, Horner SF, Jack PL and Corley RHV (1996) The application of restriction fragment length polymorphism for the genetic fingerprinting of oil palm (*E. guineensis* Jacq). *Mol. Breed.,* **2** : 175-180.

Mayes S, James C, Price Z, Groves LM, Jack PL and Corley RHV (1996) Integration of DNA markers into oil palm breeding programmes. *Proc. Int. Palm Oil Conf. Malaysia,* 49-54.

Mayes S, Jack PL, Marshall DF and Corley RHV (1997) Construction of a RFLP genetic linkage map for oil palm (*Elaeis guineensis* Jacq), *Genome,* **40** : 116-122.

Markle SA (1995) Somatic embryogenesis in Magnoliaceae, In : *Somatic embryogenesis and Synthetic Seed I, Biotechnology in Agriculture and Forestry,* (Ed. Bajaj YPS), Springer Verlag, Vol **30** : 38-403.

Meunier J and Gascon JP (1972) Le schéma général d'amelioration du palmier á huile á l'IRHO. *Oléagineux,* **27** : 1-12.

Meunier J (1975) Le palmier á huile américain *Elaeis melanococca, Oléagineux,* **30** : 51-61.

Michelmore RW, Paran I and Kesseli RV (1991) Identification of markers linked to disease resistance genes by bulked segregant analysis: a rapid method to detect markers in specific genomic regions using segregating populations. *Proc. Natl. Acad. Sci. USA,* **88** : 9828-9832.

Mohan M, Nair S, Bhawt A, Krishna TG, Yano M, Bahtia CR and Sasaki T (1997) Genome mapping, molecular markers and marker-assisted selection in crop plants. *Mol. Breed.,* **3** : 87-103.

Marcillo F, Aberlenc-Bertossi F, Trouslot P and Duval Y (1997) Characterization of 25 and 7S storage proteins in embryos of oil palm, *Plant Sci.,* **122** : 141-151.

Morcillo F, Aberlenc-Berossi F, Hamon S and Duval Y (1998) Differential accumulation of storage protein, 7S globulins, during zygotic and somatic embryos development in oil palm (*Elaeis guineensis* Jacq), *Plant Physiol. Biochem.,* **36** : 507-514.

Morcillo F, Aberlenc-Bertossi F, Noirot M, Hamon S and Duval Y (1999) Differential effects of glutamine and arginine on 7S globulins accumulation during maturation of oil palm somatic embryos. *Plant Cell Rep.,* **18** : 868-872.

Morcillo F, Hartmann C, Duval Y and Tregear JW (2001) Regulation of 7S globulin gene expression in zygotic and somatic embryo of oil palm. *Physiol. Plant.*, **112** : 233-243.

Mouyna I, Renard JL and Brygoo Y (1996) DNA polymorphism among *Fusarium oxysporum* f. sp. *elaeidis* populations from oil palm, using a repeated and dispersed sequence. *Curr. Genet.*, **30** : 174-180.

Muranti H (1996) Power of tests for quantitative trait loci detection using full-sib families in different schemes. *Heredity,* **76** : 156-165.

Muranti H (1997) Valorisation de plans de croisements pour la recherche de QTL chez les arbres forestiers, exemple d'n dialléle merisier. *Ph.D Thesis*, Centre d'Orsay,. University of Paris XI, France, 131 pp.

Nakano Y, Steward N, Sekine M, Kusano T and Sano H (2000) A tobacco *NtMET1* cDNA encoding a DNA methyltransferase : molecular characterization and abnormal phenotypes of transgenic tobacco plants. *Plant Cell Physiol.,* **41** : 448-457.

Noiret JM (1981) Application de la culture *in vitro* á l'améilioration et á la production de matériel clonal chez le palmier a huile, *Oléagineux,* **36** : 123-126.

Oakeley EJ, Podesta A and Jost JP (1997) Developmental changes in DNA methylation of the two tobacco pollen nuclei during maturation. *Proc. Natl. Acad. Sci. USA,* **94** : 11721-11725.

Ochoa I, Villegas V and Beebe S (1997) Identification of RAPD molecular marker associated with resistance to the bud rot complex in oil palm (*Elaeis guineensis* Jacq). *Palmas,* **18** : 33-38.

Paranjothy K and Othman R (1982) *In vitro* propagation of oil palm *5th. Intl. Cong. Plant Tissue and Cell Culture,* Tokyo. Abstracts p. 755-756.

Paranjothy K, Ong LM and Sharifah S (1993) DNA and protein changes in relation to clonal abnormalities. *Proc. ISOPB Symposium Malaysia,* 86-97.

Phillips RL, Kaeppler SM and Olhoft P (1994) Genetic instability of plant tissue cultures: breakdown of normal controls, *Proc. Natl. Acad. Sci. USA,* **91** : 5222-5226.

Purba AR, Noyer JL, Baudouin L, Perrier X, Hamon S and Lagoda PJL (2000) A new aspect of genetic diversity of Indonesian oil palm (*Elaeis guineensis* Jacq), revealed by isoenzyme and AFLP markers and its consequences to breeding. *Theor. Appl. Genet.,* **101** : 956-961.

Purba AR, Flori A, Baudouin L, Hamon S and Logoda PJL (2001) Prediction of oil palm (*Elaeis guineensis* Jacq.) agronomic performance using the best linear unbiased predictor (BLUP). *Theor Appl Genet.* (in Press).

Rabéchault H, Ahée J and Guénin G (1970) *Colonies* cellulaires et formes d'embryoides obtenues *in vitro* à partir de cultures d'embryons de palmiers à huile (*Elaeis* guineensis Jacq. Var. dura Becc.), *C.R.A.S.* Série D., **270** : 367-370.

Rao V and Donough CR (1990) Preliminary evidence of a genetic cause for the floral abnormalities in some oil palm ramets. *Elaeis*, **2** : 199-207.

Rashid O and Shah FH (1996) The isolation, characterization and sequencing of cDNA clones coding for stearyl-ACP desaturase gene from oil palm (*Elaeis guineensis*). *Proc. 1996 PORIM Int. Palm Oil Congress Malaysia,* 575-582.

Redenbaugh K (1993) Introduction chapter. In : *Synseeds. Applications of Synthetic Seeds to Crop Improvement.* (Ed Redenbaugh K), CRC Press, London, pp. 3-7.

Rival A, Nato A, Lavergne D and Duval Y (1994) Carboxylases (PEPc and RUBISCO) activities during *in vitro* development and acclimatization of oil palm (*Elaeis guineensis* Jacq.), *In Proc. VIIth International Congress of Plant tissue and Cell Culture, IAPTC,* Firenze, Abstract n°S 20-13, p. 261.

Rival A, Beule T, Nato A and Lavergne D (1996) Immunoenzymatic study of RubisCO in oil palm and coconut. *Plantations, Research, Development,* **3** : 55-61.

Rival A, Bernard F and Mathieu Y (1997a) Changes in peroxidase activity during *in vitro* rooting of oil palm (*Elaeis guineensis* Jacq), *Sci. Hortic.,* **71** : 103-112.

Rival A, Aberlenc-Bertossi F, Morcillo F, Tregear J, Verdeil JL and Duval Y (1997b) Scaling-up *in vitro* clonal propagation through somatic embryogenesis: the case of oil palm (*Elaeis guineensis* Jacq). *Plant Tissue Culture and Biotechnology,* **3** : 74-83.

Rival A, Beulé T, Barre P, Hamon S, Duval Y and Noirot M (1997c) Comparative flow cytometric estimation of nuclear DNA content in oil palm (*Elaeis guineensis* Jacq.) tissue cultures and seed-derived plants. *Plant Cell Rep.,* **16** : 884-887.

Rival A, Beulé T, Lavergne D, Nato A, Havaux M and Purad M (1997d) Development of photosynthetic characteristics in oil palm during *in vitro* micropropagation, *J. Plant Physiol.,* **150** : 11-26.

Rival A, Beule T, Lavergne D, Nato A and Noirt M (1998a) Growth and carboxylase activities in *in vitro* micropropagation oil palm plantlets during acclimatization : comparison with conventionally germinated seedlings. *Advances in Hort. Sci.,* **3** : 111-117.

Rival A, Tregear J, Verdeil JL, Richaud F, Beule T, Duval Y, Hartman C and Rode A (1998b) Molecular search for mRNA and genome markers of the oil palm "*mantled*" somaclonal variations. *Acta Hort.,* **461** : 165-171.

Rival A, Bertrand L, Beule T, Trouslot P and Lashermes P (1998c) Suitability of RAPD analysis for the detection of somaclonal variants in oil palm (*Elaeis guineensis* Jacq). *Plant Breed.,* **117** : 73-76.

Rival A (2000) Somatic embryogenesis in Oil Palm. In : *Somatic Embryogenesis in Woody Plants,* vol. 6, (Eds. Jain SM, Gupta PK and Newton RJ), Kluwer Academic Publishers, Dordrecht. The Netherlands, pp. 249-290.

Saghai Maroof MA, Biyashev RM, Yang GP, Zhang Q and Allard RW (1994) Extraordinarily polymorphic microsatellite DNA in barley: Species diversity, chromosomal locations, and population dynamics. *Proc. Natl. Acad. Sci. USA,* **91** : 5466-5470.

Shah FH, Rashid O, Simons AJ and Dunsdon A (1994) The utility of RAPD markers for the determination of genetic variation in oil palm (*Elaeis guineensis*). *Theor. Appl. Genet.,* **89** : 713-718.

Shah FH and Nyuk LS (1996) Use of microsatellites in the determination of genetic variation and genetic relationship between various oil palm populations. *Proc. Int. Palm Oil Conf. Malaysia,* 568-574.

Shah FH and Cha TS (2000) A mesocarp- and species-specific cDNA clone from oil palm encodes for sesquiterpene synthase. *Plant Sci.,* **154** : 153-160.

Singh R and Cheah SC (1996) Lower – specific gene expression in oil palm revealed by differential display. *Proc. PORIM Int. Palm Oil Congress Malaysia,* 583-587.

Singh R, Cheah SC and Rahman RA (1998) Generation of molecular markers in oil palm (*Elaeis guineensis* Jacq) using AFLP analysis. *Focus,* **20** : 26-27.

Singh R and Cheah SC (2000) Differential gene expression during flowering in the oil palm (*Elaeis guineensis*). *Plant Cell Rep.,* **19** : 804-809.

Smith WK and Jones LH (1970) Plant propagation through cell culture, *Chem. and Ind.,* **44** : 1399-1401.

Smith JSC, Chin ECL, Shu H, Smith OS, Wall SJ, Senoir ML, Mitchell SE, Kresovich S and Ziegle J (1997) An evaluation of the utility of SSR loci as molecular markers in maize (*Zea mays* L,) comparisons with data from RFLPs and pedigree. *Theor. Appl. Genet.*, **95** : 163-173.

Sondahl M (1991) Tissue culture of cacao, coffee and oil palm, In *: Proceedings of the fourth conference Int. Plant Biotechnology Network,* San José, Costa Rica, 14-18 January 1991. pp. 98-99.

Staritsky G (1970) Tissue culture of the oil palm (*Elaeis guineensis* Jacq) as a tool instrument for its vegetative propagation, *Euphytica,* **19** : 288-292.

Steward N, Kusano T and Sano H (2000) Expression of *ZmMET1,* a gene encoding a DNA methyltransferase from maize, is associated not only with DNA replication in actively proliferating cells, but also will altered DNA methylation status in cold-stressed quiescent cells. *Nucleic Acids Res.,* **28** : 3250-3259.

Tautz D (1989) Hypervariability of simple sequences as a general source for polymorphic DNA markers. *Nucleic Acids Res.,* **17** : 6473-6471.

Tautz D and Renz M (1984) Simple sequences are ubiquitous repetitive components of eucaryotic genomes. *Nucleic Acid Res.,* **12** : 4127-4128.

Teixeira JB, Sondhal NR, Nakamura T and Kirby E (1995) Establishment of oil palm cell suspensions and plant regeneration. *Plant Cell, Tiss. Org. Cult.,* **40** : 105-111.

Touchet (de) B, Duval Y and Pannetier C (1991) Plant regeneration from embryogenic suspension culture of oil palm (*Elaeis guineensis* Jacq), *Plant Cell Rep.,* **10** : 529-532.

Triques K, Rival A, Beule T, Puard M, Roy J, Nato A, Lavergne D, Havaux M, Verdeil JL, Sangare A and Hamon S (1997) Photosynthetic ability of *in vitro* grown coconut (*Cocos nucifera* L.) plantlets derived from zygotic embryos. *Plant Sci.*, **127** : 39-51.

Triques K, Rival A, Beule T, Verdeil JL, Hocher V and Hamon S (1997) Developmental changes in carboxylase activities in *in vitro* cultured coconut (*Cocos nucifera* L.) zygotic embryos: comparison with corresponding activities in seedlings. *Plant Cell Tiss. Org. Cult.,* **49** : 227-231.

Van Blakland R, ten Lohuis M and Meyer P (1997) Condensation of chromatin in transcriptional regions of an inactivated plant transgene : evidence for an active role of transcription in gene silencing. *Mol. Gen. Genet.,* **257** : 1-13.

Vieira J and Messing J and Messing J (1987) Production of single-stranded plasmid DNA. *Meth. Enzymol.,* **153** : 3-11.

Wade PA, Gegonne A, Jones PL, Ballestar E, Aubry F and Wolffe AP (1999) Mi-2 couples DNA methylation to chromatin remodelling and histone deacetylation. *Nature Genet.,* **23** : 62-66.

Weissenbach J, Gyapay G, Dib C, Vignal A, Morissette J, Millaseau P, Vayseix G and Lathrop M (1992) A second generation map of the human genome. *Nature.* **359** : 794-801.

Wilkins MR, Sanchez JC, Gooley AA, Appel RD, Himphery-Smith I, Hochstrasser DF and Williams KL (1995) Progress with proteome projects: why all proteins expressed by a genome should be identified and how to do it. *Biotechnol. Genet. Eng. Rev.,* **13** : 19-50.

Wong G, Tan CC and Soh AC (1996) Large scale propagation of oil palm clones – experience to date, In Abst. of International Society for Horticultural Science (ISHS) (Commission Biotechnology) *Third International Symposium on In Vitro Culture and Horticultural Breeding,* Jerusalem, Israel, 16-21 June, p. 12.

Wooi KC (1990) Oil Palm (*Elaeis guineensis* Jacq) : Tissue Culture and Micropropagation, In : *Biotechnology in Agriculture and Forestry, Legumes and Oilseed Crops* I. (Ed. Bajaj YPS), Springer-Gerlag, Berlin, Heidelberg, **10** : 569-592.

Chapter 11

DOMESTICATION OF *SALICORNIA BIGELOVII* TORR. INTO A SEAWATER IRRIGATED NEW CROP

Zhongjin Lu[1,2★], Burgund B Bassuner[1,3], and Maliyakal E John[1,4]

[1]*Monsanto Company, Agracetus Campus, 8520 University Green, Middleton, WI 53562, USA*
[2]*Current address : Seaphire International, 4500 North 32nd Street, Suite 100, Phoenix, AZ 85018, USA*
[3]*Current address : University of Wisconsin, Wisconsin Center for Space Automation and Robotics, 1415 Engineering Drive, Madison, WI 53562, USA*
[4]*Current address : 3622 Rolling Hill Drive, Middleton, WI 53562, USA*

Summary

The domestication of salicornia is a path-breaking technology as it may lead to utilization of halophytes as important food, feed and fiber. As the world's population and demand for food increase and available agricultural land decreases through degradation of soil, the search for and cultivation of sustainable plants on marginal lands becomes urgent. Successful commercial cultivation of salicornia will be a major impetus in the search for other underutilized halophytes. The increasing salinity of our irrigated productive lands makes it imperative that we consider halophytes as a new source for global agriculture. Once developed, the salicornia crop may yield as many as seven or more products. The best opportunity for profitable returns is oil followed by meal, forage, and wood fiber. Advancing salicornia to a crop status will result in many secondary product opportunities such as vegetable tips, speciality applications of biomass, forage and soil and air

---

★Corresponding author : Zhongjin Lu, Ph. D., Senior Scientist, Seaphire International, 4500 North 32nd Street, Suite 100, Phoenix, AZ 85018, USA
E-mail : zlu@ag.arizona.edu

remediations. Many of these opportunities may be pursued by coastal communities stimulating marginal land economies. In conclusion, domestication and further genetic improvement of salicornia and other halophytes is indispensable, since these plants have the potential to contribute significantly to agricultural productivity.

Keywords : Domestication, *Salicornia*, seawater irrigated oil crop

1. INTRODUCTION

Doubling of the world population, increasing loss of arable lands to development, soil erosion and salinization, as well as the need for economic revitalization of marginal lands, necessitates the development of under-utilized plant resources such as halophytes. Halophytes offer a unique opportunity to harvest food and fiber through seawater irrigation (Somers, 1979; O'Leary, 1984; Aronson, 1989; Glenn *et al.*, 1998). Halophytes are plants with a set of ecological and physiological characteristics allowing growth and reproduction in a saline environment. They can tolerate a salinity of at least 5 parts per thousand (ppt) sodium chloride in soil water whereas, glycophytes experience significant productivity loss at that salinity. Of the more than 1500 species identified as halophytes, the vast majority are wild plants (Aronson, 1989; Le Houerou, 1993; Glenn *et al.*, 1999). Attempts to domesticate useful halophytes have already come to fruition and examples include edible grains from palmer salt grass *Distichlis* (Neary, 1981), animal forage from *Atriplex* (Watson, 1990), edible kernel from Indian almond, seed oil and meal from *Salicornia bigelovii* (Charnock, 1988), aromatic resins from *Grindelia*, and specialty oils, medicine and vegetable dye from Chinese tallow (National Research Council, 1990). Of the known useful halophytes, *Salicornia bigelovii* Torr. is the premier candidate for further development because of its potential as a mass cultivated seed oil crop that can change marginal unproductive lands and coastal deserts into productive, economically viable communities (O'Leary *et al.*, 1985). Its development as a lucrative crop requires extensive breeding, selection and seawater irrigation technology. Its commercialization may be also accelerated through genetic engineering. Successful mass cultivation of salicornia will have positive implications for global economies.

This chapter provides an overview of *S. bigelovii* biology, its potential as an oilseed crop, and describes preliminary breeding results and biotechnological studies.

2. PRODUCTS FROM *S. BIGELOVII* CROP

The primary products from salicornia are considered to be seed oil for human consumption and seed meal for animal feed (Hodges, 2000; Lu *et al.*, 2000). Evaluation studies of these products have been completed by many investigators and product feasibility confirmed (Glenn *et al.*, 1991; El-Mallah *et al.*, 1994; Magboul *et al.*, 1996; Kraidees *et al.*, 1998; Abouheif *et al.*, 2000). Studies on several secondary products, such as particle board, fire-wood, carbon sequestration, animal forage and vegetable tips from biomass, saponins and sulfur containing compounds from seeds or biomass have also been completed (Glenn *et al.*, 1992; Hodges *et al.*, 1993; Clark, 1994; Olsen *et al.*, 1996; Glenn *et al.*, 1999; John unpublished results). The following description is limited to the characteristics of seed oil and meal from salicornia as a new halophytic oilseed crop on seawater irrigation.

2.1. Fatty acid composition of salicornia oil

Several investigators have reported the fatty acid profiles of salicornia oil (Glenn *et al.*, 1991; El-Mallah *et al.*, 1994; Magboul *et al.*, 1996) and a typical analysis is shown in Table 1. Like other vegetable oils, salicornia oil has a high concentration of unsaturated fatty acids (90%). It is rich in linoleic acid (70%), considered to be essential for human nutrition as it is the precursor for longer chain fatty acids like arachidonic acids. The fatty acid profile of salicornia oil is very similar to safflower or sunflower oil (Lu *et al.*, 2000).

2.2. Nutritional properties of salicornia meal

A second valuable product of the salicornia seed is its meal. On a dry weight basis, salicornia meal contains approximately 40% crude protein (CP) which is higher than that of rapeseed and linseed (Bell, 1989). Analysis of the amino acid profiles shows that salicornia meal is moderately rich in sulfur containing amino acids (methionine and cystine), but poor in lysine (Table 2). In addition, it is noteworthy that salicornia growing on seawater irrigation, does not accumulate large quantities of salts in the seed (O'Leary, 1984). The ash content in the salicornia meal appears similar to other conventional oilseed meals.

Orthophosphate is a major source of phosphorus nutrition for animals. As shown in Table 2, the orthophosphate content in salicornia meal is nearly three times of that in soybean meal. In contrast, phytates are

Table 1. Comparison of fatty acid composition of oilseed crops.

Oil crop	Botanical name	Fatty acid composition (%)[a]													Ref.[b]
		14:0	16:0	16:1	18:0	18:1	18:2	18:3	20:0	20:1	22:0	22:1	24:0	24:1	
Salicornia	*Salicornia bigelovii*	0.0	8.1	0.0	2.2	12.5	74.0	2.6	0.0	0.0	0.0	0.0	0.0	0.0	1
	cv. SOS-7	0.3	9.0	0.0	3.4	17.5	66.9	1.4	0.5	0.4	0.4	0.0	0.2	0.0	2
	cv. SOS-10	0.1	7.7	0.1	2.6	15.8	68.2	2.0	0.4	1.0	0.2	0.1	0.0	0.0	3
Soybean	*Glycine max*	0.0	15.3	0.0	4.2	23.6	48.2	8.7	0.0	0.0	0.0	0.0	0.0	0.0	4
Cottonseed	*Gossypium hirsutum*	1.0	23.4	0.8	2.5	17.9	54.2	0.0	0.0	0.0	0.0	0.0	0.0	0.0	5
Peanuts	*Arachis hypogaea*	8.0	0.0	0.0	3.7	64.3	17.3	0.0	1.5	1.2	2.7	0.0	1.6	0.0	6
Sunflower	*Helianthus annuus*	0.1	5.8	0.1	5.2	16.0	71.5	0.2	0.2	0.1	0.7	0.0	0.1	0.0	7
Brassicas															
Rapeseed	*Brassica napus*	0.0	2.9	0.3	0.9	11.4	12.9	9.4	0.7	7.2	0.8	51.9	0.2	1.1	8
Canola-type		0.0	4.2	0.2	1.5	57.7	24.6	8.4	0.6	1.3	0.3	0.2	0.2	0.0	8
Turnip rape	*Brassica rapa*	0.0	1.9	0.2	1.0	13.0	12.7	8.7	0.8	7.9	0.1	52.7	0.0	1.2	8
Canola-type		0.0	3.8	0.1	1.2	58.6	24.0	10.3	0.6	1.0	0.1	0.3	0.0	0.0	8
Mustard	*Brassica juncea*	0.0	2.5	0.3	1.2	8.0	16.4	11.4	1.2	6.4	1.2	46.2	0.7	1.9	9

Table 1. Continued

Oil crop	Botanical name	Fatty acid composition (%)[a]													Ref.[b]
		14:0	16:0	16:1	18:0	18:1	18:2	18:3	20:0	20:1	22:0	22:1	24:0	24:1	
Ethiopian mustard	*Brassica carinata*	tr	3.2	0.2	0.9	9.8	16.2	13.9	0.7	7.5	0.7	41.6	0.6	2.0	8
Corn	*Zea mays*	tr	11.5	0.0	2.2	26.6	58.7	0.8	0.2	0.0	0.0	0.0	0.0	0.0	10
Safflower	*Carthamus tinctorius*	0.0	7.6	0.0	2.0	10.8	79.6	0.0	0.0	0.0	0.0	0.0	0.0	0.0	11

[a] Fatty acids represented by carbon chain length and number of double bonds; tr, trace amounts.
[b] References: (1) Glenn *et al.* (1991), (2) El-Mallah *et al.* (1994), (3) Magboul *et al.* (1996), (4) Hymowitz *et al.* (1972), (5) Anderson and Worthington (1971), (6) Worthington and Hammons (1971), (7) Earle *et al.* (1968), (8) Downey and Rimmer (1993), (9)Appelqvist (1970), (10) Beadle *et al.* (1965), and (11) Knowles (1968).

Table 2. Comparison of major nutrients and anti-nutritional factors in salicornia and soybean meals

Seed meal	Soybean[a]	Salicornia
Nutrient composition (%)		
Crude protein	48.5	40.5
Crude fiber	3.0	7.6
Ash	5.9	8.7
Amino acid composition in protein (N x 6.25, %)		
Lysine	6.2	1.7
Methionine	1.4	2.5
Cystine	0.6	1.4
Phosphorous (μg/g)		
Orthophosphate	2.5	8.7
Phytate	2300	150
Saponins (%)	0.5	0.5

[a] All data of soybean meal were cited from Bell (1989).

anti-nutritional factors that conglomerate phosphorus and other mineral nutrients thereby reducing their intake efficiency. The content of phytates is only 150 μg/g in salicornia meal, compared to 2300 μg/g in soybean meal.

Salicornia meal contains saponins, an anti-nutritional factor. Saponins can complex with cholesterol, causing swelling of the intestinal mucosa cells. They also give salicornia fresh tissues and seed meal a bitter flavor, which affects the palatability of salicornia hay and meal. However, previous studies showed that the anti-nutritional effect of saponins in salicornia meal can be removed by treating the meal with sodium hydroxide (NaOH) or by adding the saponin-antagonist, cholesterol, to the meal (Glenn *et al.*, 1991). Like other meals, salicornia meal can be used as a protein source of feed for fish such as Nile tilapia (Belal Ibrahim and Al-Dosari, 1999), poultry such as broilers (Attia *et al.*, 1997), and ruminant animals such as Lambs (Riley *et al.*, 1994; Swingle *et al.*, 1996; Kraidees *et al.*, 1998).

3. BOTANY, TAXONOMY AND GEOGRAPHICAL DISTRIBUTION

3.1. Botanical description

The genus *Salicornia* L. was nominated by Linnaeus in 1753, and derives its name from Greek, where *sal* (or *sali*) means salt and *cornu* a horn (*i.e.* a saline plant with hornlike branches; Munz and Keck, 1970). Salicornia, commonly known as glasswort, saltwort, pickleweed, samphire, sea bean, and sea asparagus, is a dicotyledonous C_3 vascular plant from the family Chenopodiaceae.

Salicornia has succulent and articulated stems; opposite branches and short internodes. Leaves are reduced to scales and form a short sheath with rudimentary lamina and sharply pointed tips. The photosynthetic organs in salicornia are leafless stems and green inflorescences. The inflorescence (seed spike) is formed by cymes each of which contains three flowers distributed in a triangle. The middle flower in the cyme, is generally higher than the two lateral ones. Flowers are perfect and completely sunk in cavities of the segment subtended by bracts, including a four-lobed perianth, two stamens and two styles. Seeds are vertical and almost ovoid with a thin membranous testa. The endosperm is very sparse.

3.2. Taxonomy

The genus *Salicornia*, consists of about 50 species (Clapham *et al.*, 1987; Tutin *et al.*, 1993), out of which 10 are found in North America (Wiggins, 1980). The taxonomy of this genus is complex and difficult (Ball, 1964; Clapham *et al.*, 1987; Stace, 1999). Individuals of a single specimen usually exhibit large phenotypic plasticity in response to different external conditions, although they are not different in their genetic makeup based on electrophoretic mobility of enzymes (Jeffries *et al.*, 1982). *Salicornia bigelovii* (a species under domestication) is endemic to North America. Therefore, the taxonomic description of the main salicornia species distributed at this region is given below (Standley, 1916; Munz and Keck, 1970; Wiggins, 1980).

Perennial salicornia : *Salicornia subterminalis* Parish and ***Salicornia virginica*** L.

Perennial salicornia (*Salicornia subterminalis* Parish, syn. *Arthrocnemum subterminalis* Standl.) are dwarf subshrubs with woody stems found in saline soils, mostly in coastal salt marshes. Fruiting spikes bear only sterile flowers toward the apex and flowering occurs April through September. Their fresh vegetative joints are 1.5 - 3.5 mm in diameter. Seeds are glabrous, with a slight amount of endosperm.

In *Salicornia virginica* L. (syn. *S. ambigua* Michx.; *S. pacifica* Standl.), the fruiting spikes are fertile to the apex and the fresh vegetative joints are 2 - 5 mm in diameter. Their seeds are distinctly pubescent, lacking endosperm. *S. virginica* is also found in coastal salt marshes and nearby alkaline flats. Flowering occurs August through November and the seeds are 0.5-1 mm long.

Annual salicornia : *Salicornia europaea* L. and ***Salicornia bigelovii*** Torr.

The annual salicornia (*Salicornia europaea* L., syn. *S. herbacea* (L.) L.; *S. rubra* A. Nels.; *S. depressa* Standl.) is a bushy annual, branching freely from the base. These plants are 10 - 20 cm tall. The joints of the spike are longer than they are thick and the scales at the apices of joints are broadly round-ovate, slightly acute but not mucronate. The flowering spikes are 2-6 cm long, 2-3 mm thick and; the central flower is higher than the visible part of the lateral ones. Flowering occurs July through November. Seeds are pubescent and 1-1.5 mm long.

Salicornia bigelovii Torr. (syn. *S. mucronata* Bigel.) is an erect annual plant with simple-stemmed or with strongly ascending branches mostly above the middle. They are generally 10-50 cm tall and the joints of the flowering spike are mostly thicker than longer; with scales at the apices of the joints triangular-ovate, sharply mucronate; the spikes 2-10 cm long, 4-6 mm thick. The central flower is higher than the visible part of the lateral flowers and reaches nearly to the summit of the joint. Flowering occurs July through November. The seeds are pubescent and 1.5-2 mm long.

3.3. Geographical distribution

The genus *Salicornia* is distributed worldwide. In particular, *S. europaea* L. can be found in coastal salt marshes and occasionally on inland saline flats throughout the world (Duval-Jouve, 1868; Munz and Keck, 1970; Clapham *et al.*, 1987; Tutin *et al.*, 1993). However, many species of this genus appear to be locally distributed, after long term adaptation to inhabitant environments. For example, *Salicornia bigelovii* Torr., nominated by the famous American botanist Bradford Torrey (1847-

1912), is a native species in North America. *S. bigelovii* naturally grows in coastal salt marshes along the Pacific coast of northern Baja California to Los Angeles and also along the Atlantic coast of Maine, Massachusetts, Delaware, the Carolinas, Virginia, and Georgia, through Florida to the Gulf of Mexico and some of the Caribbean islands (Lu *et al.* unpublished data; Ball, 2000). Therefore, ecological niches for this species are in temperate and subtropical zones, with an approximate geographical range from 22° to 45° Northern Latitudes.

4. PHYSIOLOGY OF SALT TOLERANCE

The detrimental effects of salt stress on plant growth and development generate considerable interest and have been frequently reviewed (Flowers *et al.*, 1977; Levitt, 1980; Munns *et al.*, 1983; Bohnert *et al.*, 1995; Zhu *et al.*, 1997; Glenn *et al.*, 1999). These effects can be summarized as (1) osmotic stress arising from elevated osmotic pressure due to salts, which leads to reduced water uptake or even water efflux from plant roots (Munns 1993); (2) ion toxicity usually associated with excessive intake of sodium, chloride, boron or other ions (Sharma and Gupta, 1986); and (3) nutrient imbalance when the excess of these ions leads to a diminished uptake of potassium, nitrate or phosphate, or to impaired internal distribution of one or another of these elements (Langdale and Thomas, 1971; Grattan and Grieve, 1994). Continued severe salt stress eventually results in plant death (Neumann, 1997).

Many studies have indicated that *Salicornia* L. is a euhalophyte, *i.e.* a true salt tolerant or salt-loving plant (Halket, 1915; Webb, 1966; Waisel, 1972; L'Roy and Hendrix, 1980; Glenn and O'Leary, 1984; Stumpf and O'Leary, 1985; Ungar, 1991; Ayala and O'Leary, 1995; Ayala *et al.*, 1996). Since *S. bigelovii* originated in coastal salt marshes, it appears to have evolved the physiological mechanisms to cope with extreme saline environments. Similar to other species, *S. bigelovii* initiates osmotic adjustment in response to osmotic stress induced by saltier solution in the soil than in the root. This physiological mechanism helps increase the osmotic potential of the cell sap through the accumulation of osmolytes, such as salts and organic compounds, so that the water from the soil continues to influx into plants (Riehl and Ungar, 1982; Glenn and O'Leary, 1984; Plaut and Heuer, 1985; Stumpf and O'Leary, 1985). To cope with the toxic effects of excessive salt intake in the cells, *S. bigelovii* actively transports sodium and chloride into vacuoles that can serve as a storage area for toxic wastes. This facilitates to protect the salt-sensitive proteins and enzymes in the cytoplasm. The sodium accumulation in the vacuoles appears to be mediated by Na^+/H^+ antiporters using a pH gradient generated through vacuolar H^+-ATPase that couples the active transport of Na^+ with passive efflux of protons (Blumwald and Poole, 1987; Ayala and O'Leary, 1995; Ayala *et al.*, 1996). The vacuoles are able to achieve a concentration of more than 60 ppt salt, while pure seawater is only 32 ~ 35 ppt salt (Douglas, 1994). Table 3 indicates that *S. bigelovii* can grow and yield well up to a salinity of about 50 dS/m (nearly equal to seawater salinity) in the soil moisture (Glenn, 1984). In contrast, all conventional crops show a loss of plant productivity at the salinity levels of 1 – 8 dS/m (Mass and Hoffmann, 1977; Maro *et al.*, 1991; Francois and Maas, 1994; Pasternak and De Malach, 1994).

5. REPRODUCTIVE BIOLOGY

Recent studies by York *et al.* (2001) have revealed that *S. bigelovii* is a facultative short-day plant, although there is diversity in the sensitivity to photoperiod between ecotypes. Under a light-controlled greenhouse, mechanisms underlying the floral induction of *S. bigelovii* were investigated by using ten geographically isolated ecotypes collected across North America. The treatments of photoperiod consisted of short day (SD: 8-, 9-, and 10-hour light per day), neutral day (ND: 12-hour light per day), and long day (LD: 16-hour light per day). Results indicated that treatments with SD photoperiods significantly promoted floral initiation in all populations. In terms of days to preflowering, the main stem of 50% of the individuals showed a flower cone at the apical meristem at 23.4 ± 1.6, 22.2 ± 1.2, and 25.0 ± 1.2 days, respectively, after the onset of 8-, 9-, and 10-hour treatments (mean ± SE, n = 10). These values from SD treatments were smaller than 76.0 ± 3.5 days with the ND photoperiod at the level of $p = 0.01$. In contrast, the same ecotypes exposed to a LD photoperiod needed 107.5 ± 13.4 days to reach 50% preflowering, even significantly longer than with the ND photoperiod. Figure 1 shows dynamics of the index of floral initiation for different photoperiodic treatments. This also revealed effects of SD photoperiods on stimulating floral initiation of *S. bigelovii*.

Dadby (1962) reported that the breeding behavior of *S. europaea* is self-pollinated and even cleistogamous. In contrast, our observations showed that the mating system of *S. bigelovii* is primarily wind-pollinated and protogynous. In the field, styles and stigmas in *S. bigelovii* generally start protruding 2 - 10 days before the undehisced anthers are exerted. The degree of protogyny appears to be associated with some morphological

Table 3. Comparison of yield loss as a response to salinity among salicornia and conventional field crops.

Species	Botanical name	Yield Based on	Response to salinity		Ref.[b]
			Threshold EC_e (dS/m)	Yield loss (% per dS/m)	
Salicornia	*Salicornia bigelovii*	Seed yield	50.0	0.5	1
Barley	*Hordeum vulgare*	Grain yield	8.0	5.0	2
Wheat	*Triticum aestivum*	Grain yield	6.0	7.1	2
Durum wheat	*Triticum durum*	Grain yield	5.9	3.8	3
Rice	*Oryza sativa*	Grain yield	3.0	12.0	2
Corn	*Zea mays*	Grain yield	1.6	9.6	2
		Ear FW[a]	1.7	12.0	2
Sorghum	*Sorghum bicolor*	Grain yield	6.8	16.0	4
Cotton	*Gossypium hirsutum*	Seed yield	7.7	5.2	2
Soybean	*Glycine max*	Seed yield	5.0	20.0	2
Peanut	*Arachis hypogaea*	Seed yield	3.2	29.0	2
Sugar beet	*Beta valgaris*	Storage root	7.0	5.9	2
Sugarcane	*Saccharum officinarum*	Shoot DW	1.7	5.9	2
Potato	*Solanum tuberosum*	Tuber yield	1.7	12.0	2
Alfalfa	*Medicago sativa*	Shoot DW	2.0	7.3	2
Bermuda grass	*Cynodon dactylon*	Shoot DW	6.9	6.4	2
Pea	*Pisum sativum*	Seed FW	3.4	10.6	5
Mung bean	*Vigna radiata*	Seed yield	1.8	20.7	6

Table 3. Continued

Species	Botanical name	Yield Based on	Response to salinity		Ref.[b]
			Threshold EC_e (dS/m)	Yield loss (% per dS/m)	
Onion	*Allium cepa*	Bulb yield	1.2	16.0	2
Onion	*Allium cepa*	Seed yield	1.0	8.0	7
Garlic	*Allium sativum*	Bulb yield	1.7	10.0	8
Carrot	*Daucus carota*	Storage root	1.0	14.0	2
Radish	*Raphanus sativus*	Storage root	1.2	13.0	2
Cabbage	*Brassica oleracea*	Head FW	1.8	9.7	2
Turnip	*Brassica rapa*	Storage root	0.9	9.0	9
Tomato	*Lycopersicon lycopersicum*	Fruit yield	2.5	9.9	2
Tomato	*L. cerasiforme*	Fruit yield	1.7	9.1	10
Spinach	*Spinacia oleracea*	Top FW	2.0	7.6	2
Strawberry	*Fragaria × Ananassa*	Fruit yield	1.0	33.0	2
Asparagus	*Asparagus officinalis*	Spear yield	4.1	2.0	11

[a]FW stands for fresh weight and DW for dry weight.

[b]References: (1) Glenn and O'Leary (1984), (2) Maas and Hoffman (1977), (3) Francois *et al.* (1986), (4) Francois *et al.* (1984), (5) Cerda *et al.* (1982), (6) Minhas *et al.* (1990), (7) Mangal *et al.* (1989), (8) Mangal *et al.* (1990), (9) Francois (1984), (10) Maro *et al.* (1991), and (11) Francois *et al.* (1987).

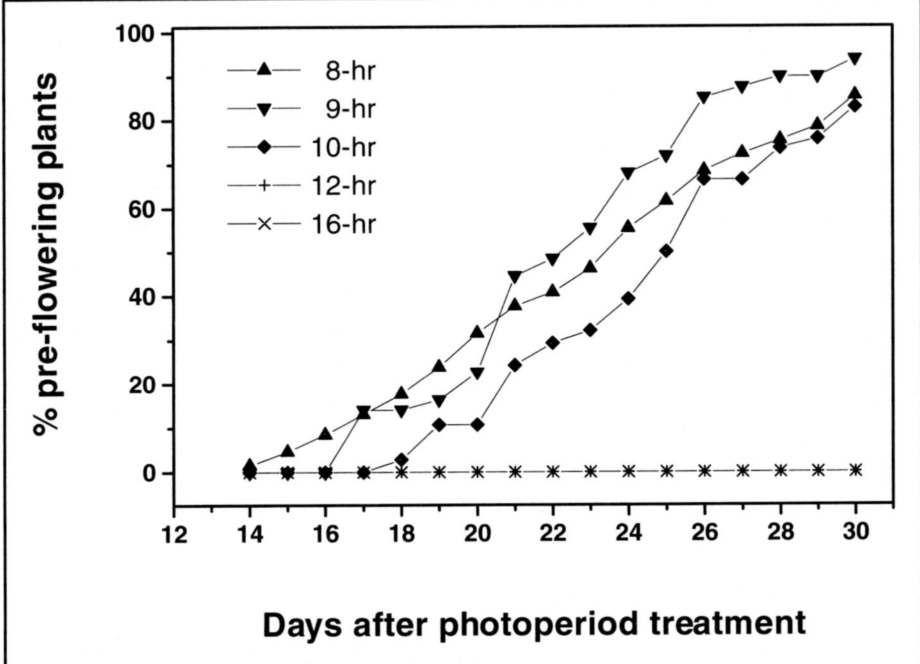

Figure 1 : Effects of different photoperiodic treatments on floral initiation in *S. bigelovii* Torr. Photoperiodic treatments were started with 42-day-old seedlings. The morphological changes at the apical meristem of a main stem were observed daily after the experimental start, and the date of floral initiation for each individual was recorded. Each point represents the mean for more than 120 individuals.

characteristics. Plants with spreading architectures usually display a stronger protogynous property, while ones with upright and compact types are feebly protogynous, even protandrous (*i.e.,* anthesis occurs before the appearance of pistils). Environmental factors such as temperature and humidity also play a role in controlling protogyny of *S. bigelovii* flowers. Low temperature and high relative humidity were found to induce a comparatively longer interval between the first appearance of styles and anthers. For an individual plant, flowering begins at the basal part of each inflorescence and continues upward. At each cymule, the stigmas protruded earlier in the central floret than in the two lateral florets. Microscopic inspection indicated that the stigma matures about 1-2 days before its emergence, suggesting that the pistils of *S. bigelovii* become functional before flowering. The anthesis occurs mainly in the morning, but continues throughout the day. The maximum anthesis takes place between 7 a.m. and 12 p.m. with the peak at 9 - 11 a.m. In addition, our pollination experiments showed that pollens stored at 4°C maintained viability for up to 15 days.

The inflorescent stage of individual plants of *S. bigelovii* ecotypes varies from 5 to 30 days. For example, under a greenhouse in Tucson, Arizond, USA (31.20° N.L.), populations from Alabama (USA), though flowering late, showed a short inflorescent duration (about 5 days). Populations from Florida (USA) and Texas (USA) took 12 – 20 days to complete the flowering process. However, populations along the Pacific coast such as California (USA) and Tijuana (Mexico) displayed a longer duration (16-40 days), although they usually began flowering much earlier. Additionally, it was observed that the flowering period was shortened by high temperature, but was extended by high relative humidity in the air.

6. AGRONOMY AND CULTIVATION PRACTICES

6.1. Biomass, seed and oil yield

The precise seed yield of salicornia is not known, but projections place it as high as that of soybeans. Yield estimates of salicornia grown under experimental conditions with undiluted seawater irrigation at Puerto Penacso, Sonora, Mexico for five years suggested a seed yield of 1.99 Mt/ha, as shown in Table 4 (Glenn *et al.*, 1991). The highest seed yield in Saudi Arabia was reported to be 3.5 Mt/ha (Clark, 1994). These seed yields are comparable to soybeans in the United States (2 Mt/ha, USDA 1990). Soybeans have mean oil content of ca. 20% with a range from 16% to 24%, and produces about 0.4 Mt oil/ha. In comparison, salicornia on seawater irrigation may yield 0.6 Mt oil/ha with the range of 0.5 to 0.9 Mt/ha. However, to date little is known about its full potential for seed yield improvement. Salicornia seems to have a very low harvest index (HI). The HI is 50 – 60% in cereal crops, 30 – 32% in soybean, and 35% in oilseed brassicas, respectively (Austin *et al.*, 1985; Borojevic, 1990). Even under drought conditions, wheat can achieve 35 ~ 40% of harvest index (Austin, 1989). In contrast, the HI of salicornia, averaged 12.2% under seawater irrigation and ranged from 9.47% to 14.96% (Table 4)

6.2. Water requirements and irrigation management

The irrigation management of salicornia with seawater requires special considerations (Pasternak and De Malach, 1994; Glenn *et al.*, 1998).

Table 4. Biomass, seed, and oil yields of *S. bigelovii* grown on seawater irrigation. (Modified from Glenn *et al.*, 1991).

Year	Month sown	Biomass (Mt/ha)	Seed	Oil yield[a]	Harvest index[b] (%)
1984	December	24.6	2.33	0.68	9.47
1985	April	13.9	2.08	0.60	14.96
1986	February	15.1	1.93	0.56	12.78
	April	12.7	1.77	0.51	13.94
1987	February	19.8	2.46	0.71	12.42
1988	February	14.4	1.39	0.40	9.65
Mean ± SE		16.8 ± 1.8	1.99 ± 0.16	0.58 ± 0.05	12.20 ± 0.91

Note : [a] Oil yield is a product of seed yield per unit area and oil content in seeds. [b] Harvest index (HI) is percentage of seed yield in relation to the above-ground biomass.

Due to a continuous process of upward capillary water movement coupled with water evaporation, the concentration of salts accumulated in the upper soil layers can be as high as 100 ppt (Noriega, 1996), beyond the salinity limit of plant survival in salicornia (Glenn and O'Leary, 1984). According to our observations, salicornia is a crop with a very shallow root system and its main root is 10 – 20 cm long. This suggests that salicornia may suffer more than deep-rooted halophytes for a given mean soil salinity. Therefore mass cultivation of salicornia requires very frequent, even daily, irrigation of seawater to leach excess salts away from the upper soil layers, thereby controlling salinity levels in the root zone. Glenn *et al.* (1997) used full-strength seawater to grow salicornia in drainage lysimeters placed within open fields during two years of field trials, and estimated water balances and salt balances for growth under seawater irrigation. It was found that the biomass yield of salicornia responded well to the amount of seawater applied. Additionally, a leaching fraction (excess water application above crop needs) of 35%, appeared to be necessary for keeping salt levels below 80 ppt in the soil surface. Salicornia can be cultivated in a variety of irrigation systems. If hand labor is available, it can be grown in small beds or plots and irrigated by hand or flood (*e.g.* in China and India). On a larger scale, it can be irrigated using a central pivot sprinkler system (*e.g.* in Saudi Arabia and the USA).

6.3. Nutrient requirements and fertilizer management

As compared with irrigation with freshwater in traditional agriculture, seawater irrigation supplies plants with certain essential nutrients, such as potassium, sulfur, calcium, and magnesium, in adequate concentrations for plant growth (National Research Council, 1990; Grasshoff *et al.*, 1999). However, seawater still lacks the two primary macronutrients, nitrogen and phosphorus, and several micronutrients (*e.g.* iron, copper, zinc, manganese, molybdenum). To date, the actual nutrient requirements of salicornia plants under seawater irrigation have not been thoroughly investigated, although some studies have been done on nitrogen, phosphorus and a combination of both.

6.3.1. Nitrogen

In order to investigate the response of salicornia biomass and seed yields to nitrogen supplementation under seawater irrigation, urea-based nitrogen fertilizer was applied to give final N treatment levels of 0, 110, 220, and 330 kg N/ha over the growing season (Mota-Urbina, 1996). Results showed that additions of nitrogen significantly increased the seed yield. Compared to the control (0 kg N/ha), supplementations of 110, 220 and 330 kg N per hectare increased seed yields by 145.06%, 275.69%, and 368.72%, respectively. Glenn *et al.* (1991) reported optimum nitrogen fertilizer applications of 200 kg/ha or more under experimental conditions. A nitrogen source/level experiment by Riley and Abdal (1993) revealed that ammonium nitrate appeared to be the optimal source of nitrogen. Furthermore, 200 and 400 kg/ha of nitrogen, yielded similar mean air-dried biomass. In this study, the experimental field plots were pre-treated with a phosphorus fertilizer (triple super phosphate; 100 kg of P_2O_5/ha). Nitrogen is one of the "mobile" nutrients (like phosphorus), that can be remobilized from mature to immature tissue. Thus, its deficiency symptoms in plants can be seen first in mature tissues. Based on our observations in the field, nitrogen deficiency symptoms include reddish branches and stunting of plants. Excess of nitrogen in plants often leads to lodging, susceptibility to water deficit, stimulation of vegetative growth, delay in development of flower primordia and/or shift of conic floral meristems back to a flat vegetative state.

6.3.2. Phosphorus

Al Saeedi (1997) examined the response of salicornia to fertilizer treatments under seawater irrigation. Results indicated that phosphorus (100 kg P_2O_5/ha) had a doubling effect on mean plant height and growth rate, as compared to treatments without phosphorus. Hawke and Maun (1987) observed similar positive correlation between concentrations of nitrogen and phosphorus in plant growth of three dune species. Al Saeedi (1997) concluded that this interaction might result from the otherwise strong fixation of phosphorus by seawater. In general, crop yield decreases with enhanced salinity level. This fixation of specific phosphorus ions is determined by the pH-value of saline soil and seawater. At a pH-level of 8.2, 90% of the present phosphorus species is HPO_4^{2-}. However, for nutrient uptake, plants prefer the phosphorus species $H_2PO_4^-$ (Dixon *et al.*, 1989). To estimate critical phosphorus status in plant tissue, Al Saeedi and Elprince (2000) grew *S. bigelovii* hydroponically in the greenhouse. The solution medium was seawater enriched with a half-strength modified Hoagland's solution at pH 6.2, and P treatments ranged from 0 to 120 mg/L. The concentration of phosphorous critical for growth in semiwoody internode tissue was found to be 480 mg soluble P per kg dry biomass. In addition, P deficiency symptoms observed were stunting of plants, dark green and purple color, and shriveled seeds (Al Saeedi and Elprince, 2000).

6.3.3. Ratio of N to P

The optimum ratio between applied amounts of nitrogen, phosphorus, and potassium fertilizer is crop specific and has thus to be determined for optimal growth and development. In general, conventional crops require a N:P:K fertilizer ratio of 3:1:5. For salicornia plants, potassium is already supplied with a certain amount from irrigated seawater. Al Saeedi (1997) showed an optimum N:P ratio of 4:1 with a pots experiment. Field studies by Riley and Abdal (1993) indicated an optimum N:P ratio of 2-4:1. These data indicate that Salicornia growth requires the N:P ratio similar to conventional crops on freshwater irrigation. However, under mass cultivation of salicornia, the optimal ratio of N to P additions that would produce the maximal seed yield has not yet been determined.

6.4. Disease, pest, and weed control

S. bigelovii grows in its native habitats and on test farms without being significantly affected by disease or pest, although Stanghellini *et al.* (1988, 1990, 1992) reported instances of salicornia being attacked by *Metachroma* larvae and *Macrophomina phaseolina*. It is assumed that the saponins in salicornia may deter fungal infestations and discourage pests from feeding on plants (Bowyer *et al.*, 1995; Osbourn, 1996).

Table 5 lists common diseases, pests and weeds that have been

Table 5. Diseases, pests, and weeds of *S. bigelovii.*

Biotic stress	Farm field (Mexico, Saudi Arabia, United Arab Emerus)	Greenhouse (ERL in Tucson, USA)	Reference
Fungi	*Macrophomina phaseolina* (Tassi) Goidanich *Pythium aphanidermatum* (Edson) Fitzp. *Rhizoctonia solani* Kuhn *Fusarium* spp.	*Macrophomina phaseolina* (Tassi) Goidanich *Pythium aphanidermatum* (Edson) Fitzp. - -	Stanghellini *et al.* (1988, 1990, 1992)
Bacteria	*Bacillus subtilis*	*Bacillus subtilis*	Stanghellini and Rasmussen (1989)
Virus	Curly top virus	Curly top virus	Brown (1996)
Nematodes	*Heterodera* ssp. (Cyst nematode)	*Heterodera* ssp. (Cyst nematode)	Baldwin *et al.* (1997); Ferris *et al.* (1999)
	Trichodorus spp. (Stubby-root nematode)	-	
	Longidorus spp. (Needle nematode)	-	
Insects	*Lepidotera* (Leaf caterpillar; Spike larvae)	*Lepidotera* (Leaf caterpillar; Fruit larvae)	
	Linothrips (Thrips)	*Linothrips* (Thrips)	Riley and Abdal (1993)

Table 5. Continued

Biotic stress	Farm field (Mexico, Saudi Arabia, United Arab Emerus)	Greenhouse (ERL in Tucson, USA)	Reference
	Hemiptera (True Bugs)	-	
	Coleoptera (Beetle larvae)	-	Stanghellini *et al.* (1988); Proudfoot (1993)
	Homoptera (Leafhopper)	-	
	Aphis (Aphids)	*Aphis* (Aphids)	
Weeds	*Cressa truxilensis* L.	*Cressa truxilensis* L.	Lu *et al.* (unpublished data)
	Suaeda	-	

observed in *S. bigelovii* grown on farms and in greenhouses. The fungus *Macrophomina phaseolina* (Tassi) Goidanich is a soilborne pathogen widely distributed in habitat of arid and semi-arid lands. It has been reported to cause root rot on mature salicornia plants in both wild and cultivated environments (Stanghellini *et al*., 1992). On pilot seawater farms at Puerto Penasco, Sonora, Mexico, *M. phaseolina* caused rotting of the roots and led to stand mortality rates as high as 80% in 1989 and 30% in 1990. The disease occurrence is attributed to the plant undergoing environmental and/or physiological stresses. Stress from water deficit, sub-optimal salinity (less than 5 ppt), high temperature, and flowering are associated with the onset of *M. phaseolina* root rot. Therefore, avoidance of stress is one of effective control methods for *M. phaseolina* root rot. Timely irrigation will prevent water deficit stress. Early planting should help avoid temperature and reproductive stress by initiating flowering during relative cooler weather. In a greenhouse at the Environmental Research Laboratory (ERL), University of Arizona, Tucson, USA, *Pythium aphanodermatum* appeared to be a major seedling pathogen and caused severe stand loss. The source of the pathogen was found to be the river washed sand used in the pots (Stangellini, personal communication). This pathogen can be controlled with the broad-spectrum fungicide Bavistine™ or Ridmil™.

With regard to pests in salicornia, Stangellini *et al.* (1988) reported that small beetle larvae of the genus *Metachroma* caused a stand loss of up to 35% in the field at Puerto Penasco, Mexico. They cut off the seedling roots 3 – 5 cm below the soil surface. Treatment of the soil with Diazonin™ or Vapam™ stopped plant damage by these beetle larvae. In a greenhouse, three pests have been identified on salicornia plants. Aphids (*Aphis gossypii*) were observed during the entire crop season (March through November). They feed on the fresh tips of young branches and spikes, and become a serious pest for salicornia grown in a greenhouse. Thrips (*Linothrips cerealium* Haliday) are very tiny pests on salicornia plants, usually appearing in May through September. They attacked the basal part of young joints. This leads to woody and shriveled symptoms and then death of apical meristems. In addition, very limited quantities of caterpillars (*Spodoptera latebrosa*) were found during the flowering stage (June – August). They ate fresh tissues of plants, but did not cause significant crop damage. These insect pests in a greenhouse can be controlled by overhead spraying Diazonin™ at a concentration of about 0.25%.

7. GENETICS AND BREEDING

7.1. Genetic variability

Nearly all programs of conventional crop improvement benefit from the genetic variability available among wild germplasm (Borojevic, 1993). As shown in Table 6, twenty natural populations of *S. bigelovii* were collected from seven locations in North America, which covered the Atlantic coast, the Gulf of Mexico, the Gulf of California and the Pacific coast. Evaluation of phenological, morphological and agronomic performances indicated large genetic variation among geographically distant accessions. For instance, *S. bigelovii* population (SBP-07) collected from Indian River Lagoon, Merrit Island, Florida, USA had an upright and compact plant type with a plant height of $57.^{37} \pm 1.^{19}$ cm at harvest (mean ± SE, n = 144). In contrast, *S. bigelovii* plants of population SBP-04 from Tijuana Salt Marsh, Mexico, uniformly displayed a bushy architecture with a much branching property. The plant height was $9.^{95} \pm 0.^{82}$ cm, only one sixth time that of SBP-07. In addition, spike morphology and color, seed size and weight, seedling resistance to diseases were remarkably different between populations. The genetic variation in *S. bigelovii* is also confirmed at the molecular level by random amplified polymorphic DNA (RAPD) markers as shown in Figure 2. By using RAPD techniques, tribe *Salicornieae* has been differentiated into three genera: *Salicornia* L., *Sarcocornia* A.J., and *Arthrocnemum* Moq.

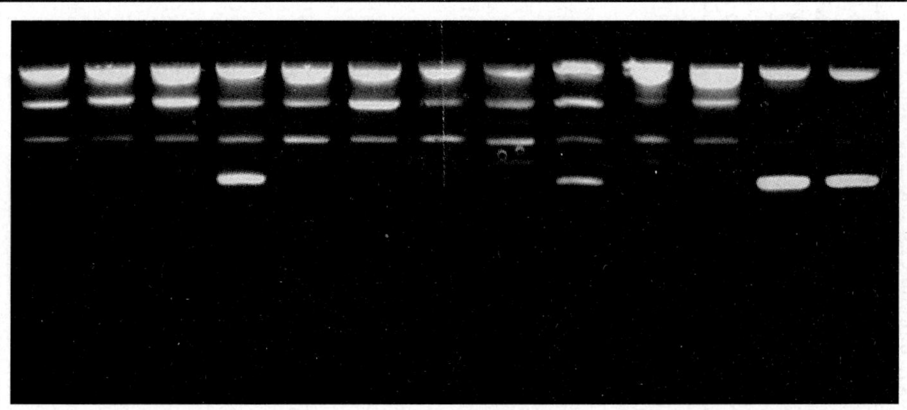

Figure 2 : Genetic variation detected with RAPD techniques. The sequence of a RAPD primer (P3) used for the PCR amplification was GATTCGACGAT. Amplified products were electrophoresed on a 1.2% agarose gel at 90 V for 1.5 h. Each panel represented different genotypes in *Salicornia bigelovii* Torr.

Table 6. Phenology, morphology and productivity variation among wild accessions of *S. bigelovii* Torr.

Region	Location	Population	Crop[a] cycle (day)	Plant[b] height (cm)	Stem[c] diameter (cm)	Inflorescence		Plant productivity[d]		
						Length (cm)	Width (cm)	Biomass (g)	Seed (g)	HI[e] (%)
Atlantic coast	Florida, Merrit Island, Mosquito Lagoon # 1	SBP-05	260.4	15.72	1.0	2.9	3.3	124.9	5.4	4.3
	Florida, Merrit Island, Mosquito Lagoon # 2	SBP-06	269.5	20.33	0.9	2.6	3.5	121.2	7.2	5.9
	Florida, Merrit Island, Indian River Lagoon	SBP-07	248.7	57.37	1.3	7.8	3.4	294.5	30.3	10.3
	Georgia, Sapelo Island Airport	SBP-08	265.9	25.37	1.1	8.1	3.8	124.4	18.6	15.0
	Georgia, Sapelo Island Cabretta	SBP-09	282.1	17.26	0.9	6.8	4.2	82.9	11.9	14.4
	Georgia, Sapelo Island Site ML	SBP-10	272.5	18.04	0.8	8.4	3.7	119.3	16.1	13.5
Gulf of Mexico	Alabama, Point Aux Pins, East	SBP-01	278.6	35.19	1.3	6.6	3.4	210.3	28.1	13.4
	Alabama, Point Aux Pins, West	SBP-02	275.1	36.79	1.3	6.4	3.3	222.1	31.5	14.2
	Texas, Padre Island, Site A	SBP-18	265.5	32.81	1.3	6.5	3.6	310.6	25.6	8.2

Table 6. Continued

Region	Location	Population	Crop[a] cycle (day)	Plant[b] height (cm)	Stem[c] diameter (cm)	Inflorescence		Plant productivity[d]		
						Length (cm)	Width (cm)	Biomass (g)	Seed (g)	HI[e] (%)
	Texas, Padre Island, Site B	SBP-19	260.1	33.17	1.3	6.7	3.5	301.9	26.5	8.8
	Texas, Padre Island, Site C	SBP-20	263.2	32.99	1.3	5.9	3.6	316.8	25.0	7.9
Pacific coast	California, Sweet water Marsh	SBP-03	261.1	17.50	0.7	7.6	2.9	63.1	2.9	4.6
	Tijuana Salt Marsh, Tijuana, Mexico	SBP-04	247.0	9.95	0.4	4.2	3.1	12.4	0.5	4.0
Gulf of Upper California	Puerto Penasco, Mexico, Cholla Bay, Puerto Penasco, Mexico, Estero Morua	SBP-15	261.5	10.23	0.4	5.0	3.6	35.5	7.1	20.0
		SBP-16	262.6	13.46	0.4	3.2	3.8	75.4	5.4	7.2
	Puerto Penasco, Mexico, La Pinta Estuary	SBP-17	259.8	13.60	0.5	3.1	3.5	39.8	2.3	5.8
Gulf of Lower California	La Paz Bay, Mexico, Site C	SBP-11	256.0	25.01	1.0	7.8	3.3	180.1	22.8	12.5
	La Paz Bay, Mexico, Site M	SBP-12	267.8	26.81	1.0	8.5	3.0	157.5	17.7	11.2

Table 6. Continued

Region	Location	Population	Crop[a] cycle (day)	Plant[b] height (cm)	Stem[c] diameter (cm)	Inflorescence		Plant productivity[d]		
						Length (cm)	Width (cm)	Biomass (g)	Seed (g)	HI[e] (%)
	La Paz Bay, Mexico, Site 3	SBP-13	260.0	17.10	0.8	6.4	3.1	126.6	15.3	12.1
	La Paz Bay, Mexico, Site 4	SBP-14	259.5	22.24	1.0	9.4	3.0	196.9	25.0	12.7

Note : [a]Crop cycle stands for the duration of salicornia from seed sowing to seed maturity.
[b]Plant height is measured at harvest as the distance from the ground to the top of the main stem.
[c]Stem diameter is measured at harvest as the maximal section at the basal part of salicornia plants.
[d]Both above-ground biomass and seed yields are determined after drying in the air.
[e]HI stands for harvest index and calculates as a ratio of seed to above-ground biomass yield.

(Luque *et al.*, 1995). Moreover, within the genus *Salicornia*, inter-specific differences between *S. europaea* and *S. ramosissima* can be detected with this PCR-based marker system. Large genotypic variation within *Salicornia* L. is also found by the use of other molecular marker systems such as restriction fragment length polymorphism (RFLP; Davy *et al.*, 1990) and nuclear rDNA variation (Noble *et al.*, 1992).

7.2. Breeding

7.2.1. Breeding objectives

Salicornia is a wild plant and requires extensive breeding and selection to remove undesirable negative traits and enhance positive ones before it can be mass cultivated as a profitable seawater agricultural crop. Table 7 shows a list of major undesirable traits that limit its potential. Therefore, our fundamental objectives for salicornia breeding, are to develop commercial cultivars with high yield, short crop cycle, high oil content, low saponin content, and disease resistance.

Table 7. Undesirable traits of *Salicornia bigelovii* as an oilseed crop.

Undesirable trait	Comments
Long crop cycle	Currently salicornia crop cycle is in the range of 7 to 10 months, which leads to the increased production cost. Reducing the crop cycle to 5-6 months will significantly reduce farming costs.
Seed shattering	About 10 – 20% seeds are lost in the field due to shattering.
Uneven maturity	Plant maturity in a given field varies significantly. This leads to a difficulty in determining the harvest time in the field.
Small seed size	Salicornia seeds are ~1mm in size causing loss during harvesting and cleaning. Due to not uniform maturity, seed shattering, and tiny size, *S. bigelovii* seed recovery is about 60-70%, but crop species more than 90%.
Low seed oil content	Increasing the oil content from 22-29% to 35 - 40% will increase the profitability.
High saponin content	Saponins are main anti-nutrient factors present in seed meal and biomass. Salicornia meal without saponins will benefit end users.
Low seed germinability	Storage of seeds at room temperature results in loss of germination. In addition, seeds germinate much better at low salinity than on seawater.

7.2.2. Seed yield

According to the morphological and physiological characteristics of salicornia, two breeding strategies may improve the seed yield. One strategy is to employ an ideotypic approach that enhances the harvest index, thereby maximizing the yield potential. The second strategy is to implement defensive breeding to increase the resistance of salicornia plants to primary stresses (abiotic and biotic), thereby minimizing the loss of final seed yield.

7.2.2.1. Ideotypic approach : In conventional crops, ideotype breeding has proven effective in achieving a high harvest index, resulting in increased seed yield (Borojevic, 1990; Austin, 1989, 1994). Austin *et al.* (1985) compared seed yield and harvest index between old and modern wheat varieties under field conditions. Improvement in plant height and leaf architecture, resulted in the increased harvest index (HI) of 32.3% in old cultivars and 49.5% in modern ones. The improvement in HI of modern cultivars corresponds to a genetic gain of 50% the grain yield. The HI of wild salicornia is about 10%, less than one third that of other conventional oilseed crops, such as soybean and rapeseed. Thus, improving the HI of salicornia plants would be economically important. Previous observations (Glenn *et al.*, 1991) and our studies indicated statistically significant positive correlation between the HI and seed yield of salicornia in the field as well as under greenhouse conditions. It is likely that ideotypic breeding may increase the HI of salicornia from the present level of 12% to about 25%, without a change in the above-ground biomass (18 Mt/ha). Therefore, on a theoretical basis, the seed yield could increase from the present level of 1.99 Mt/ha to 4.50 Mt/ha, achieving a gain of 126% in seed yield.

In 1998 and 1999, the salicornia accessions listed in Table 6 were grown with saline water irrigation (10 ppt salt) in eight repeats. Some morphological and physiological traits conducive to enhancing the seed yield, have been identified within a wide range of salicornia germplasms. In the vegetative growth phase, seedling survivability, plant height and stem diameter at the base of plants showed significant positive correlation with the final seed yield (r = 0.9138★★★★, 0.8486★★★ and 0.8894★★★, respectively). However, seed germination appeared not to correlate with the seed yield (r = -0.1545). In the pre-flowering phase, the duration (as described in section 5) and spike length significantly correlated with the seed yield, but inflorescent diameter did not (r = 0.0516). In the seed filling phase, shoot biomass on the basis of either fresh or dry weight had excellent correlation with the seed yield at the level of $p = 0.001$. These findings suggest that seedling survival, plant height, stem diameter, spike length, low numbers of secondary branches, compact and upright plant type, uniform maturity and seed size might be used as valuable criteria for breeding selection of salicornia seed yield in early generations.

7.2.2.2. ***Defensive approach :*** Primary stresses are defined as biotic and abiotic ones that significantly decrease the seed yield of salicornia plants in the field. According to our observations, seed losses occur due to various stresses: seed senescence during storage (poor germination), seedling dehydration (loss of plant stand), pests and diseases [estimated combined yield loss of 20 - 30%], plant lodging [5 - 20%], and seed shattering [10 - 20%]. In addition, about 10% loss of seed yield results from non uniformity of ripening due to immature green seeds and 10-20% loss from mechanical threshing due to small seed size. A defensive approach with the assistance of PCR-based markers is, therefore, employed to improve salicornia resistance to primary stresses, with an attempt to protect the yield potential.

7.2.3. Crop cycle

As salicornia requires frequent irrigation with seawater (O'Leary, 1984), energy for pumping accounts for 70 ~ 80% of its production cost. Thus, a short crop cycle would reduce irrigation costs and increase the profitability of the crop. Under greenhouse conditions in Tucson, USA, salicornia populations showed significant diversity in the crop cycle. The entire duration averaged 263.8 ± 8.8 days (mean ± SD, $n = 20$) and ranged from 247 – 282 days. As shown in inset of Figure 3, the life cycle of salicornia plants (about 9 months) consists of the three stages: vegetative growth phase (4 months), pre-flowering phase (1 month); and seed-filling phase (4 months). Regression analyses indicated that the duration of vegetative growth phase correlated with the crop cycle very well ($r = 0.7057^{\star\star}$), but not significantly with the final seed yield ($r = -0.1741$, Figure 3). Pre-flowering duration, although comparatively short (only one month), appeared to have a good correlation with both the crop cycle and the final seed yield ($r = 0.6711^{\star\star}$ and $r = 0.5368^{\star}$, respectively). In contrast, the duration of seed filling phase did not significantly correlate with either the crop cycle ($r = -0.3766$) or the final seed yield ($r = -0.0135$). Thus selection for a short crop cycle (about 6 - 7 months), might be feasible through shortening the duration of both vegetative and reproductive growth of salicornia plants.

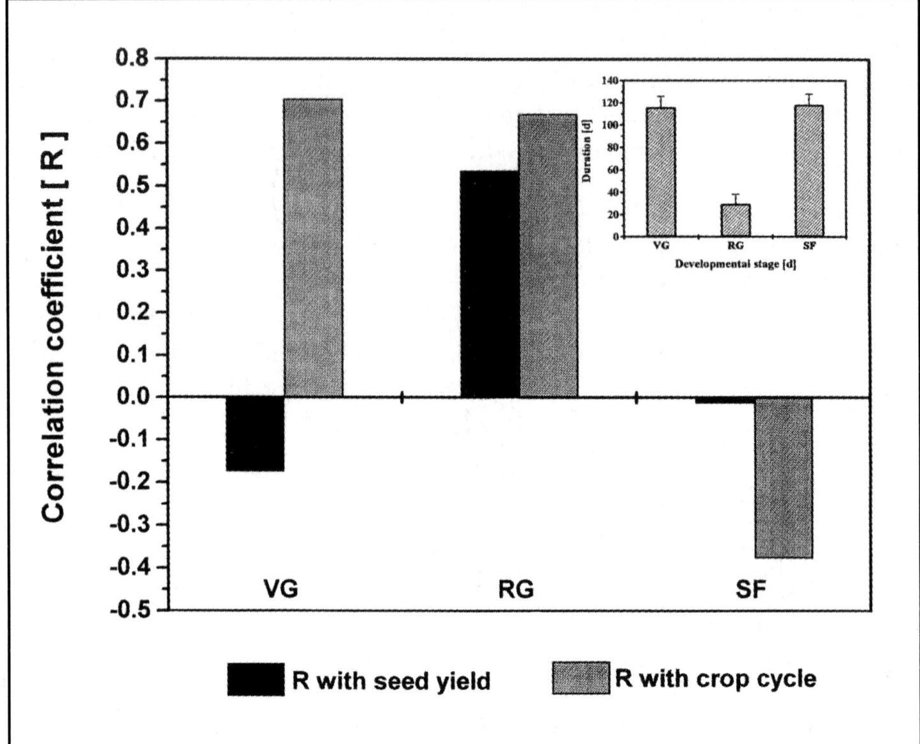

Figure 3 : Developmental phases in salicornia and their correlation with both duration of the crop cycle and the final seed yield. VG stands for vegetative growth phase (from sowing to floral initiation); PF for preflowering phase (from floral initiation to blossom); SF for seed filling phase (from blossom to maturity), and R for correlation coefficient respectively. Inset shows the duration of each phase in units of days.

7.2.4. Oil content

Salicornia accessions were screened for their oil content by accelerated solvent extraction (ASE; Association of Official Analytical Chemists, 1984) and near-infrared spectroscopy (NIR), a non-destructive method. As shown in Table 8, the results obtained by NIR were in agreement with results by the ASE for the same accession. On the average, the oil content of assayed samples was 24.96 ± 0.50% (mean ± SE, n =20) by the NIR method and 24.98 ± 0.55% by ASE method. This suggests that NIR can be used as a reliable and rapid method in salicornia, as in oilseed brassicas (Downey and Rimmer, 1993), for screening a large number of seed samples of early generations in a non-destructive fashion.

Table 8. Comparisons of oil contents in salicornia seeds analyzed by accelerated solvent extraction assay (ASE) and near-infrared spectroscopy (NIR).

Location	Accession ID	Oil content (%)		
		ASE	NIR	Difference
Alabama	SBP010508	25.8	25.7	0.1
	SBP010608	22.2	22.3	0.0
	SBP011205	22.8	23.0	- 0.2
	SBP011107	23.4	23.5	- 0.1
	SBP011707	24.5	24.2	0.3
California	SBP031108	26.8	26.4	- 0.4
Florida	SBP051005	21.0	21.3	- 0.3
	SBP051407	18.2	18.2	0.0
Georgia	SBP080905	21.8	21.9	- 0.1
	SBP081105	24.7	24.5	0.2
	SBP081204	25.0	24.8	0.2
	SBP081806	24.7	24.8	- 0.1
La Paz, Mexico	SBP110407	26.9	25.9	1.0
	SBP110705	26.3	25.9	0.4
	SBP111306	22.1	22.6	- 0.5
	SBP111408	25.3	25.1	0.2
Texas	SBP180606	22.3	22.9	- 0.6
	SBP181007	23.4	23.4	0.0
	SBP181504	29.2	29.0	0.0
	SBP181707	23.9	23.7	0.2
Mean ± SE	N = 20	23.98 ± 0.55	23.96 ± 0.50	0.02 ± 0.08

Our studies showed that the seed oil content of salicornia accessions from the wild averaged 22.49 ± 2.98% (mean ± SD, n = 236) and ranged from 11.5% to 30.6%. In the greenhouse, a good positive correlation was found between the duration of the seed-filling phase (from flowering to ripening) and the seed oil content in salicornia (r = 0.5138, p = 0.02). Regression analysis indicated that the seed oil content increases by about 1.5% when the seed filling phase is prolonged by 10 days. The seed oil content is also positively correlated with the length of spikes (r = 0.6508, p = 0.0025). This revealed that seeds from genotypes with longer spikes might contain more oil as spike length increases from 2.6 to 9.4 cm. In addition, the oil content is associated with the size of seeds (Figure 4). However, the correlation between them is negative (r = - 0.7262, p = 0.0009), suggesting that for a 0.1-mg increase in seed weight, the oil content would decrease by 1.1%. These findings would be helpful in formulating an effective breeding strategy for the oil-content improvement in salicornia.

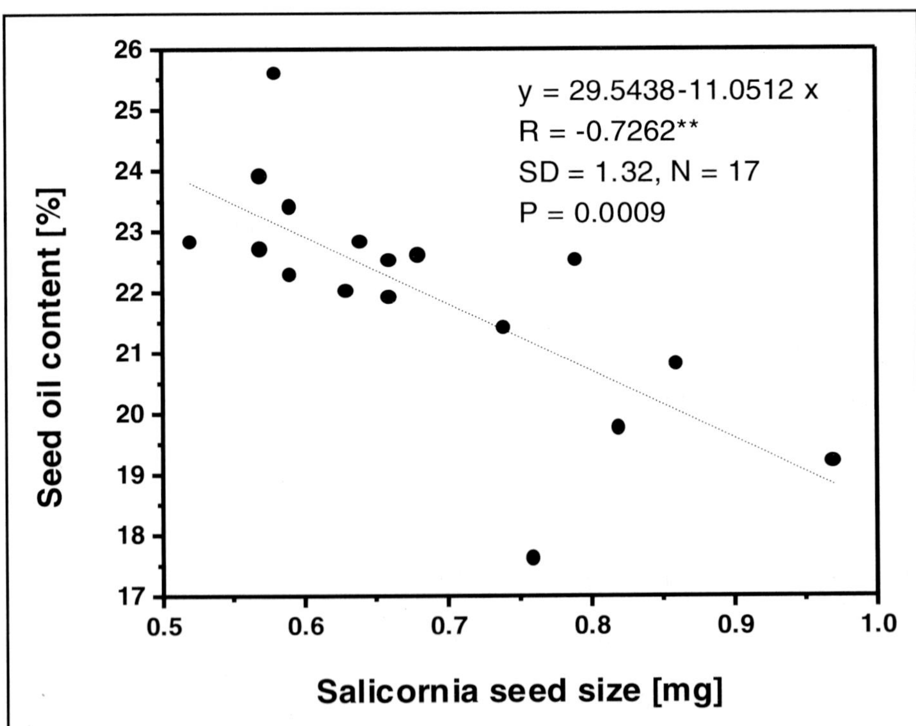

Figure 4 : Relation between seed size and seed oil content in salicornia. Seed size was measured in terms of dry weight. The oil content in seeds was determined in a non-destructive fashion with near-infrared spectroscopy (NIR).

7.2.5. Saponin content

Saponins are a type of glycosides widely distributed in plants (Hostettmann and Marston, 1995). They represent high-molecular-weight compounds, harboring a sugar part linked to a triterpene or steroid aglycone. Saponins have been identified in more than 400 plant species, including soybean, alfalfa, spinach and sugar beet. Plants, such as soapwort (*Saponaria officinalis*), soaproot (*Chlorogalum pomeridianum*) and soapbark (*Quillaja saponaria*), contain large quantities of saponins and have been used in the manufacture of soaps. Many saponins have detergent surfactant properties because they consist of water-soluble and fat-soluble components. However, saponins may reduce seed germination (Bowyer *et al.*, 1995). Furthermore, they are considered to be a main anti-nutritional factor in food and feed stuffs, as discussed in section 2.2. Thus, selection for seeds with a low content of saponins would increase the quality and monetary value of the salicornia meal.

The saponin content of salicornia germplasm has been qualitatively

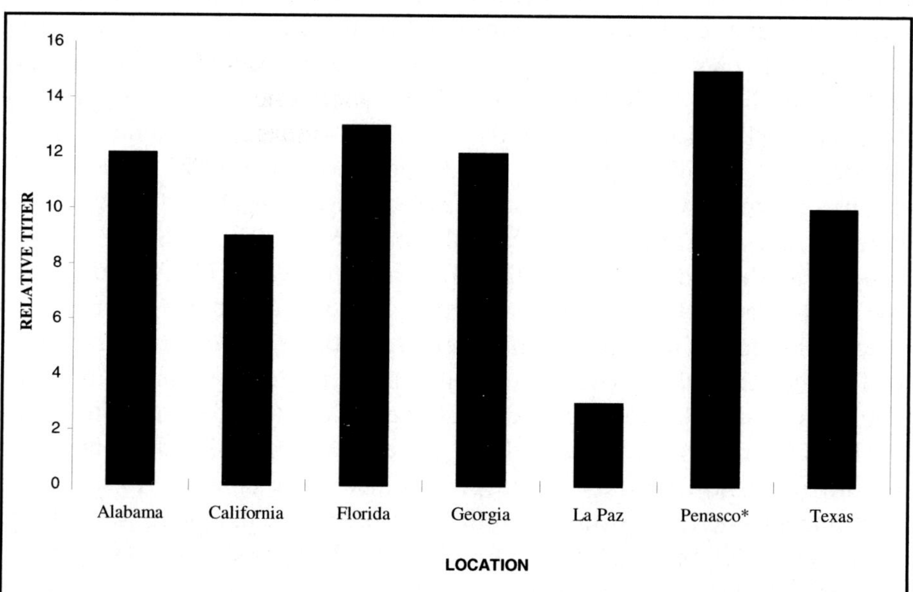

Figure 5 : Diversity in saponin content of seeds among geographically isolated ecotypes in *S. bigelovii*. Twenty populations of wild salicornia accessions were collected from seven sites across North America. The content of saponins in seeds, in units of relative titers, was determined by a homelysis assay with rabbit red blood cells.

determined by a hemolysis assay, as described in Coxworth *et al.* (1969). Preliminary results indicated a large genetic variation in the saponin content of salicornia seeds among geographically isolated populations. As shown in Figure 5, accessions collected from La Paz Bay, Mexico showed the lowest saponin content in seeds, with a mean of 2.8 relative titers. In contrast, seeds from Puerto Penasco, Mexico contained 15.2 relative titers, representing the highest content of saponins. It is noteworthy that the saponin content appears to be unrelated to the oil content in the seed. Coefficient of correlation between saponin and oil content is low ($r = -0.2169$) and is not statistically significant ($p = 0.33$). In addition, saponins are considered as one of the most effective defense substances in plants which serve as non-specific natural toxins to fungi and pests (Bowyer *et al.*, 1995; Osbourn, 1996). In *S. bigelovii*, our studies also indicated an apparent negative correlation between disease incidence (%) and saponin content (relative titer).

7.3. PCR-based markers and breeding consideration

In breeding programs of conventional crops, molecular marker techniques, especially polymerase chain reaction (PCR)-based marker systems, have been widely employed in the identification and rapid selection of traits that are simply inherited (Quarrie, 1997). The primary PCR-based DNA marker systems include random amplified polymorphic DNA (RAPD), sequence tagged sites (STS), simple sequence repeats (SSR), and amplified fragment length polymorphism (AFLP). RAPD is a fairly inexpensive technique and has been used for DNA fingerprinting (Welsh and McClelland, 1990; Williams *et al.*, 1990). Short random primers are hybridized to the DNA and the intervening DNA is amplified by PCR. Products from this reaction are separated by agarose gel electrophoresis. The size and number of bands are indicative of genetic differences in DNAs. This method is simple to perform and only small amounts of DNA are necessary. Luque *et al.* (1995) confirmed that RAPD techniques are powerful for generic and specific discrimination in the tribe *Salicornieae*. Our preliminary results also revealed the usefulness of RAPD system in salicornia breeding. A marker with a size of about 250 bp generated by the RAPD primer P3 (sequence: GATTCGACGAT), appeared to be associated with a low saponin content in seeds, but not with other traits, such as seed size and the oil content in seeds.

Apart from the RAPD system, microsatellite or SSR techniques are also useful for salicornia breeding. SSRs are highly mutable loci comprised of short tandem repeats that are present at many sites throughout the genome. The SSR system is very attractive since SSRs are co-dominant markers (in contrast with RAPDs and AFLPs) and usually represent a single locus that is often multi-allelic. These PCR-based marker techniques would assist to enhance the breeding efficiency of simply inherited traits in salicornia, such as stress resistance and saponin content, thereby accelerating the development of superior genotypes.

8. BIOTECHNOLOGY

The gentic manipulation of salicornia offers unique opportunities to complement trait improvements through conventional breeding and selection. For example new transgenic lines with desirable traits, such as reduced saponins, reduced seed-shattering, improved yield, larger seed size, better drought tolerance, insect resistance and higher oil quality can be introduced into the market at a faster pace than possible with conventional breeding alone. Many of the genes responsible for these traits have been identified from other plants, animals or microorganisms as shown in Table 9. However, their effectiveness in salicornia can not be assured and requires production of transgenic plants. In this regard the first step is to develop a procedure to introduce genes into salicornia. Genetic transformation of many conventional crops is now in routine, as discussed in other chapters. Protocols to introduce and express diverse foreign genes in plants that were first described for tobacco (De Block *et al.*, 1984; Horsch *et al.*, 1984; Paszkovsky *et al.*, 1984) have been extended to over 120 plant species in at least 35 families (Birch, 1997).

Plant transformation of foreign genes may be accomplished through biological methods (*Agrobacterium tumefaciens* T-DNA mediated) or through physical methods (particle bombardment with DNA-coated microprojectiles; whisker mediated gene transfer; microinjection; electroporation) or chemical methods (polyethylene glycol treatment; lypofectin treatment). Our preliminary studies to transform salicornia were focused on *Agrobacterium* and particle bombardment mediated protocols. The disarmed soil bacterium *A. tumefaciens* is employed widely to transform genes of interest into many plant species. The virulence genes located on the Ti plasmid of the bacterium make it possible to transfer genes into the genome of most plants. However, some plants are not susceptible or have a very low susceptibility to *Agrobacterium*. An efficient tissue culture and regeneration system is necessary for *Agrobacterium* mediated transformation. Different growth factor combinations (BA and NAA, TDZ, 2ip and IAA) were applied to test

Table 9. Traits and genes that may be useful in salicornia biotechnology.

Traits of interest		Genes
Herbicide resistance	1)	Glyphosate (N-[phosphonomethyl]glycine) is a foliar-applied selective postemergence herbicide. Glyphosate resistance is conferred by 5-enolpyruvylshikimate-3-phosphate synthase (EPSPS: Shah *et al.* 1986; Shyr *et al.* 1992)
	2)	Glufosinate is a non-selective herbicide. Bialaphos is a natural analog of glufosinate produced by *streptomyces*. Glufosinate resistance is conferred by phosphinothricin acetyltransferase (PAT: Strauch *et al.* 1988).
	3)	2,4-D(2,4-dichlorophenoxyacetic acid) is an effective herbicide for dicot weeds. 2,4-D resistance is conferred by 2,4-D monooxygenase (*tfdA*: Bayley *et al.* 1992).
Insect resistance	*1)*	*Bacillus thuringiensis* (Bt) is a bacterial delta-endotoxin effective against insects (Schnepf and Whiteley 1987).
	2)	Protein toxins that interfere with digestion and absorption of nutrition. Examples include, polyphenol oxidase, proteinase inhibitors, alpha-amylase inhibitors (Felton *et al.* 1992; Huesing *et al.* 1991).
	3)	Cholesterol oxidase (Purcell *et al.* 1993) and scorpion venom peptides (Bougis *et al.* 1989) are also found to be effective against certain insects.
Yield improvement	1)	Adenosine diphosphoglucose pyrophosphorylase (*sh2*: Giroux *et al.* 1996).
	2)	Cytokinin gene inhibiting senescence and PHY A gene that increases harvest index (Gan *et al.* 1995; Robson 1996).
Reduced saponin		Avenacinase or tomatinase are fungal enzymes that detoxify saponins. Corresponding gene may be useful in reducing saponins in salicornia.
Oil modifications		Several genes useful in oil modifications are known. Examples include fatty acid hydroxylase, delta-6-fatty acid saturase, lauroyl ACP thioesterase, C_{10} Acyl-[ACP]thioesterase, Antisense- ω3 and ω6 fatty acid desaturase, sn-2 acyltransferase, sn-1 and sn-3 acyltransferase, and palmitoyl-ACP desaturase.
Anti-seed shattering		Gene expressed in the dehiscence zone and valve opening. (University of California, San Diego)
Drought resistance		Gene responsible for drought, cold and salt resistance (Cornell Research Foundation).

the regeneration ability of salicornia cotyledons and hypocotyls. Preliminary results showed that in salicornia while it was possible to induce callus formation on hypocotyls and cotyledons isolated from two to seven day old seedlings and on immature embryos through the application of growth factors (BA, NAA, TDZ and IAA) or silver nitrate and 2,4-D, embryogenesis was not observed. A more promising result of regeneration via organogenesis was obtained with cotyledons isolated from three-day-old seedlings. These cotyledons placed on S1 medium with 4 mg/l 2ip + 1 mg/l IAA indicated a regeneration rate of 16% after 4 weeks.

Different wild-type *Agrobacterium tumefaciens* strains (such as A348 an octopine type, A208 nopaline type, and the supervirulent strain A281) were tested for their abilities to infect salicornia plants. *Agrobacterium* culture was applied to wounded tissue and was maintained in a growth chamber (16 hours light, 8 hours dark, 23°C) for three months. Young sunflower plants were used as a positive control. Tumor tissue was found on sunflower after two months, whereas no callus tissue was observed in salicornia plants. These studies indicate that salicornia has a very low susceptibility to the *Agrobacterium* strains tested. Further attempts to transform salicornia were conducted with disarmed strain ABI4750, harboring the selection genes *nptII* for kanamycin resistance, the cp4 for the glyphosate resistance and the marker gene, β-glucuronidase (*uidA*) on binary vector. Salicornia explants (cotyledons, hypocotyls) and seedlings were wounded by sonication or acupuncture needle or a scalpel in the presence of marker genes. The explants were co-cultured with *Agrobacterium* for 12 hrs, placed on selective MS medium with antibiotics and growth factors. Transient expression of a GUS reporter gene was observed for all wounding methods, but none of the methods gave stable transformed tissue. *Agrobacterium* mediated transformation of the floral dip was also attempted. The floral dip method is described as being successful for flowering *Arabidopsis* plants (Clough and Bent, 1997). The labor intensive vacuum infiltration process was eliminated in favor of a simple dipping of floral tissue into a solution containing *Agrobacterium* (ABI4750 containing *nptII* and cp4 genes), sucrose and a surfactant. The harvested seeds were surface sterilized and placed on selective MS medium containing 200 mg/l kanmycin and 0.1 mM glyphosate. About 2000 seeds were screened but none of them were transformed.

A second gene delivery protocol, the gene gun was also tested for salicornia transformation. The gene gun is a very versatile devise to introduce DNA molecules into plant cells. The basic design of a high voltage electric discharge device that accelerates DNA-coated gold particles into living cells has been reported by Christou *et al.* (1990). Cotyledons, hypocotyls, meristem explants isolated from two- to four-day-old seedlings were used for bombardment of the marker gene *uidA*, selectable markers cp4 and *nptII*. Transient expression of the GUS marker gene was observed in the bombarded tissue for up to 8 days post bombardment. However, in these initial experiments stable transformation was not observed, as judged by histochemical staining for GUS after 10 days post bombardment. Further research need still remains in this field.

REFERENCES

Abouheif MA, Al-Saiady M and Kraidees M (2000) Influence of inclusion of *Salicornia* biomass in diets for rams on digestion and mineral balance. *Asian Austral Journal of Animal,* **13** : 967-973.

Al Saeedi AH (1997) The response of *Salicornia* to fertilizers under irrigation with seawater: A pots experiment. *Alexandria J. Agric. Res.,* **42** : 187-192.

Al Saeedi AH and Elprince AM (2000) Critical phosphorus levels for *Salicornia* growth. *Agronomy J.,* **92** : 336-345.

Anderson OE and Worthington RE (1971) Boron and manganese effects on protein, oil content, and fatty acid composition of cottonseed. *Agronomy J.,* **63** : 566-569.

Appelqvist LA (1970) Lipids in Cruciferae. VI. The fatty acid composition of seeds of some cultivated *Brassica* species and of *Sinapis alba* L. *Fette Seifen Anstrichm,* **72** : 783-792.

Aronson JA (1989) *HALOPH: A Data Base of Salt Tolerant Plants of the World.* Arid Land Studies, University of Arizona, Tucson, AZ, USA.

Association of Official Analytical Chemists (1984) *Official Methods of Analysis.* 14th edn. AOAC, Washington DC.

Attia FM, Alsobayel AA, Kriadees MS, Al-Said MY and Bayoumi MS (1997) Nutrient composition and feeding value of *Salicornia bigelovii* Torr. meal in broiler diets. *Animal Feed Sci. and Technol.,* **65** : 257-263.

Austin RB (1989) Prospects for improving crop production in stressful environments. In: *Plants under Stress: Biochemistry, Physiology and Ecology and their Application to Plant Improvement.* Jones HG, Flowers TJ, Jones MB, eds. SEB Seminar Series: 39. Cambridge University Press, UK. pp. 235-248.

Austin RB (1994) Plant breeding opportunities. In: *Physiology and Determination of Crop Yield.* Madison, WI. Am Soc Agron; Crop Sci Soc Am; Soil Sci Soc Am. pp. 567-585.

Austin RB, Morgan CL and Ford MA (1985) Grain and straw yields of old and new wheat varieties. *Annual Report 1984.* Plant Breeding Institute, Cambridge (UK). pp. 106-107.

Ayala F and O'Leary JW (1995) Growth and physiology of *Salicornia bigelovii* Torr. at suboptimal salinity. *Intl. J. Plant Sci.,* **156** : 197-205.

Ayala F, O'Leary JW and Schumaker KS (1996) Increased vacuolar and plasma membrane-ATPase activities in *Salicornia bigelovii* Torr. in response to NaCl, *J. Exp. Bot.,* **47** : 25–32.

Baldwin JG, Mundo-Ocampo M and McClure MA (1997) *Cactodera salina* n. sp. from the estuary plant, *Salicornia bigelovii* in Sonora, Mexico. *J. Nematology,* **29** : 465-473.

Ball PW (1964) A taxonomic review of *Salicornia* in Europe. *Feddes Repertorium,* **69** : 1-8.

Ball PW (2000) *Salicornia* L. In : *Flora of North America.* Vol. 4. Chenopodiaceae, (Ed Shultz L) Flora of North America Association, USA.

Bayley C, Trolinder N, Ray C, Morgan M, Quisenberry JE and Ow DW (1992) Engineering 2,4-D resistance into cotton. *Theor. Appl. Genet.,* **83** : 645-662.

Beadle JB, Just DE, Morgan RE and Reiners RA (1965) Composition of corn oil. *J. Am. Oil Chem. Soc.,* **42** : 90-95.

Belal Ibrahim EH and Al-Dosari M (1999) Replacement of fish meal with salicornia meal in feeds for Nile tilapia *Oreochromis niloticus. J. World Aquaculture Soc.,* **30** : 285-292.

Bell JM (1989) Nutritional characteristics and protein uses of oilseed meals. In: *Oil Crops of the World,* (Eds. Robbelen G, Downey RK, Ashri A). McGraw-Hill Publishing Company, New York. pp. 192-207.

Birch RG (1997) Plant Transformation: Problems and strategies for practical application. *Ann. Rev. Plant Physiol. Plant Mol. Biol.,* **48** : 297-326

Blumwald E and Poole R (1987) Salt-tolerance in suspension cultures of sugar beet. I. Induction of Na^+/H^+-antiport activity at the tonoplast by growth in salt. *Plant Physiol.,* **83** : 884-887.

Bohnert HJ, Nelson DE and Jensen G (1995) Adaptations to environmental stresses. *Plant Cell,* **7** : 1099-1111.

Borojevic S (1990) *Principles and Methods of Plant Breeding.* Elsevier, New York, USA.

Bougis PE, Rochat H and Smith LA (1989) Precursors of an *Androctonus australis* nurotoxins. Structure of precursors, processing outcomes and expression of a functional recombinant toxin II. *J. Biol. Chem.,* **264** : 19259-19265.

Bowyer P, Clarke BR, Lunness P, Daniels MJ and Osbourn AE (1995) Host-range of a plant-pathogenic fungus determined by a saponin detoxifying enzyme. *Science,* **267** : 371-374.

Brown JK (1996) First report of beet curly top germinivirus isolated from diseased *Salicornia* in Mexico. In: *Proceedings of the First International Technical Seminar for Salicornia,* September 25 –29, Phoenix, Arizona USA. pp. 50-58.

Cerda A, Caro M and Fernandez FG (1982) Salt tolerance of two pea cultivars. *Agronomy J.,* **74** : 796-798.

Charnock A (1988) Plants with a taste for salt. *New Scientist,* **20** : 41-45

Christou P, MaCabe D, Martinell B and Swain W (1990) Soybean genetic-engineering - Commercial production of transgenic plants. *Trends Biotechnol.,* **8** : 145-151

Clapham AR, Tutin TG and Moore DM (1987) *Salicornia* L. In: *Flora of the British Isles,* 3rd edn, Cambridge University Press. Cambridge. pp. 161-163.

Clark A (1994) Samphire: from sea to shining seed. *Aramco World,* Novemeber/ December. pp. 2-9.

Clough S and Bent A (1997) Floral dip: A simplified method for *Agrobacterium*-mediated transformation of *Arabidopsis thaliana. Plant J.,* **16** : 735-743

Coxworth EC, Bell JM and Ashford A (1969) Preliminary evaluation of Russian thistle, kochia, and garden atriplex as potential high protein content seed crops for semiarid areas. *Can. J. Plant Sci.,* **49** : 427-435.

Dadby DH (1962) Chromosome number, morphology and breeding behavior in the British Salicorniae. *Watsonia,* **5** : 150-162.

Davy AJ, Noble SM and Oliver RP (1990) Genetic variation and adaptation to flooding in plants. *Aquatic Botany,* **38** : 91-108.

de Block M, Herrera-Estrella L, van Montagu M, Schell J and Zambryski P (1984) Expression of foreign genes in regenerated plants and their progeny. *EMBO J.,* **3** : 1681-1689

Dixon JB, Weed SB and Dinauer RC (1989) *Minerals in Soil Environment.* Madison, Wisconsin: Soil Science Society of America.

Douglas J (1994) A rich harvest from halophytes. *Resource* (May): 15-18.

Downey RK and Rimmer SR (1993) Agronomic improvement in oilseed brassicas. *Adv. Agronomy,* **50** : 1-66.

Duval-Jouve MJ (1868) Des *Salicornia* de l'Herault. *Bulletin de la Societe Bototanique de France,* **15** : 132-140 and 164-178.

Earle FR, van Etten CH, Clark TF and Wolff IA (1968) Compositional data on sunflower seed. *J. Am. Oil Chem. Soc.,* **45** : 876-879.

El-Mallah MH, Murui T and El-Shami S (1994) Detailed studies on seed oil of *Salicornia* SOS-7 cultivated at the Egyptian border of red sea. *Grasas y Aceites,* **45** : 385-389.

Felton G, Donato K, Broadway R and Duffey S (1992) Impact of oxidized plant phenolics on the nutritional quality of dietary protein to a noctuid herbivore *Spodoptera exigua. J. Insect Physiol.,* **38** : 277-285.

Ferris VR, Krall E, Faghihi J and Ferris JM (1999) Phylogenetic relationships of *Globodera millefolii, G. artemisiae,* and *Cactodera salina* based on ITS region of ribosomal DNA. *J. Nematology,* **31** : 498-507.

Flowers T, Troke PF and Yeo AR (1977) The mechanism of salt tolerance in halophytes. *Ann. Rev. Plant Physiol.,* **28** : 89-121.

Francois LE (1984) Salinity effects on germination, growth, and yield of turnips. *HortScience,* **19** : 82-84.

Francois LE (1987) Salinity effects on asparagus yield and vegetative growth. *J. American Soc. Horticult. Sci.,* **112** : 432-436.

Francois LE, Donovan TJ and Maas EV (1984) Salinity effects on seed yield, growth and germination of grain sorghum. *Agronomy J.,* **76** : 741-744.

Francois LE and Maas EV (1994) Crop response and management on salt-affected soils. In: *Handbook of Plant and Crop Stress,* (Ed Pessarakli M) Marcel Dekker Inc. New York. pp. 149-182.

Francois LE, Maas EV, Donovan TJ and Youngs VL (1986) Effects of salinity on grain yield, quality, vegetative growth, and germination of semi-dwarf and durum wheat. *Agronomy J.,* **78** : 1053-1058.

Gan S and Amasino RM (1995) Inhibition of leaf senescence by auto regulated product of cytokinin. *Science,* **270** : 1986-1988.

Giroux MJ, Shaw J, Barry G, Cobb G, Greene T, Okita T and Hannah C (1996) A single gene mutation that increases maize seed weight. *Proc. Natl. Acad. Sci. USA*, **93** : 5824-5829

Glenn EP, Brown JJ and Blumwald E (1999) Salt tolerance and crop potential of halophytes. *Crit. Rev. Plant Sci.,* **18** : 227-255.

Glenn EP, Brown JJ and O'Leary JW (1998) Irrigating crops with seawater. *Scientific American,* **279** : 76-81.

Glenn EP, Coates WE, Riley JJ, Kuehl RO and Swingle RS (1992) *Salicornia bigelovii* Torr.: A seawater-irrigated forage for goats. *Animal Feed Sci. Technol.,* **40** : 21-30.

Glenn EP, Miyamoto S, Moore D, Brown JJ, Thompson TL and Brown P (1997) Water requirements for cultivating *Salicornia bigelovii* Torr. with seawater on sand in a coastal desert environment. *J. Arid Environ.,* **36** : 711-730.

Glenn EP and O'Leary JW (1984) Relationship between salt accumulation and water content of dicotyledonous halophytes. *Plant, Cell and Environ.,* **7** : 253-261.

Glenn EP, O'Leary JW, Watson MC, Thomspon TL and Kuehl RO (1991) *Salicornia bigelovii* Torr.: An oilseed halophyte for seawater irrigation. *Science,* **251** : 1065-1067.

Grasshoff K, Kremling K and Ehrhardt M (1999) *Methods of Seawater Analysis.* Wiley-VCH, Weinheim, Germany.

Grattan SR and Grieve CM (1994) Mineral nutrient acquisition and response by plants grown in saline environments. In: *Handbook of Plant and Crop Stress,* (Ed Pessarakli M) Marcel Dekker Inc. New York. pp. 203-226.

Halket AC (1915) The effect of the salt on the growth of *Salicornia. Ann. Bot.,* (London) **29** : 149.

Hawke MA and Maun MA (1987) Some aspects of nitrogen, phosphorus, and potassium nutrition of three colonizing beach species. *Can. J. Bot.,* **66** : 1490-1496.

Hodges CN, Thompson TL, Riley JJ and Glenn EP (1993) Reversing the flow: Water and nutrients from the sea to the land. *AMBIO,* **22** : 483-490.

Hodges RM (2000) *Salicornia bigelovii:* An oilseed crop irrigated with full-strength seawater. *INFORM,* **11** : 95-96.

Horsch R, Fraley BT, Rodgers SG, Sander PR and Lloyd A (1984) Inheritance of functional foreign genes in plants. *Science,* **223** : 496-498.

Hostettmann KK and Marston AA (1995) *Saponins.* Cambridge University Press, Cambridge, UK. pp. 1-548.

Huesing JE, Shade RE, Chrispeels MJ and Murdock LL (1991) Alpha amylase inhibitor, not phytohemagglutin explains the resistance of common bean seeds to cowpea weevil. *Plant Physiol.,* **96** : 993-996.

Hymowitz T, Palmer RG and Hadley HH (1972) Seed weight, protein, oil and fatty acid relationships within the genus *Glycine. Tropical Agriculture Guildford,* UK, **49** : 245-250.

Jeffries RL and Gottlieb LD (1982) Genetic differentiation of the microspecies *Salicornia europaea* L. (sensu stricto) and S. *ramosissima* J. Woods. *New Phytol.,* **92** : 123-129.

Kraidees MS, Abouheif MA, Al-Saiady MY, Tag-Eldin A and Metwally H (1998) The effect of dietary inclusion of halophyte *Salicornia bigelovii* Torr. on growth performance and carcass characteristics of lambs. *Animal Feed Sci. and Technol.,* **76** : 149-159.

Langdale GW and Thomas JR (1971) Soil salinity effects on absorption of nitrogen, phosphorus, and protein synthesis by coastal bermudagrass. *Agronomy J.* **63** : 708-711.

Le Houerou HN (1993) Salt-tolerant plants for the arid regions of the Mediterranean isoclimatic zone. In: *Towards the Rational Use of High Salinity Tolerant Plants*, Vol. 1. (Eds Lieth H, Al-Masoom A), Kluwer Academic Publishers, Dordrecht. pp. 403-422.

Lee CW, Glenn EP and O'Leary JW (1992) *In vitro* propagation of *Salicornia bigelovii* by shoot-tip cultures. *HortScience,* **27** : 472.

Levitt J (1980) *Responses of Plants to Environmetnsal Stresses: Water Radiation, Salt and Other Stresses.* 2nd edn. Academic Press, New York, USA.

L'Roy A and Hendrix DL (1980) Effect of salinity upon cell membrance potential in the marine halophyte, *Salicornia bigelovii* Torr. *Plant Physiol.,* **65** : 544-549.

Lu Z, Glenn EP and John ME (2000) *Salicornia bigelovii* Torr.: An overview. *INFORM,* **11** : 418-423.

Luque T, Ruiz C, Avalos J, Calderon IL and Figueroa ME (1995) Detection and analysis of genetic variation in *Salicornieae* (Chenopodiaceae) using random amplified polymorphic DNA (RAPD) markers. *Taxon.,* **44** : 53-63.

Knowles PF (1968) Registration of UC-1 safflower. *Crop Sci.,* **8** : 641.

Maas EV and Hoffman GJ (1977) Crop salt tolerance – Current assessment. *Journal of Irrigation and Drainage Division, ASCE,* **103** : 115-134.

Magboul AM, Al-Saeedi A and Al-Dossary M (1996) Seawater utilization for food production and coastal desert greening. In: *Proceedings of the First International Technical Seminar for Salicornia,* September 25 –29, Phoenix, Arizona USA. pp. 71-86.

Mangal JL, PS Hooda and Lal S (1989) Salt tolerance of the onion seed crop. *J. Hort. Sci.,* **64** : 475-477.

Mangal JL, Singh RK, Yadav AC, Lal S and Pandey UC (1990) Evaluation of garlic cultivars for salinity tolerance. *J. Hort. Sci.,* **65** : 657-658.

Maro M, Cruz V, Cuartero J, Estan MT and Bolarin MC (1991) Salinity tolerance of normal-fruited and cheey tomato cultivars. *Plant and Soil,* **136** : 249-255.

Minhas PS, Sharma DR and Khosla BK (1990) Mungbean response to irrigation with waters of different salinities. *Irrigation Science,* **11** : 57-62.

Mota-Urbina JC (1996) Experiments from the Kino Bay Farm 1987-1988. In: *Proceedings of the First International Technical Seminar for Salicornia,* September 25 –29, Phoenix, Arizona USA. pp. 18-40.

Munns R (1993) Physiological processes limiting plant growth in saline soils: some dogmas and hypotheses. *Plant, Cell and Environ.,* **16** : 15-24.

Munns R, Greenway H and Krist GO (1983) Halotolerant eukaryotes. In : *Encyclopedia of Plant Physiology* Vol. 12, (Eds Lange OL, Osmond CB, Nobel PS, Ziegler H) Springer-Verlag, Heidelberg. pp. 59-135.

Munz PA and Keck DD (1970) *Salicornia* L. In: *A California Flora.* University of California Press, Berkeley, CA, USA. p.382.

National Research Council (1990) *Saline Agriculture: Salt-Tolerant Plants for Developing Countries,* National Academy Press, Washington DC, USA. pp. 1-143.

Neary J (1981) Pickleweed, Palmer's grass and saltwort: Can we grow tomorrow's food with today's salt water? *Science,* **81** : 39-43.

Neumann PM (1997) Salinity resistance and plant growth revisited. *Plant, Cell and Environ.,* **20** : 1193-1198.

Noble SM, Davy AJ and Oliver RP (1992) Ribosomal DNA variation and population differentiation in *Salicornia* L. *New Phytol.,* **122** : 553-565.

Noriega JR (1996) Soil salt dynamics in a sea water irrigated farm, Kino Bay, Mexico. In : *Proceedings of the First International Technical Seminar for Salicornia,* September 25-29, 1996. Phoenix, Arizona, USA. pp. 41-49.

O'Leary JW, Glenn EP and Watson MC (1985) Agricultural production of halophytes irrigated with seawater. *Plant and Soil,* **89** : 311-321.

O'Leary JW (1984) The role of halophytes in irrigated agriculture. In: *Salinity Tolerance in Plants: Strategies for Crop Improvement*, (Eds Staples RC, Toenniessen GH). John Wiley & Sons, New York, USA. pp. 285-301.

Olsen MW, Frye RJ and Glenn EP (1996) Effect of salinity and plant species on CO_2 flux and leaching of dissolved organic carbon during decomposition of plant residue. *Plant and Soil,* **179** : 183-188.

Osbourn A (1996) Saponins and plant defence - A soap story. *Trends Plant Sci.,* **1** : 4-9.

Pasternak D and de Malach Y (1994) Crop irrigation with saline water. In: *Handbook of Plant and Crop Stress,* (Ed Pessarakli M) Marcel Dekker Inc. New York. pp. 599-622.

Paszkovski J, Shillito RD, Saul M, Mandak V and Hohn T (1984) Direct gene transfer to plants. *EMBO J.,* **3** : 2717-2722

Proudfoot AM (1993) Relationships between *Coleophora atriplicis* and its host plants on a salt marsh. Ph. D. Thesis, University of East Anglia, Norwich.

Purcell JP, Greenplate JT, Jennings MG, Ryerse JS, Pershing JC, Sims SR, Prinsen MJ, Corbin DR, Tran M, Sammons RD and Stonard RJ (1993) Cholesterol oxidase: A potent insecticidal protein active against boll weevil larvae. *Biochem. Biophys. Res. Comm.,* **196** : 1406-1413.

Quarrie SA (1997) New molecular tools to improve the efficiency of breeding for increased drought resistance. In: *Drought Tolerance in Higher Plants: Genetical, Physiological and Molecular Biological Analysis,* (Ed Belhassen E) Kluwer Academic Publishers, The Netherlands. pp 89-100.

Riehl TE and Ungar IA (1982) Growth and ion accumulation in *Salicornia europaea* under saline field conditions. *Oecologia* (Berlin), **54** : 193-199.

Riley JJ, Glenn EP and Mota-Urbina CU (1994) Small ruminant feeding on the Arabian peninsula with *Salicornia bigelovii* Torr. In: *Halophytes as a Resource for Livestock and for Rehabilitation of Degraded Lands.* (Eds Squires VR, Ayoub AT) Kluwer Academic Publishers. The Netherlands. pp. 273-276.

Riley JJ and Abdal M (1993) Preliminary evaluation of *Salicornia* production and utilization in Kuwait. In: *Towards the Rational Use of high Salinity Tolerant Plants, Vol. 2.* Lieth H, Al-Masoom A eds. Dordrecht, Holland: Kluwer Academic Publishers. pp. 319-329.

Robson PRH, McCormick AC, Irvine AS and Smith H (1996). Genetic engineering of harvest index in tobacco through expression of phytochrome gene. *Nature (Biotechnology)* **14** : 995-998.

Schnepf HE and Whiteley HR (1987) Cloning and expression of the *Bacillus thuringiensis* crystal protein gene in *E. coli. Proc. Natl. Acad. Sci. USA*, **78** : 2893-2897.

Shah DM, Horsch RB, Klee HJ, Kishore GM, Winter JA, Tumer NE, Hironaka CM, Sanders PR, Gasser CS, Aykent S, Siegel NR, Rogers SG and Fraley RT (1986) Engineering herbicide tolerance in transgenic plants. *Science,* **233** : 478-481.

Shyr YYJ, Hepburn AG and Widholm JM (1992) Glyphosate selected amplification of the 5 enolpyruvylshikimate-3-phosphate synthase gene in cultured carrot cells. *Mol. Gene. Genet.,* **22** : 377-382.

Somers GF (1979) Natural halophytes as a potential resource for new salt-tolerant crops for food and forage. *Environmental Sci. Res.,* **14** : 101-115.

Stace C (1997) *Salicornia* L. In: *New Flora of the British Isles*, 2nd edn. Cambridge University Press, Cambridge, UK. pp. 147-150.

Standley PC (1916) *Salicornia* L. In: *North America Flora*, vol. 21. New York, USA. pp. 82-85.

Stanghellini ME, Mihail JD, Rasmussen SL and Turner BC (1992) *Macrophomina phaseolina*: A soilborne pathogen of *Salicornia bigelovii* in a marine habitat. *Plant Disease,* **76** : 751-752.

Stanghellini ME and Rasmussen SL (1989) Two new diseases of *Salicornia* sp. caused by *Bacillus subtilis* and *Macrophomina phaseolina. Phytopathology,* **79** : 912.

Stanghellini ME, Rasmussen SL and Turner BC (1990) *Macrophomina phaseolina* associated with mortality of *Salicornia* in an estuary on the seacoast of the State of Sonora, Mexico. *Phytopathology,* **80** : 892.

Stanghellini ME, Werner EG, Turner BC and Watson MC (1988) Seedling death of *Salicornia* attributed to *Metachroma* larvae. *The Southwestern Entomologist,* **13** : 305.

Strauch E, Wohlleben W and Puhler A (1988) Cloning of the phosphinothricin N-acetyltransferase gene from *Streptomyces viridochromo* genes Tu494 and its expression in *Streptomyces lividans* and *Escherichia coli. Gene,* **63** : 65-74.

Stumpf DK and O'Leary JW (1985) The distribution of Na+, K+ and glycinebetaine in *Salicornia bigelovii. J. Exp. Botany,* **36** : 550-555.

Swingle RS, Glenn EP and Squires V (1996) Growth performance of lambs fed mixed diets containing halophyte ingredients. *Animal Feed Sci. Technol.,* **63** : 137-148.

Troyo-Dieguez E, Ortega-Rubio A, Maya Y and Leon JL (1994) The effect of environmental conditions on the growth and development of the oilseed halophyte *Salicornia bigelovii* Torr. in arid Baja California Sur, Mexico. *J. Arid Environ.,* **28** : 207-213.

Tutin TG, Burges NA, Chater AO, Edmondson JR, Heywood VH, Moore DM, Valentine DH, Walters SM and Webb DA (1993) *Arthrocnemum* Moq - *Salicornia* L. In: *Flora Europaea* Vol. 1, Psilotaceae to Platanaceae, 2nd edn. Cambridge University Press, Cambridge, UK. pp. 121-123.

Ungar IA (1991) *Ecophysiology of Vascular Halophytes.* CRC Press, Boca Raton, USA.

Waisel Y (1972) *Biology of Halophytes.* Academic Press, New York, USA.

Watson MC (1990) *Atriplex* species as irrigated forage crops. *Agriculture, Ecosystems and Environment,* **32** : 107-118.

Webb KL (1966) NaCl effects on growth and transpiration in *Salicornia bigelovii*, a salt-marsh halophyte. *Plant and Soil,* **24** : 261-268.

Welsh J and McClelland M (1990) Fingerprinting genomes using PCR with arbitrary primers. *Nucleic Acids Res.,* **18** : 7213-7218.

Wiggins IL (1980) *Salicornia* L. In : *Flora of Baja California* Stanford University Press, Stanford, California, USA.

Williams JGK, Kubelik AR, Livak KJ, Rafalski JA and Tingey SV (1990) DNA polymorphisms amplified by arbitrary primers are useful as genetic markers. *Nucleic Acids Res.,* **18** : 7213-7218.

Worthington RF and Hammons RO (1971) Genotypic variation in fatty acid composition and stability of *Arachis hypogaea* L. oil. *Oleagineux,* **26** : 695-700.

York J, Lu Z, John ME and Glenn EP (2001) Photoperiod affects floral induction in *Salicornia bigelovii* Torr. *American J. Economic Botany* (in review).

Zhu JK, Hasegawa PM and Bressan RA (1997) Molecular aspects of osmotic stress in plants *Crit. Rev. Plant Sci.,* **16** : 253-277.

AUTHOR INDEX

SUBJECT INDEX